INTERPRETING
BODIES

Interpreting Bodies

Classical and Quantum Objects
in Modern Physics

ELENA CASTELLANI, EDITOR

PRINCETON UNIVERSITY PRESS

PRINCETON, NEW JERSEY

Library of Congress Cataloging-in-Publication Data

Interpreting bodies : classical and quantum objects in modern physics
/ Elena Castellani, editor.
p.   cm.
Includes bibliographical references.
ISBN 0-691-01724-7 (cloth : alk. paper) ISBN 0-691-01725-5 (pbk : alk. paper)
1. Physics—Philosophy.   2. Reality.   3. Physical measurements.
I. Castellani, Elena, 1959–   .
QC6.I637   1998
530′.01—dc21   98-17178

Contents

PART THREE: OBJECTS AND MEASUREMENT

Preface

As its title indicates, this volume is concerned with the question of the nature of physical objects—that is, the objects or "bodies" constituting the furniture of our world—when considered in the light of the results of modern physics. To be more precise, the subject matter of the book comprises the particular and additional problems posed by contemporary physical theories—relativistic as well as quantum—as they relate to the philosophical discussion on the nature of physical objects and, in particular, on traditional issues such as their "constitution," "identity," and "individuality."

The present volume is the outcome of a project I started and first developed in 1993–94 while visiting the Department of Philosophy of Princeton University. The project was to provide an overview on the foregoing subject by collecting a number of articles that, when put together, could give the reader a good idea of the cluster of problems involved and of the various approaches and solutions proposed.

Why such a project? Here are some of the motivations. First of all, consider the absence (to my knowledge) of an organic presentation of what could be called the "foundational" debate on objects: a growing area of inquiry relative to the new and particular problems raised by the developments in physics for the traditional philosophical issues associated with the nature of objects. Second, there exists a deplorable gap between such a developing area of research and the contemporary philosophical debate on physical objects—an important and well-established field of philosophical inquiry. An attempt to bridge such a gap by assembling "philosophical" and "foundational" studies on the nature of objects seemed to be of some utility. Finally, the situation characterizing the so-called foundations of physics, namely the very lively debate between the different alternative interpretations of quantum mechanics, is without a correspondingly accurate examination of the ontological options involved.

This volume consists of a collection of essays: writings by authors who kindly agreed to participate in my project by submitting either articles specially written for the collection or recently published papers on the subject of the book; and some classic works by "old masters"—Reichenbach, Born, Heisenberg, and Schrödinger—that I decided to insert in the volume in order to give an idea of the sources and the continuity of the problems debated.

The writings collected in the volume are from philosophers, philosophers of science, philosophers of physics, logicians, and physicists. Most of the authors who agreed to contribute are well-known scholars—some of them the leading experts in their field—and all have recently done significant and frontline work on the basic themes of the volume (as is also apparent from the bibliography). The "old masters" need no introduction.

The collection has no pretense of completeness or conclusiveness. Its main purpose is to introduce the reader to the subject matter being taken into consideration and to offer a "map" of related discussions, not to present final solutions and definitive accounts. In this sense, full agreement on the arguments and solutions proposed was not a leading criterion in the selection of the material.

The volume is therefore structured to help the readers orient themselves in the ensemble of the debated questions, by following as much as possible a progression from the general to the specific, and from the statements of the problems to the proposed solutions. The collection as a whole is organized into three parts, corresponding to the three main topics around which the articles have been selected. The nature of this division and the reasons for it are illustrated in my introduction to the volume.

To whom is this collection addressed? Actually, to anyone who is generally interested in both philosophical and scientific issues. Details of the questions discussed in the volume will indeed be more appreciated by a reader who is familiar with the arguments considered in contemporary philosophy and foundations of physics. However, a good number of the essays are not too technical and their introductory parts are usually of a very general and accessible character. Moreover, many of the articles are accompanied by a rich collection of bibliographical notes, which can be used as convenient guides to further reading on specific topics.

In bringing together this volume, I have also collected many debts that I should like to acknowledge here. I wish to express my gratitude to all those who have encouraged my project in some way and given helpful comments. But above all I am indebted to the authors who have contributed to the collection, thus making its realization possible. I wish to thank them for their understanding and patience, as well as for their important suggestions. To two of them, in particular, I owe a special debt: to Maria Luisa Dalla Chiara, who provided not only constant encouragement but also concrete help on several occasions, and to Bas van Fraassen, who has supported my project since its very beginning with precious comments and advice. In fact, I can say that this book would not have come into existence without his generous assistance.

Acknowledgments

The editor is grateful for permission to include the following copyrighted material.

David Lewis, "Many, but Almost One": from *Ontology, Causality, and Mind: Essays in Honor of David Armstrong*, edited by John Bacon, Keith Campbell, and Lloyds Reinhardt. © Cambridge University Press 1993. Reprinted with the permission of Cambridge University Press.

Hans Reichenbach, "The Genidentity of Quantum Particles": from H. Reichenbach, *The Direction of Time*, translated and edited by Maria Reichenbach, published by the University of California Press. © 1956 renewed 1984 Maria Reichenbach. Reprinted with the permission of the University of California Press.

Bas C. van Fraassen, "The Problem of Indistinguishable Particles": from *Science and Reality: Recent Work in the Philosophy of Science*, edited by James T. Cushing, C. F. Delaney, and Gary M. Gutting. © 1984 by the University of Notre Dame Press. Used by permission of the publisher.

Max Born, "Physical Reality": from *Philosophical Quarterly* 3 (1953): 139–149. © The Management Committee of The Philosophical Quarterly. Reprinted with the permission of Blackwell Publishers.

Peter Mittelstaedt, "The Constitution of Objects in Kant's Philosophy and in Modern Physics": from *Kant and Contemporary Epistemology*, edited by Paolo Parrini. © 1994 Kluwer Academic Publishers. Reprinted with kind permission from Kluwer Academic Publishers.

Erwin Schrödinger, "What Is an Elementary Particle?": from *Collected Papers*, vol. 4. © 1984 by Österreichische Akademie der Wissenschaften. Reprinted with the permission of the Austrian Academy of Science.

Werner Heisenberg, "The Nature of Elementary Particles": from *Physics Today* 29, no. 3 (1976): 32–39. Reprinted with permission from American Institute of Physics Journals. © 1976 American Institute of Physics.

INTERPRETING
BODIES

Introduction

Elena Castellani

1. The Philosophical Question of Physical Objects and the Objects of Modern Physics

In the philosophical literature, as well as in ordinary usage, physical objects are most frequently referred to as "bodies," "material things," "material beings," "objects in space and time." With respect to the totality of existing objects, they are usually (but not always) distinguished from "persons." Whichever are the particular descriptions employed, *having mass*, *being located in space and time*, and *persisting through time* seem to constitute the fundamental features required for something to qualify as a "physical object."

Such is more or less the common picture that is abstracted from the experience of everyday physical "things," which indeed appear to be material, to occupy space, and to endure through time. But how are physical objects to be identified on the basis of their mass and spatiotemporal determinations? Do we need any further identification criterion? This is the sort of question that is at the heart of the philosophical debate on the nature of objects.

Now what about the objects of physics? Are those entities, whose nature and behavior it is the aim of physical theories to describe, "physical objects" in the preceding sense? Let us consider, for example, the entities at the very center of the developments of contemporary physics: the so-called elementary particles, the microscopic "objects" supposed to be the ultimate constituents of the physical world. To begin with, the particles of contemporary physics are not always material: think of photons, which are massless particles. Moreover, it is not certain that we can always know their exact positions, at any instant of time. Finally, it is not at all clear what "enduring through time" means in the case of particles that have, for example, no "spatiotemporal continuity," or whose existence can only be inferred through indirect observations (as in the case of quarks). Are such entities to be called *objects* after all?

Philosophers seem to have different and not always clear positions on this point. They typically feel a little uncomfortable about what to do with such "objects."[1] It is quite clear that entities like microphysical particles do represent a very special kind of physical objects (if they are to be included in the category) and that the arguments developed with regard to everyday material things do not generally apply to such a peculiar class of beings. How then are we to deal with this sort of entity when considering the question of the nature of physical objects?

One possibility is simply to ignore the problem: we can take physical particles to be only "borderline cases" of objects,[2] or—to quote another expression—"not clear cases of material objects,"[3] and accordingly dismiss the specific problems arising in connection with these entities as not being of real concern for the philosophical question of physical objects. But even if we consider microparticles to be just borderline cases of objects, we usually do admit that they are the parts of which every material object is made: which makes it difficult not to consider such "parts" when speaking of the "wholes."

An opposite attitude is to search for a very general concept of a physical object. Take the view according to which a more "liberal notion of physical object" should be adopted, as suggested by Quine in his "Whither Physical Objects?", thus including into the category of physical objects also such entities as physical processes or events and, of course, elementary particles.[4] What can result in this way? Quine gives us a good example: on his attempt to arrive at a very comprehensive notion, what he obtains is a progressive "evaporation" of physical objects, from the "bodies" he starts with, to "space-time regions," to the "pure sets of numerical coordinates" with which he ends.[5]

The problem of defining a more comprehensive notion of a physical object is one of the many issues arising when the object question is addressed in the light of the results of modern physics. Bewildering features of contemporary physical theories—such as the relativistic space-time structure, the wholeness (nonseparability) of the quantum world, and the peculiarities of so-called quantum statistics—challenge the traditional ways of conceiving objects in space and time. How to divide the world in parts and wholes, in which way to confer individuality upon a object, how to relate objective and measurable properties to a physical entity: similar issues acquire new and unexpected aspects when the developments of modern physics are taken into account.

The implications of contemporary physical theories for the debate on the nature of objects constitute the central theme of this collection of essays. The underlying idea is that ontological questions could be more thoroughly and profitably investigated by bringing together the philosophical and the foundational debates on objects. The volume therefore focuses on those aspects of the world of modern physics that are particularly significant from both points of view. As will be illustrated in more detail, this area of inquiries can be organized around three main issues: namely the question of *individuality*, the significance of *invariance*, and the problem of *measurement*. The present collection is correspondingly structured into three parts, each of which is specifically devoted to one of these prominent issues.

2. Objects and Individuality

How are we able to identify a physical object as an individual—that is, how are we able to claim that it is *this* physical object and no other? How can we divide the world into individual objects? Part One is centered on these kinds of questions.

The first contribution to the volume, Giuliano Toraldo di Francia's "A World of Individual Objects?" (chapter 1), provides a historical survey of the general problematic. Starting from the conceptions of the philosophers of antiquity and arriving to the more recent achievements in the foundations of physics, Toraldo di Francia gives us a good idea of the evolution of most basic issues—for example, the legitimacy of conceiving the world as a collection of separate objects. His chapter may thus be taken as an introduction to the themes more specifically explored in the contributions that follow.

The individuality issue is at the center of a problem cluster that, following current philosophical usage, can be investigated by focusing on two main points of discussion: (1) the *part/whole relation*, whose formal theory is the so-called *mereology*; and (2) the search for some *individuation principle*, that is some principle telling us how to ascribe individuality to a physical object.

With respect to the first point, two sorts of questions can be distinguished depending on whether the individual taken into account plays the role of a "whole" or of a "part." Accordingly, one has to deal, on the one hand, with the fundamental question of how an individual as a "whole" can be identified on the basis of its "parts"; and, on the other hand, with the questions arising in relation to the individuality of physical objects that are "parts" of compounds or aggregates.

Many of the problems and puzzles regarding the identity of material objects—"artifacts" like tables and statues, or natural things such as stones and cats—just arise because those objects are composed of parts. Think of the much debated "problem of the ship of Theseus," that is, the problem of the identity of a ship whose parts are gradually all replaced by new ones.[6] But replacement or variation of parts is only one side of the question. What does it mean for something to be a *part* of a material object? And what about things that have "questionable parts," parts "not definitely included and not definitely not included,"[7] as in the case of the water droplets composing a cloud (which is the example illustrating Peter Unger's much debated "problem of the many")?[8] Are things completely determined by their constituters, or is there a difference between the constitution and the identity of physical objects? A thorough investigation of this sort of question is provided by David Lewis's "Many, but Almost One," which is reprinted in this volume (chapter 2). *Vagueness* of objects, doctrines of *relative* and *partial* identity, and the semantic method of *supervaluations* are some of the points described and evaluated in this influential work, which ends up with Lewis's own proposal for addressing the problem of the many.

It is commonly accepted that all material objects are ultimately composed of elementary particles, the microphysical objects with which this volume is particularly concerned. "Elementary" is usually taken to mean "without (proper) material parts." But parts are not only *material parts*. There can be *temporal parts*, to begin with. Moreover, on the view that objects are sums of properties, *properties* may be taken to play the role of "parts," and it is worthwhile recalling that "property parts" could also be modal, dispositional, or counterfactual.[9] Questions about the identity of individuals composed of parts thus regard macroscopic as well as microscopic—even "elementary"—physical objects.

The implications of microphysics for the relation between part and whole are spelled out in Tim Maudlin's "Part and Whole in Quantum Mechanics" (chapter 3). In the quantum domain, because of the existence of entangled physical states for a collection of systems—one of the most striking features of quantum theory—it is not always clear what kind of properties can be attributed to the "individual" physical systems that are "parts" of compounds. What may be the individuality of physical objects that are "parts" of compounds or aggregates is therefore a question of particular significance in the case of quantum objects. By exploring the real roots of the ineliminable "holism" of quantum theory and the problems that thus arise for metaphysical doctrines such as "reductionism," Maudlin's chapter shows us how the world cannot be seen just as "a set of separately existing localized objects, externally related only by space and time."[10]

To focus on the nature of the relation between part and whole is one way of dealing with the problem of individuality. The problem can also be addressed by directly concentrating on the search for some *principle of individuation*—that is, some principle that establishes what confers individuality upon a physical object.

In this latter approach, one usually starts from the consideration of the problem of *identity through time* (or identity through change in time). Physical objects are "in time"—that is, they are supposed to exist over some period of time. This raises the problem of establishing in what the identity of a physical object consists that allows one to think that it is the same individual object that has persisted over time. Note that this problem of identity through time (or "reidentifiability") can also be understood in the sense of searching for "something" that unites the successive parts or momentary *stages* of an object. In the literature, this "something" is usually called *genidentity*, after Hans Reichenbach's use of the term for characterizing the relation connecting different states of the same physical object at different times.[11]

In today's debate on physical objects, we can distinguish between two main theories of individuation, depending on whether the role of conferring individuality is ascribed to properties of the object, or to something "transcending" the object's set of properties, such as some kind of persisting substantial substratum or essence ("transcendental individuality," "haecceity," or "primitive thisness" are some of the denominations employed for this latter view).[12]

Within the first viewpoint, a prominent place is usually assigned to space-time properties. The most diffused conception of individuality is founded on the space-time location of physical objects. *Space-time location*, together with a condition of continuity upon the space-time path corresponding to an object's career (the so-called *spatiotemporal continuity*)[13] and the further condition given by the *impenetrability assumption* (two distinct individual physical objects cannot occupy the same position at the same time), constitute the basic ingredients of such a view.[14]

In the philosophical literature, conditions for individuation are usually discussed with regard to ordinary physical objects. What happens when we turn to the domain of microscopic objects? Take the preceding example of spatiotemporal continuity and impenetrability assumption, which are both reasonable conditions for warranting the reidentifiability of macroscopic objects, but no more at hand in the case of microphysical entities. As stated by Hans Reichenbach in his fundamental essay on "the genidentity of quantum particles," reprinted in this volume (chapter 4), "the usual methods of identification break down in the atomic domain"; what the situation suggests is to "replace an individual examination of particles by inferences based on statistical properties of an assemblage of particles." Which are the implications of these statistical properties for the genidentity of microphysical entities? Reichenbach's investigation on this point is seminal as regards the contemporary discussion on quantum objects.[15]

As Reichenbach points out, the shift to statistical considerations raises new problems in the debate on the individuality of physical objects, first of all the "problem of identical particles" (or "problem of indistinguishable particles"). The historical roots of this problem are to be found in some physical developments in the 1920s that were of decisive importance for quantum physics: namely, the appearance of new statistics (the Bose-Einstein and Fermi-Dirac statistics) for aggregates of similar physical systems, and the connection between those "quantum statistics" and the principle of permutation invariance for quantum particles of the same kind (the so-called identical particles).[16] Permutation invariance, seen as a condition of physical indistinguishability of identical particles,[17] is at the core of a very lively discussion about the significance of the notions of identity, individuality, and indistinguishability in the quantum domain. The "problem of indistinguishable particles" (the existence of entities that are physically indistinguishable although "numerically distinct") is indeed the center of a whole cluster of problems: from the problem raised by statistical correlations between identical particles apparently defying a causal explanation, to the question as to whether Leibniz's principle of the identity of indiscernibles should be regarded as violated in quantum physics.

A detailed account of the sources, nature, and development of this debate is offered in the essay by Bas van Fraassen, "The Problem of Indistinguishable Particles," which is reprinted in this volume (chapter 5). Van Fraassen's contribution, by providing a systematic description and clarification of the various problems involved as well as a discussion of some of the most significant attempts (including his own) at a solution, is of the utmost utility for a real understanding of the field.

The last three chapters of Part One are concerned with the more recent developments in this debate.

Does quantum physics force us to consider microscopic particles as nonindividuals? Steven French, in his "On the Withering Away of Physical Objects" (chapter 6), explores the impact of the fundamental underdetermination existing between physics and metaphysics on similar attempts to draw ontological conclusions from physical theories.[18] In this respect, French's chapter critically discusses the attempts to break the underdetermination by appeal to philosophical principles (such as Leibniz's principle) as well as recent work on quantum individuality, rigid designation, and nonsupervenient relations.

In Paul Teller's "Quantum Mechanics and Hacceities" (chapter 7) the basic question of the identity and distinctness of quantum entities is addressed in terms of "haecceities." Teller's suggestion is that "belief in haecceities plays a crucial role in the felt puzzles about quantum statistics." The purpose of Teller is then to show how quantal facts give good reasons for rejecting any aspect of quantum entities that might be thought to do the job of "haecceity" (such as the "particle labels" in the traditional formulations of quantum theories).[19] If we abandon particles with haecceities, are we left with fields? Teller rejects the dichotomy of particles and fields, arguing that both particle and field concepts are "imprecise conglomerates of more general component concepts."[20]

If quantum entities are nonindividuals after all, how are we to provide a logicomathematical framework for dealing with these objects? From a logical point of view, the indistinguishability of "identical particles" has suggested revisions of semantics for quantum entities, such as, for example, the proposal of a theory of "quasisets" (or "quasets") for describing collections of objects having cardinality but no order type—that is, objects to which the concept of identity of classical logic does not apply.[21] The essay by Maria Luisa Dalla Chiara, Roberto Giuntini, and Decio Krause, "Quasiset Theories for Microobjects: A Comparison" (chapter 8), compares the basic ideas and intuitive aspects of two different approaches to the notion of *quasiset*: namely the "semiextensional" approach developed by da Costa, French and Krause (the theory S^{**}), and the "intentional" approach by Dalla Chiara and Toraldo di Francia (the theory of quasets QST).

3. Objects and Invariance

The chapters in Part Two are specifically concerned with the relevance of the notion of invariance to the problem of defining physical objects.

As we have seen, invariance understood as permanence (permanence through change in time) does not constitute any new argument in relation to the identity of objects. The specific point considered here is how this "permanence argument" can be exploited by using the *mathematical* notion of invariance—that is, the notion of "invariance with respect to a group of transformations" (or "symmetry"). This notion has acquired a special significance in contemporary theories that describe the characteristics and behavior of physical "particles." This is due to some very important results in both mathematics and physics, the core of which can be traced back to the introduction of the concept of *group of transformations* in nineteenth-century mathematical thought and the subsequent development of the theory of transformation groups and their invariants.[22]

The use of the group-theoretic notion of invariance in relation to the object question is basically grounded on the idea that the possibility of speaking in terms of "objects" in a given context is connected with the possibility of individuating invariants with respect to the symmetry group of the context. This idea was first introduced by Felix Klein with regard to "geometrical objects," as a corollary of his new conception of geometry proposed in the famous 1872 *Erlanger Programm*.[23] With the subsequent application of group theory to other domains of science and in particular to physics, this view could then be extended to other sorts of "objects" and, in particular, to "physical objects" (the objects of physical theories as well as the objects of our common perception).[24]

In contemporary physics, a decisive support to this way of thinking was provided by the interpretation of the theory of relativity as a theory of physical invariance with respect to the group of transformations of reference frames (or "observers"). It is a common opinion that what counts in defining an object should not depend upon the particular perspective under which the object is taken into consideration. In the light of the foregoing interpretation of relativity, such an opinion could then be reformulated in the following terms: *objective* is what is invariant with respect to the group of transformations of the spatiotemporal frame or, in the words of Hermann Weyl, "objectivity means invariance with respect to the group of automorphisms [of space-time]."[25]

The implications of relativity theory as regards the object question are very clearly pointed out in Max Born's "Physical Reality," which is reprinted in this volume (chapter 9). In this classic essay, the significance of the notion of invariance and its group-theoretic treatment for our conception of objects and more generally of "reality" is thoroughly explored. Born's analysis of the basic features of an approach to the object question that is grounded on the idea of invariance is seminal. Let us just recall the following two main points he sets forth: first, the possibility of dealing in an unitary way—that is, within the same approach—with very different kinds of objects (from ordinary material things to the microphysical particles of contemporary physics); second, the view according to which objects (*all* objects, classical as well as quantum) are essentially "sets of invariants," that is, sets of quantities that are invariant with respect to the relevant transformation groups.

In physics, the main invariants are quantities like mass (rest-mass), spin, charge—those properties of physical particles which are usually known as "intrinsic properties." The important point is that the application of group-theoretic methods to contemporary physics has made it possible, in some way, to derive these invariant properties on the ground of symmetry considerations.[26] In other words, we are provided with a general procedure for "constructing" or "constituting" the objects of physical theories as sets of invariants. This possibility supplies the basic motivation for what may be called the *group-theoretic approach* to the problem of defining physical objects, a recently developing area of inquiry centered on the invariance idea and its exploitation by using the results of the application of group theory to contemporary physical theories.

Within such an approach, can a relation be established between the constitution of the objects of physics and the constitution of objects in Kantian philosophy? The essay by Peter Mittelstaedt, "The Constitution of Objects in Kant's Philosophy and in Modern Physics," which is reprinted in this volume (chapter 10), explores this issue. Mittelstaedt argues that the Kantian way of constituting objects of experience by means of categories is actually the one to follow in constituting the objects of classical as well as of quantum physics. But whereas in classical mechanics objects can be "completely determined," in quantum mechanics we can at most obtain either "incomplete" objects or objects that are "unsharp" and only approximately constituted.

Elena Castellani's "Galilean Particles: An Example of Constitution of Objects" (chapter 11) enters into the details of the group-theoretic approach to the object question. Because the chapter's main purpose is to provide an introduction to this recent area of research, it illustrates the basic ideas as well as the historical background and offers a brief survey of its key concepts. Then the chapter presents a concrete example of object constitution within the group-theoretic approach, namely the constitution of classical and quantum particles in the nonrelativistic (i.e., "Galilean") case.

4. Objects and Measurement

In dealing with physical objects, one has inevitably to face the problem of how to take into account the results of "experience." In fact, in addressing such issues as those concerning the identity, individuality, and constitution of physical objects, we are always more or less implicitly confronted with the problem of how to relate our experience of "external things" to the concept of a physical object we are elaborating. The relation between *physical objects* (in particular, microphysical objects) and *experience* is the ground theme of the papers of Part Three.

What the term "experience" exactly means and where the boundary is—if there is any—between experience and theory are traditional philosophical questions, and we shall not enter into the subject here. It will be sufficient to recall that, in considering the objects of physical theories, one usually understands "experience" in the sense of what we can *measure* or, in other words, in the sense of the *results of experimental physics*. The question is then, What has experimental evidence to tell us when we tackle the problem of the nature of the objects of physical theories? More specifically, does it make sense to speak of "particles" in the light of the experimental results in quantum and relativistic physics?

As is well known, quantum physics raises specific difficulties with regard to the concept of *measurement*. Let us just mention the problem generated by the presence of an ineliminable interference between the measuring apparatus and the measured system at the microphysical level. This problem, known in the literature as the *measurement problem*, was first brought out in connection with the formulation of the famous Heisenberg's uncertainty relation. Aspects connected with this result and related consequences of the quantum description—such as, for example, the impossibility of attributing a definite trajectory to a particle in the same sense that classical mechanics does and the "wave-particle duality" of quantum entities—have been regarded as seriously challenging the legitimacy of the "classical" concept of an individual particle.

The implications of the results of quantum theory for the concept of an elementary particle are extensively analyzed in Erwin Schrödinger's "What Is an Elementary Particle?" which is reprinted in this volume (chapter 12). Schrödinger's specific purpose is to demonstrate how we are forced by "observed facts" to deny microphysical particles the "dignity of being an absolutely identifiable individual." Much time has gone by (and many new physical results have been achieved) since this essay was written, but Schrödinger's masterful piece remains an illuminating example of how to proceed with regard to these kinds of issues.

More than two decades later, Werner Heisenberg's "The Nature of Elementary Particles," reprinted in this volume (chapter 13), addresses the same question as Schrödinger by taking into account the developments in contemporary particle physics and its more significant experimental results. Such "facts" as the "discovery" of *antimatter* (with the consequence that the number of particles is not a constant—particles can be created and annihilated), the transformation of matter into energy, the proliferation of the "observed" particles, and the fundamental role of symmetry groups need to be accounted for in dealing with the objects of microphysics. What can be concluded in the light of these results? Heisenberg's position is that "words such as 'divide' or 'consist of' have to a large extent lost their meaning." Our task is therefore "to adapt our thinking and speaking... to the new situation created by experimental evidence."

The measurement problem is at the center of the foundational debate on contemporary physics. Still today, this problem constitutes the crucial point in the proposal and discussion of an interpretation of quantum mechanics. Many of the issues connected with the problem, such as the nonlocal behavior of quantum entities, the localizability of quantum particles, and the possibility of attributing objective properties to an individual physical (quantum) system, have fundamental implications for the discussion concerning the objects of physics. The foundational problems of quantum physics and their ontological implications are treated in both "The Entity and Modern Physics: The Creation-Discovery View of Reality" by Diederik Aerts (chapter 14) and "Dynamical Reduction Theories as a Natural Basis for a Realistic Worldview" by Gian Carlo

Ghirardi's (chapter 15). By addressing the basic question of what an adequate view of reality can be in the light of the most recent physical achievements, both chapters propose new approaches to the interpretative problems of quantum mechanics, thus contributing as well to the general foundational debate on quantum physics.

The "creation-discovery view" proposed by Aerts is a realistic interpretation of quantum theory—in the sense that it considers the quantum entity as existing in the outside world—incorporating two essential aspects: an aspect of "discovery" (referring to the properties that the entity already had before the measurement) and an aspect of "creation" (referring to the new properties that are created during the act of measurement). In which sense this "view" differs from both the de Broglie-Bohm interpretation and the Copenhagen interpretation of quantum mechanics, and how the paradoxes of orthodox quantum theory can be revisited in assuming this new perspective, are the central items treated in the chapter.

The starting point of Ghirardi's article is the question as to whether it is possible to elaborate a worldview that can accommodate our knowledge about natural phenomena. How can we obtain a satisfactory and objective description of reality at the macroscopic level within the quantum mechanical framework? The fundamental problem of the relation between macroscopic and microscopic objects, closely connected with the measurement problem, is here analyzed from the point of view of *dynamical reduction theories*.[27] The chapter is aimed to show that it is possible to work out an interpretation of the formalism—more specifically, an interpretation of the wave function—allowing a "satisfactory description of the world in terms of the values taken by an appropriately defined mass density function in ordinary configuration space."

Finally, in considering the nature of microphysical objects it is still to be clarified how exactly the particles of physical theories are related to what is effectively "observed" in physical laboratories. Giulio Peruzzi's "Microphysical Objects and Experimental Evidence" (chapter 16) focuses on this point, by analyzing in the framework of contemporary particle physics (i.e., quantum field theory) the experimental techniques used for detecting physical particles. A brief survey of the scattering techniques is provided, in which the crucial role of the notion of "cross-section" is particularly stressed. What is actually measured in a typical scattering experiment? And what is the image of microphysical objects resulting from today's experimental physics? Peruzzi's conclusion is that, as far as experimental evidence is concerned, microphysical objects are "cross-sectional entities."

Notes

1. As is well stated by Peter van Inwagen: "Few philosophers would be perfectly happy about calling a quark or a proton or even a large organic molecule a material object." P. van Inwagen, *Material Beings* (Ithaca, N.Y.: Cornell University Press, 1990), 17.
2. "Borderline cases of material things" is, for instance, the expression used for electrons and other elementary particles in R. Coburn, "Identity and Spatiotemporal Continuity," in M. K. Munitz, ed., *Identity and Individuation* (New York: New York University Press, 1971), 85.
3. Van Inwagen, *Material Beings*, 19.
4. W.V.O. Quine, "Whither Physical Objects?" in R. S. Cohen, P. K. Feyerabend, and M. W. Wartofsky, eds., *Essays in Memory of Imre Lakatos* (Dordrecht: Reidel, 1976).
5. See Quine, "Whither Physical Objects?" 502–504. Note that it has become quite usual, in searching for a more liberal ontology of physical objects, to attribute special importance to the spatiotemporal (i.e., four-dimensional) character of physical objects, as in the case of the Quinean "space-time regions" or "portions of space-time." Take, for example, the ontology of "four-dimensional hunks of matter" that Mark Heller proposes instead of the "standard ontology" of physical objects, in his *The Ontology of Physical Objects: Four-Dimensional Hunks of Matter* (Cambridge: Cambridge University Press, 1990). And see also, for a strenuous defense of a four-dimensional "Minkowskian view" of physical objects (in opposition to a view of objects as three-dimensional things that endure through time), J.J.C. Smart, "Space-Time and Individuals," in R. Rudner and I. Scheffler, eds., *Logic and Art* (Indianapolis: Bobbs-Merrill Company, 1972).
6. In modern philosophy, the case was first treated by Hobbes (*De Corpore* **II**, 11). The discussion of the puzzle of the ship of Theseus in the contemporary literature on physical objects is due especially to David Wiggins, who discussed it in his *Identity and Spatio-Temporal Continuity* (Oxford: Blackwell, 1967). On this problem see also by the same author the more recent *Sameness and Substance* (Oxford: Blackwell, 1980), 92–94.
7. The expressions are from D. Lewis, "Many, but Almost One," in J. Bacon, K. Campbell, and L. Reinhardt, eds., *Ontology, Causality, and Mind: Essays in Honor of David Armstrong* (Cambridge: Cambridge University Press, 1993), which is reprinted here as chapter 3.
8. That is, the problem of how a given cloud can be identified with one of the many aggregates of water droplets differing in composition only in respect of some few droplets, and therefore equally good candidates to be that cloud. See P. Unger, "The Problem of the Many," *Midwest Studies in Philosophy* 5 (1980): 411–467. The problem is quite general, if one thinks, following Lewis, that all material objects are ultimately composed of "swarms of particles." Of the same kind is the "Geach's paradox of 1.001 cats," also discussed in the essay by Lewis (chapter 2).
9. It is open to question whether such kind of "parts" (modal, dispositional, and counterfactual properties) should be taken into account in establishing

the identity of individual physical objects. In this respect, see for example M. Johnston, "Constitution Is Not Identity," *Mind* 101. 401 (1992): 89–105, and, in response to Johnston, H. W. Noonan, "Constitution Is Identity," *Mind* 102. 405 (1993): 133–145. About the different meanings and functions of "parts," see in particular P. Simons, *Parts: A Study in Ontology* (Oxford: Clarendon Press, 1987).

10. On Maudlin's position with regard to quantum holism and its consequences, see also his recent book *Quantum Non-Locality and Relativity* (Oxford: Blackwell, 1994).

11. This term, introduced by K. Lewin, was first used by Reichenbach in his *The Philosophy of Space and Time* (New York: Dover, 1957; original German edition published in 1928). With regard to the meaning of the term, see also the essay of Reichenbach on the genidentity of quantum particles, which is reprinted here as chapter 4. For a recent discussion of Reichenbach's notion of genidentity, see, for example, B. van Fraassen, *An Introduction to the Philosophy of Time and Space* (New York: Columbia University Press, 1985; originally published 1970).

12. A thorough discussion of the different views of individuality as regards the objects of physics can be found, for example, in S. French, "Identity and Individuality in Classical and Quantum Physics," *Australasian Journal of Philosophy* 67 (1989): 432–446. For the use of "transcendental individuality" (the terminology is due to H. R. Post) in the actual debate on the identity of physics objects, see, in particular, S. French and M. Redhead, "Quantum Physics and the Identity of Indiscernibles," *British Journal for the Philosophy of Science* 39 (1988): 233–246, and French, "Identity and Individuality." A first discussion of the meaning of "primitive thisness" and "haecceity" in relation to the contemporary debate on objects can be found in R. M. Adams, "Primitive Thisness and Primitive Identity," *Journal of Philosophy* 76 (1979): 5–26. To the question of "haecceities" in contemporary physics, Paul Teller has recently devoted much attention. See his "Quantum Mechanics and Haecceities" (chapter 7) and his recent book, *An Interpretive Introduction to Quantum Field Theory* (Princeton: Princeton University Press, 1995).

13. "Spatiotemporal continuity" is a central notion in the contemporary debate on physical objects (ordinary material beings as well as objects of physical theories). For a detailed discussion of the meaning and function of spatiotemporal continuity in relation to the problem of the identity of individual physical objects, see for instance E. Hirsch, "Essence and Identity," in M. K. Munitz, ed., *Identity and Individuation* (New York: New York University Press, 1971).

14. Of such kind is, for example, the view of individuality that Reichenbach calls "material genidentity," grounded on the conditions of "continuity of change" and "spatial exclusion." See Reichenbach, "The Genidentity of Quantum Particles" (chapter 4).

15. A detailed and critical discussion of the arguments defended in these pages by Reichenbach is provided by Bas van Fraassen in "The Problem of Indistinguishable Particles," in J. T. Cushing, C. F. Delaney, and G. M. Gutting, eds., *Science and Reality: Recent Work in the Philosophy of Science* (Notre

Dame, Ind.: University of Notre Dame Press, 1984), which is reprinted here as chapter 5.

16. That is, particles having the same "intrinsic" or state-independent properties, such as mass, charge, spin, and so on. A critical discussion of the connection between quantum statistics and the principle of permutation invariance can be found in Bas van Fraassen's recent book *Quantum Mechanics: An Empiricist View* (Oxford: Clarendon Press, 1991), whose chapter 11 is entirely devoted to the problem of identical particles.

17. From a historical point of view, a useful account of the development of the indistinguishability concept in microphysics is provided in A. Kastler, "On the Historical Development of the Indistinguishability Concept for Microparticles," in A. van der Merwe, ed., *Old and New Questions in Physics, Cosmology, Philosophy, and Theoretical Biology* (New York: Plenum, 1983).

18. French is one of the authors who has most contributed to the recent debate on the significance of the notions of identity, individuality, and indistinguishability in modern physics. Among his contributions to the field, see, for instance, French, "Identity and Indidividuality"; S. French, "Individuality, Supervenience and Bell's Theorem," *Philosophical Studies* 55 (1989): 1–22; S. French, "Why the Principle of the Identity of Indiscernibles Is Not Contingently True Either," *Synthese* 78 (1989): 141–166; and S. French, "Hacking Away at the Identity of Indiscernibles: Possible Worlds and Einstein's Principle of Equivalence," *Journal of Philosophy* 92 (1995): 455–466.

19. Teller has recently devoted much attention to this point. In this regard, see, for instance, M. Redhead and P. Teller, "Particles, Particle Labels, and Quanta: The Toll of Unacknowledged Metaphysics," *Foundations of Physics* 21 (1991): 43–62, and, of the same authors, "Particle Labels and the Theory of Indistinguishable Particles in Quantum Mechanics," *British Journal for the Philosophy of Science* 43 (1992): 201–218.

20. This crucial issue is extensively treated in Teller's *An Interpretive Introduction to Quantum Field Theory*. Apart from the collection edited by H. R. Brown and R. Harré, *Philosophical Foundations of Quantum Field Theory* (Oxford: Clarendon Press, 1988), Teller's book is one of the very few works in today's literature that are specifically concerned to explore the philosophical implications of quantum field theory.

21. The notion of quaset was introduced in the essay by M. L. Dalla Chiara and G. Toraldo di Francia, "Individuals, Kinds and Names in Physics," in E. Agazzi and M. Mondadori, eds., *Logica e Filosofia della Scienza, oggi,* Proceedings Soc. Italiana di Logica e Filos. delle Scienze, San Gimignano, 1983 (Bologna: Clueb, 1986).

22. Broadly speaking, a *group* is a set of elements (here, operations or "transformations") having the property that the combination of any two elements is an element that also belongs to the set.

23. That is, the conception according to which each *geometry* is characterized by a given group of transformations and the *geometrical properties* are those properties that do not change (i.e., that are *invariant*) with respect to the transformations of the group.

24. On the significance of the developments of the theory of transformation groups in relation to the problem of characterizing "objects," it is worthwhile signaling the seminal contribution of Ernst Cassirer, "The Concept of Group and the Theory of Perception," *Philosophy and Phenomenological Research* 5 (1944): 1–35, where the group concept is applied to the object question in the context of geometry, physics, and perception theory. See also, by the same author, "Reflections on the Concept of Group and the Theory of Perception" (1945), in *Symbol, Myth and Culture: Essays and Lectures of Ernst Cassirer, 1935–1945* (New Haven: Yale University Press, 1979).
25. H. Weyl, *Symmetry* (Princeton: Princeton University Press, 1982; originally published in 1952), 132 (automorphism = symmetry transformation).
26. Whence, for example, the possibility of regarding physics objects as "nomological objects," "knots of (invariant) properties prescribed by physical laws," as stressed on many occasions by Toraldo di Francia. On this point, also signaled in Toraldo di Francia's "A World of Individual Objects?" (chapter 1), see especially G. Toraldo di Francia, "What Is a Physical Object?" *Scientia* 113 (1978): 57–65, and, by the same author, *Le cose e i loro nomi* (Rome: Laterza, 1989).
27. This very important approach to the foundational problems of quantum mechanics developed in the past decade by Ghirardi and others is clearly and extensively illustrated in Ghirardi's contribution to this volume (chapter 15).

PART ONE

OBJECTS AND INDIVIDUALITY

1

A World of Individual Objects?

Giuliano Toraldo di Francia

1. Things and Bodies

For many centuries philosophers—as well as the man on the street—seem to have taken for granted that the world consists of individual *objects*, alternatively called *things* or *bodies*. From Lucretius to Telesio and Spinoza, the nature of the world was currently referred to as *rerum natura*. It was understood that everybody knew the meaning of the term "thing" (*res*). Consequently no definition was attempted; it was a primitive term.

Scholars did not bother to question the *legitimacy* of dividing the world into individual things. They were rather struck by Zeno's paradoxes, which shed a grim light on the very opposite notion of *continuity*. Possibly this distrust prompted Leucippus, Democritus, and their followers to introduce the atoms. In their view, the apparent continuity of matter was only a delusion due to the extremely small size of the atoms.

Be that as it may, the philosophers of antiquity went on conceiving the world as a collection of separate things; moreover, they assumed that everybody shared that tenet. Incidentally, it will be noted that a clear understanding of the terms *subject* and *object* in the modern sense had to wait till about the end of the seventeenth century. At that time, a sort of semantic inversion had taken place. Formerly, the Latin *subjectum* and the greek ὑποκείμενον hinted at what *underlies* the perceptible appearance; consequently they were to some extent closer to the modern understanding of "object" than to that of "subject."

The essential multiplicity of what exists appeared to be so obvious that the Greek philosophers more often than not avoided even the use of a special word for "things" and resorted to a simple plural neuter, as for example in the celebrated "πάντα ῥεί" of Heraclitus. Aristotle (*Physica*) says: "τῶν γὰρ ὄντον τὰ μὲν ἔστι φύσει, τὰ δὲ δί ἄλλας αἰτίας," which in the Loeb Classical Library is translated as: "Some *things* exist, or come into existence, by nature; and some otherwise."[1]

However, Plato and Aristotle sometimes use also the expressions σώματα or σώματα αἰσθητά, referring to physical (i.e., sensible) bodies.

What are physical bodies made of? The notion of matter seems in some way to have been straightforward, as well as the distinction between two kinds of things—namely, intelligible and sensible (or ideal and material). But it is worthwhile to recall Plato (*Timaeus*), who wanted to add a third kind (τρίτον ἄλλο γένος): namely, the "ever-existing *place*— or *space*—which admits not of destruction, and provides room for all things that have birth" (τρίτον δὲ αὖ γενος ὄν τὸ τῆς χώρας ἀεί, φθορὰν οὐ προσδεχόμενον, ἑδραν δὲ παρέχον ὅσα ἔχει γένεσαν πᾶσιν). J. Derrida (in his booklet *Khôra*) has recently called attention to this outstanding intuition of Plato. In this connection a physicist may even be reminded of the modern conception of *vacuum*, which—at variance with the Kantian notion—turns out to be something instead of nothing (or some *thing*, instead of no *thing*).

2. A Human Construction

It is surprising to realize how long it took philosophers to suspect that the notion of "thing" may not be forced on us by nature, but on the contrary may represent a construction imposed on nature by our mind. Saint Augustine (*Confessions*), when describing how he learned to speak from adults, says: "Cum ipsi appellabant *rem* aliquam et cum secundum eam vocem corpus ad aliquid movebant, videbam et tenebam hoc ab eis vocari *rem* illam, quod sonabant, cum eam vellent ostendere." (When they named a certain thing and after that utterance moved their body toward something, I saw and believed that in that way they were calling the thing they wanted to show.)

Wittgenstein (*Philosophical Researches*) rightly remarks that Augustine describes the acquisition of human language as if the baby were already able to think in terms of things. It is, so to speak, a question of translating from one language into another.

Perhaps the reason why our dividing the world into individual objects appears to be necessary may reside in the circumstance that our way of thinking has become *logocentric*, as is maintained by Derrida (*De la Grammatologie*). According to this philosopher, language is but a form of writing (*écriture*), in that it substitutes a set of conventional signs for our thought. Now, verbal language does not consist of a continuous sound emission—as is the case for many animals—but of a sequence of distinct words (or phonemes). This system, adopted by our early ancestors, is probably the best one from the standpoint of communication and information theory. But there is a danger: one may end up attributing to the *signified* (our thought) the structure of the *significant* (our language). Incidentally, this view is supported by the well-known confusion inherent in the word λόγος.

It would be pointless to set up a long list of those philosophers who have taken for granted that the world is by its own nature divided into a plurality of objects. A couple of quotations will suffice. In his *Discours préliminaire* to the great *Encyclopédie*, d'Alembert says: "La première chose que nos sensations nous apprennent... c'est notre existence. La seconde connaissance que nous devons à nos sensations, est l'existence des *objets* extérieurs.... La nature, nous ne saurions trop le repeter, n'est composée que d'*individus*." (The first thing our sensations teach us is our existence. The second notion we owe to our sensations is the existence of the external objects. Nature, we cannot repeat it too often, is only made up of individuals.)

Kant (*Prolegomena*) states: "Nature, considered *materialiter*, is the totality of all the *objects* of experience." Further, in the *Critique of Pure Reason* he speaks of the objects that "strike our senses."

It has been repeatedly asserted that those who think of the world in terms of things and use a "thing-language" are necessarily *realists*. Such belief is not correct. Berkeley (*Essay concerning Human Understanding*) writes: "For as to what is said of the absolute existence of unthinking *things* without any relation to their being perceived, that seems perfectly unintelligible. Their *esse* is *percipi*." Clearly, the father of absolute idealism believed it necessary to speak of things. Therefore the question of *legitimacy* (of dividing the world into things) is to be distinguished from the question of *reality*. Incidentally, it will be recalled that even in the context of transcendental idealism, Kant thought it expedient to introduce the "thing in itself" (*Ding an sich*), although he presented it only as a "limit-concept" (*Grenzbegriff*).

3. Encoding Sensible Data

Why physical objects? Quine has repeatedly argued that man has no other evidence of the existence of physical objects than the fact that assuming them helps us to organize experience. In other words, introducing objects is a convenient way of *encoding* the sense data into something manageable by our mind (or language). Indeed could our logic dispense with individual constants? Of course, the code we have used for centuries is most suitable for dealing with the situations of everyday life. The mental operation of encoding the sense data into separate objects was termed *objectuation* by the present author.[2]

Remark now that animal life was born and is still going on at the surface of the Earth, namely in a very peculiar corner of the universe, where temperature is comparatively close to absolute zero; consequently, in our environment intermolecular binding energies are greater than kT

(Boltzmann's constant multiplied by absolute temperature). This situation is conducive to the formation of the solid state of matter. In fact, when we think of a physical object, we generally have in mind a *solid body*. Accordingly, our code has been developed to suit the situation. If, for the sake of the argument, we made the absurd assumption that another mankind is living on the surface of the Sun, we ought to grant them a way of encoding the sensible world utterly different from our own. They would have no idea of a solid object!

It will be stressed that our code is not immutable, and must evolve along with the evolution of everyday objects. For millennia we had to deal mostly with *natural* ojects, rather than with human artifacts. The code used to represent them (e.g., in palaeolithic caves) was elaborated in accordance. Paul Cézanne once asked, Is there a single straight line in nature? The negative answer to this question can explain why straight lines had virtually no part in the primitive representations.

However, straight geometry gradually gathered importance with the increasing use of artificial objects. Eventually the time came—with the Renaissance—when even architecture largely relied on the simple model of straight lines, right angles, planes, circles, spheres. It is therefore not surprising that an architect, namely Brunelleschi, was one of the great founders of *perspective*, a new code based on the confidence that the observer knew what an artificial environment should be like. With the design of "Italian style" gardens—plane-cut hedges, rectangular flower beds, and the like—the appropriate geometry was introduced into the very heart of nature.

4. Invariants

Is *permanence* or *identity* a prerequisite of our notion of a physical object? Remember that Heraclitus challenged that notion, remarking that "you cannot step twice into the same river." Why did he say the *same* river? Strictly speaking, his sentence seems to imply a contradiction in terms. We can be sure that Heraclitus knew very well—though he did not aknowledge it in modern words—that there were some topological *invariants* in the landscape, such as the two banks of the river, the flow of water in between, the meadows on both sides, that he could still find the next day. Those invariants were more important than the changing "substance" (i.e., the water) to establish the identity of the river. As a matter of fact, the notion that a physical object is essentially a *lump of invariants* has been received in modern times with increasing favor.

Early in the present century a celebrated analysis due to Bertrand Russell (*Our Knowledge of the External World*) was directed to the problem of how an observer constructs a (solid) thing out of the set of different perspectives it can show. "Given an object in one perspective, form the system of all the objects correlated with it in all the perspectives: that system may be identified with the momentary common-sense thing. Thus an aspect of a *thing* is a member of the system of aspects which is the *thing* at that moment." It is intriguing to note that cubist painters (Braque, Picasso) had already found it expedient to present "all" (i.e., many) aspects of the thing simultaneously.

In his definition Russell had relied on the physical law that a rigid body is invariant under any transformation of the rototranslatory group. He soon realized that there was no reason to stop at that particular law, and eventually came to the definition: "*Things* are those series of aspects which obey the laws of physics." This is tantamount to stating that a *thing* preserves the invariants that underlie any physical law.

5. Continuous Bodies

Russell's definition by abstraction—which, incidentally, is somewhat circular—may be all right for solid bodies. As early as the middle seventeenth century, however, physics took a decisive step in a different direction, when Torricelli showed that air has a weight and consequently is a material body in its own right. The mechanics of liquids too was given new attention by Galileo, Torricelli, Bernoulli, and many others. The ground was prepared for the advent of the physics of *continuous* bodies, which was destined to gain great impetus in the nineteenth century, with Fourier, Poisson, Cauchy, and others. A standard method was elaborated with very satisfactory results.

The method consisted in measuring the *initial* and *boundary conditions* of a region of space containing matter, and then predicting its future evolution by applying the differential equations of physics. The next and natural step was the introduction of the notion of *field* by Faraday and Maxwell. In this connection, it is a little surprising that Einstein, when introducing relativity, still spoke of *Elektrodynamik bewegter Körper*; today we would rather speak of moving *systems*. But we must consider that Einstein had mainly in mind the riddle of a wire moving in a magnetic field.

Whatever may be the actual procedure used by humans to pinpoint the bodies of the outer world, it is undoubted that—apart from a few "optical illusions"—the code worked for centuries in a useful and blameless way.

An important landmark was represented by the advent of the telescope and microscope. It is well known that Galileo experienced some difficulty when trying to convince his contemporaries that what they saw in the telescope were no "fallacies" but real objects. In other words, he had to persuade his opponents that the sense data arriving via the telescope could be encoded exactly in the same way as in direct observation. Galileo's opinion was nonetheless to win in the long run: eventually people looking through an optical instrument started to apply with confidence the good old code. But there was a pitfall: encouraged by the success of the telescope and microscope, scholars were tempted to conclude that the code could be extended without limit, both upward to the infinitely large and downward to the infinitely small. That conclusion might have been right—but it was wrong! Such extrapolation turned out to be a failure.

6. Nomological Objects

A turning point of tremendous importance was represented by the discovery of the *nomological* objects of physics—atoms, electrons, protons, and other particles—whose characteristic properties were fixed by *law*. A macroscopic object is *contingent* in his properties: its form is not necessary; its volume, weight, temperature, electric charge, magnetization can take any values. On the contrary, an elementary particle can have only well-determined values for mass, charge, spin, magnetic moment. Incidentally, it will be stressed that in microphysics the problem of the "three-legged" tiger, which has lately worried students of ordinary semantics, simply cannot exist. When physicists discovered an "electron" weighing about 200 times an ordinary electron, they stipulated that it was *not* an electron and termed it *muon*.

About the turn of the century, physicists tried hard to apply the old code to the newly discovered entities. At first there was indeed a strong bias in favor of the construction of "tiny balls" of some sort. As an example, one can mention J. J. Thomson, who is currently credited with having discovered the electron. But, strictly speaking, he did not demonstrate that cathode rays consisted of a swarm of separate particles; he did not go beyond measuring the ratio m/e of the cathode ray "fluid," which could be interpreted as density of mass divided by density of charge.

In the 1920s there came the upsetting discovery of the wave aspects of particles. The dilemma "particles or waves," as well as the struggle that arose about it, can today be discarded as senseless. Neither the code applied to construct billiard balls nor the code applied to ordinary waves could work without leading to inconsistencies and contradictions. Consequently, we cannot escape the conclusion that the new objects are *neither* particles *nor* waves. There is nothing weird in this statement. Indeed we must humbly *learn*, rather than *prescribe*, what nature is made of. Moreover it is really naive to believe that anything obeying a certain differential equation is a physical wave like sound.

There would be no point to describe here the birth and development of quantum mechanics from Planck to de Broglie, to Schrödinger, to Heisenberg, to Dirac. We take for granted that—by and large—the story is known to the reader. The fact that the theory is indeterministic—in the sense that all we can predict is a *probability*—does not per se impair the legitimacy of dividing the world into separate objects.

Still better, one may be tempted to declare that modern science has at long last reached *the* real objects of nature and demonstrated that the physical world *is* constituted by those individual objects. A deeper insight, however, brings us to a far less optimistic conclusion. There are some bewildering features lurking in the theory of modern physics, that are likely to upset any reassuring belief.

7. Identical Particles

Nomological objects of a given kind (say, electrons) are by definition all exactly equal to one another. Now, according to a classical identity principle stressed by Leibniz, two things having exactly all the same properties are one and the same thing; but, surely, two electrons are not one electron!

Of course, if we talk of electrons, we must take into account Pauli's exclusion principle: two or more electrons cannot occupy exactly the same quantum state. For instance, in a nonexcited *He* atom the two electrons, though being both in a $1s$ state, have their spins pointing in opposite directions; however, one cannot tell which points in which direction. Interchanging the two particles has no effect whatsoever on the state of the atom and nobody can tell whether the interchange has taken place.

This is a general feature of identical particles, and the success of quantum statistics (Bose-Einstein and Fermi-Dirac) confirms beyond any doubt such impossibility. All this contradicts Leibniz's statement that *eadem sunt quorum unum potest substitui alteri salva veritate* (if one thing can be substituted to another, without violating truth, they are one and the same thing). What is then the sense (if any) of *genidentity* in microphysics?

What is an electron (or any other elementary particle)? From an *extensional* point of view, an electron is an element of the set of particles we call electrons. But here we stumble on a serious problem: can one deal with *sets* of identical particles in any classical sense? No, because a collection of identical particles has only a *cardinal* number; but its elements cannot be ordered and cannot be named individually. In order to treat adequately the collections of identical particles, M. L. Dalla Chiara and the present author have introduced the notion of *quaset* (quasiset). For further details on this subject, the reader is referred to the literature.[3]

Anyhow, it seems that in subatomic physics it is difficult to preserve the long-standing logical tool of extensional semantics. An *intensional* definition, based on the description of invariant properties, would be better to the point. But there is still some drawback. Intensional logic is so far not very well developed and, as Quine once remarked, intensional and extensional ontologies are like oil and water. There is still much work to be done in this field.

8. The Puzzle of Inseparability

Albert Einstein is generally credited with having introduced in physics a number of prodigious new ideas, like special and general relativity, photons, the theory of brownian motions. But few people seem to have realized that he was a giant even when he was wrong. This is beautifully examplified by the so-called EPR (Einstein, Podolski, and Rosen) paradox, a sort of *Gedankenexperiment*, originally devised in order to disprove the reliability of orthodox (Copenhagen) interpretation of quantum theory. According to the authors, the results of the experiment predicted by quantum mechanics would have been absolutely unacceptable; consequently the theory seemed to be at fault (or to be *incomplete*, as Einstein preferred to say).[4]

For a few decades, lengthy and fierce discussions went on about EPR, till a fundamental theorem was given by J. Bell (1964) and then a *real* (not *Gedanken*) experiment was set up by A. Aspect and others (1982), which eventually decreed the victory of orthodox quantum mechanics.[5]

Without attempting to give a full coverage of the subject—which would involve such fundamental issues as locality and instantaneous propagation—we will limit ourselves to stating the main result of interest in the present context. Two particles, having interacted in the past, form an inseparable whole, even if at present they are far apart. They share a common state vector, consequently a measurement made on one of them entails an instantaneous influence on the possible result of a measurement made on the other.

This finding has a tremendous implication for our discussion. Since any particle has certainly interacted with other particles in the past, the world turns out to be *nonseparable* into individual and independent objects. The world is in some way a single object. But what operational meaning may have that statement? How could one *observe* such an object? Could we stand outside the world and oserve the world?

We are really in a predicament, and it seems impossible not to agree with B. d'Espagnat when he says: "The existence of nonseparability clearly reveals at least a *dissonance* between quantum mechanics and the very notion of space. This is undeniably an important philosophical conclusion. It is surprising that philosophers of science have so far given so little attention to it."[6]

Ironically, there are some comparatively modest developments of science that seem to have deserved the pompous name of *revolutions*, whereas the frightening notion of nonseparability tends to be ignored. Probably, it upsets too much our long-standing and cherished beliefs.

Notes

1. We warn the reader that such terms as "thing," "object," "body," and the like, whenever appearing in a quotation, are printed in italics, regardless of the original.
2. See G. Toraldo di Francia, *Le cose e i loro nomi* (Rome: Laterza, 1986).
3. See M. L. Dalla Chiara and G. Toraldo di Francia, "Individuals, Properties and Truth in the E.P.R. Paradox," in P. Lahti and P. Mittelstaedt, eds., *Symposium on the Foundations of Modern Physics, 1985* (Singapore: World Scientific, 1985), and M. L. Dalla Chiara and G. Toraldo di Francia, "Individuals, Kinds and Names in Physics," in G. Corsi, M. L. Dalla Chiara, and G. C. Ghirardi, eds., *Bridging the Gap: Philosophy, Mathematics, and Physics* (Dordrecht: Kluwer, 1993).
4. A. Einstein, B. Podolski, and N. Rosen, "Can Quantum-Mechanical Description of Physical Reality Be Considered Complete?" *Physical Review* 47 (1935): 777–780.
5. J. S. Bell, "On the Einstein-Podolsky-Rosen Paradox," *Physics* 1 (1964): 195–200; A. Aspect, P. Grangier, and G. Roger, "Experimental Realization of Einstein-Podolsky-Rosen-Bohm *Gedankenexperiment*: A New Violation of Bell's Inequalities," *Physical Review Letters* 48 (1982): 91–94.
6. B. D'Espagnat and E. Klein, *Regards sur la matière* (Paris: Fayard, 1993), 174.

2

Many, but Almost One

David Lewis

The Problem of the Many

Think of a cloud—just one cloud, and around it clear blue sky. Seen from the ground, the cloud may seem to have a sharp boundary. Not so. The cloud is a swarm of water droplets. At the outskirts of the cloud the density of the droplets falls off. Eventually they are so few and far between that we may hesitate to say that the outlying droplets are still part of the cloud at all; perhaps we might better say only that they are near the cloud. But the transition is gradual. Many surfaces are equally good candidates to be the boundary of the cloud. Therefore many aggregates of droplets, some more inclusive and some less inclusive (and some inclusive in different ways than others), are equally good candidates to be the cloud. Since they have equal claim, how can we say that the cloud is one of these aggregates rather than another? But if all of them count as clouds, then we have many clouds rather than one. And if none of them count, each one being ruled out because of the competition from the others, then we have no cloud. How is it, then, that we have just one cloud? And yet we do.

This is Unger's "problem of the many."[1] Once noticed, we can see that it is everywhere, for all things are swarms of particles. There are always outlying particles, questionably parts of the thing, not definitely included and not definitely not included. So there are always many aggregates, differing by a little bit here and a little bit there, with equal claim to be the thing. We have many things or we have none, but anyway not the one thing we thought we had. That is absurd.

Think of a rusty nail, and the gradual transition from steel, to steel with bits of rust scattered through, to rust adhering to the nail, to rust merely resting on the nail. Or think of a cathode, and its departing electrons. Or think of anything that undergoes evaporation or erosion or abrasion. Or think of yourself, or any organism, with parts that gradually come loose in metabolism or excretion or perspiration or shedding

---

[1] P. Unger, "The Problem of the Many", *Midwest Studies in Philosophy* 5 (1980): 411–467.

of dead skin. In each case, a thing has questionable parts, and therefore is subject to the problem of the many.

If, as I think, things perdure through time by having temporal parts, then questionable temporal parts add to the problem of the many. If a person comes into existence gradually (whether over weeks or over years or over nanoseconds doesn't matter for our present purpose) then there are questionable temporal parts at the beginning of every human life. Likewise at the end, even in the most sudden death imaginable. Do you think you are one person?—No, there are many aggregates of temporal parts, differing just a little at the ends, with equal claim to count as persons, and equal claim to count as you. Are all those equally good claims good enough? If so, you are many. If not, you are none. Either way we get the wrong answer. For undeniably you are one.

If, as some think but I do not,[2] ordinary things extend through other possible worlds, then the problem of the many takes on still another dimension. Here in this world we have a ship, the *Enigma*; there in another world is a ship built at about the same time, to plans that are nearly the same but not quite, using many of the same planks and some that are not the same. It is questionable whether the ship in that other world is *Enigma* herself, or just a substitute. If *Enigma* is a thing that extends through worlds, then the question is whether *Enigma* includes as a part what's in that other world. We have two versions of *Enigma*, one that includes this questionable other-worldly part and one that excludes it. They have equal claim to count as ships, and equal claim to count as *Enigma*. We have two ships, coinciding in this world but differing in their full extent. Or else we have none; but anyway not the one ship we thought we had.

The Paradox of 1001 Cats

Cat Tibbles is alone on the mat. Tibbles has hairs $h_1, h_2, \ldots, h_{1000}$. Let c be Tibbles including all these hairs; let c_1 be all of Tibbles except for h_1; and similarly for $c_2, \ldots, c_{1000}$. Each of these c's is a cat. So instead of one cat on the mat, Tibbles, we have at least 1001 cats—which is absurd. This is P. T. Geach's paradox of 1001 cats.[3]

Why should we think that each c_n is a cat? Because, says Geach, "c_n would clearly be a cat were the hair h_n plucked out, and we cannot reasonably suppose that plucking out a hair *generates* a cat, so c_n must already have been a cat" (p. 215). This need not convince us. We can

[2] See D. Lewis, *On the Plurality of Worlds* (Oxford: Blackwell, 1986): 210–220.

[3] P. T. Geach, *Reference and Generality*, 3rd ed. (Ithaca, N.Y.: Cornell University Press, 1980): 215–216.

reply that plucking out h_n turns c_n from a mere proper part of cat Tibbles into the whole of a cat. No new cat is generated, since the cat that c_n becomes the whole of is none other than Tibbles. Nor do c_n and Tibbles ever become identical *simpliciter*—of course not, since what's true about c_n's past still differs from what's true about Tibbles's past. Rather, c_n becomes the whole of cat Tibbles in the sense that c_n's post-plucking temporal part is identical with Tibbles's post-plucking temporal part. So far, so good; except for those, like Geach, who reject the idea of temporal parts. The rest of us have no paradox yet.

But suppose it is spring, and Tibbles is shedding. When a cat sheds, the hairs do not come popping off; they become gradually looser, until finally they are held in place only by the hairs around them. By the end of this gradual process, the loose hairs are no longer parts of the cat. Sometime before the end, they are questionable parts: not definitely still parts of the cat, not definitely not. Suppose each of $h_1, h_2, \ldots, h_{1000}$ is at this questionable stage. Now indeed all of $c_1, c_2, \ldots, c_{1000}$, and also c which includes all the questionable hairs, have equal claim to be a cat, and equal claim to be Tibbles. So now we have 1001 cats. (Indeed, we have many more than that. For instance there is the cat that includes all but the four hairs h_6, h_{408}, h_{882}, and h_{907}.) The paradox of 1001 cats, insofar as it is a real paradox, is another instance of Unger's problem of the many.

To deny that there are many cats on the mat, we must either deny that the many are cats, or else deny that the cats are many. We may solve the paradox by finding a way to disqualify candidates for cathood: there are the many, sure enough, but the many are not all cats. At most one of them is. Perhaps the true cat is one of the many; or perhaps it is something else altogether, and none of the many are cats. Or else, if we grant that all the candidates are truly cats, we must find a way to say that these cats are not truly different from one another. I think both alternatives lead to successful solutions, but we shall see some unsuccessful solutions as well.

Two Solutions by Disqualification: None of the Many Are Cats

We could try saying that not one of the c's is a cat; they are many, sure enough, but not many cats. Tibbles, the only genuine cat on the mat, is something else, different from all of them.

One way to disqualify the many is to invoke the alleged distinction between things and the parcels of matter that constitute them. We could try saying that the *c*'s are not cats. Rather, they are cat-constituting parcels of matter. Tibbles is the cat that each of them constitutes.[4]

This dualism of things and their constituters is unparsimonious and unnecessary. It was invented to solve a certain problem, but a better solution to that problem lies elsewhere, as follows. We know that the matter of a thing may exist before and after the thing does; and we know that a thing may gain and lose matter while it still exists, as a cat does, or a wave or a flame. The dualists conclude that the matter is not the thing; constitution is not identity; there are things, there are the parcels of matter that temporarily constitute those things; these are items of two different categories, related by the special relation of constitution. We must agree, at least, that the temporally extended thing is not the temporally extended parcel of matter that temporarily constitutes that thing. But constitution may be identity, all the same, if it is identity between temporal parts. If some matter constitutes a cat for one minute, then a minute-long temporal segment of the cat is identical to a minute-long temporal segment of the matter. The cat consists entirely of the matter that constitutes it, in this sense: The whole of the cat, throughout the time it lives, consists entirely of temporal segments of various parcels of matter. At any moment, if we disregard everything not located at that moment, the cat and the matter that then constitutes it are identical.[5]

So only those who reject the notion of temporal parts have any need for the dualism of things and constituters. But suppose we accept it all the same. At best, this just transforms the paradox of 1001 cats into the paradox of 1001 cat-constituters. Is that an improvement? We all thought there was only one cat on the mat. After distinguishing Tibbles from her constituter, would we not still want to think there was only one cat-constituter on the mat?

Further, even granted that Tibbles has many constituters, I still question whether Tibbles is the only cat present. The constituters are cat-like in size, shape, weight, inner structure, and motion. They vibrate and set the air in motion—in short, they purr (especially when you pat them).

---

[4] This is the solution advanced in E. J. Lowe,"The Paradox of the 1,001 Cats", *Analysis* 42 (1982): 27–30.

[5] The dualism of things and their constituters is also meant to solve a modal problem: Even at one moment, the thing might have been made of different matter, so what might have been true of it differs from what might have been true of its matter, so constitution cannot be identity. This problem too has a better solution. We should allow that what is true of a given thing at a given world is a vague and inconstant matter. Conflicting answers, equally correct, may be evoked by different ways of referring to the same thing, e.g., as cat or as cat-constituter. My counterpart theory affords this desirable inconstancy; many rival theories do also. See Lewis, *On the Plurality of Worlds*: 248–263.

Any way a cat can be at a moment, cat-constituters also can be; anything a cat can do at a moment, cat-constituters also can do. They are all too cat-like not to be cats. Indeed, they may have unfeline pasts and futures, but that doesn't show that they are never cats; it only shows that they do not remain cats for very long. Now we have the paradox of 1002 cats: Tibbles the constituted cat, and also the 1001 all-too-feline cat-constituters. Nothing has been gained.

I conclude that invoking the dualism of cats and cat-constituters to solve the paradox of 1001 cats does not succeed.

A different way to disqualify the many appeals to a doctrine of vagueness in nature. We could try saying that cat Tibbles is a vague object, and that the *c*'s are not cats but rather alternative precisifications of a cat.

In one way, at least, this solution works better than the one before. This time, I cannot complain that at best we only transform the paradox of 1001 cats into the paradox of 1001 cat-precisifications, because that is no paradox. If indeed there are vague objects and precisifications, it is only to be expected that one vague object will have many precisifications.

If the proposal is meant to solve our paradox, it must be meant as serious metaphysics. It cannot just be a way of saying "in the material mode" that the words "Tibbles" and "cat" are vague, and that this vagueness makes it indefinite just which hairs are part of the cat Tibbles. Rather, the idea must be that material objects come in two varieties, vague and precise; cats are vague, the *c*'s are precise, and that is why none of the *c*'s is a cat.

This new dualism of vague objects and their precisifications is, again, unparsimonious and unnecessary. The problem it was made to solve might better be solved another way. It is absurd to think that we have decided to apply the name "Tibbles" to a certain precisely delimited object; or that we have decided to apply the term "cat" to each of certain precisely delimited objects. But we needn't conclude that these words must rather apply to certain *im*precisely delimited, vague objects. Instead we should conclude that we never quite made up our minds just what these words apply to. We have made up our minds that "Tibbles" is to name one or another Tibbles-precisification, but we never decided just which one; we decided that "cat" was to apply to some and only some cat-precisifications, but again we never decided just which ones. (Nor did we ever decide just which things our new-found terms "Tibbles-precisification" and "cat-precisification" were to apply to.) It was very sensible of us not to decide. We probably couldn't have done it if we'd tried; and even if we could have, doing it would have been useless folly. Semantic indecision will suffice to explain the phenomenon of vagueness.[6] We need no vague objects.

Further, I doubt that I have any correct conception of a vague object. How, for instance, shall I think of an object that is vague in its spatial extent? The closest I can come is to superimpose three pictures. There is the *multiplicity* picture, in which the vague object gives way to its many precisifications, and the vagueness of the object gives way to differences between precisifications. There is the *ignorance* picture, in which the object has some definite but secret extent. And there is the *fadeaway* picture, in which the presence of the object admits of degree, in much the way that the presence of a spot of illumination admits of degree, and the degree diminishes as a function of the distance from the region where the object is most intensely present. None of the three pictures is right. Each one in its own way replaces the alleged vagueness of the object by precision. But if I cannot think of a vague object except by juggling these mistaken pictures, I have no correct conception.[7]

I can complain as before that we end up with a paradox of 1002 cats: Tibbles the vague cat, and also the 1001 precise cats. Once again, the cat-precisifications are all too cat-like. More so than the cat-constituters, in fact: The precisifications are cat-like not just in what they can do and how they can be at a moment, but also over time. They would make good pets—especially since 1001 of them will not eat you out of house and home!

Don't say that the precisifications cannot be cats because cats cannot be precise objects. Surely there could be cats in a world where nature is

[6] Provided that there exist the many precisifications for us to be undecided between. If you deny this, you will indeed have need of vague objects. See P. van Inwagen, *Material Beings* (Ithaca, N.Y.: Cornell University Press, 1990): 213–283.

[7] I grant that the hypothesis of vague objects, for all its faults, can at least be made consistent. If there are vague objects, no doubt they sometimes stand in relations of "vague identity" to one another. We might think that when a and b are vaguely identical vague objects, the identity statement $a = b$ suffers a truth-value gap; but in fact this conception of vague identity belongs to the theory of vagueness as semantic indecision. As Gareth Evans showed, it doesn't mix with the idea that vague identity is due to vagueness in nature. For if a and b are vaguely identical, they differ in respect of vague identity to a; but nothing, however peculiar it may be, differs in any way from itself; so the identity $a = b$ is definitely false. See G. Evans, "Can There be Vague Objects?" *Analysis* 38 (1978): 208 (reprinted in *Collected Papers* [Oxford: Oxford University Press, 1985]). (Evans's too-concise paper invites misunderstanding, but his own testimony confirms my interpretation. See D. Lewis, "Vague Identity: Evans Misunderstood", *Analysis* 48 [1988]: 128–130.) To get a consistent theory of vague objects, different from the bastard theory that is Evans's target, we must disconnect "vague identity" from truth-value gaps in identity statements. Even if $a = b$ is definitely false, a and b can still be "vaguely identical" in the sense of sharing some but not all of their precisifications.

so much less gradual that the problem of the many goes away. It could happen that cats have no questionable parts at all, neither spatial nor temporal. (In this world, when cats shed in the spring, the hairs *do* come popping off.) So it is at least possible that cat-like precise objects are genuine cats. If so, how can the presence of one vague cat spoil their cathood?

I conclude that invoking the dualism of vague objects and their pre-cisifications to solve the paradox of 1001 cats does not succeed.

A Better Solution by Disqualification: One of the Many Is a Cat

Since all of the many are so cat-like, there is only one credible way to deny that all of them are cats. When is something very cat-like, yet not a cat?—When it is just a little less than a whole cat, almost all of a cat with just one little bit left out. Or when it is just a little more than a cat, a cat plus a little something extra. Or when it is both a little more and a little less.

Suppose we say that one of our many is exactly a cat, no more and no less; and that each of the rest is disqualified because it is a little less than a cat, or a little more, or both more and less. This invokes no unparsimonious and unnecessary dualisms; it disqualifies all but one of the many without denying that they are very cat-like; it leaves us with just one cat. All very satisfactory.

The trouble, so it seems, is that there is no saying which one is a cat. That is left altogether arbitrary. Settling it takes a semantic decision, and that is the decision we never made (and shouldn't have made, and maybe couldn't have made). No secret fact could answer the question, for we never decided how the answer would depend on secret facts. Which one deserves the name "cat" is up to us. If we decline to settle the question, nothing else will settle it for us.[8]

We cannot deny the arbitrariness. What we can deny, though, is that it is trouble. What shall we do, if semantic indecision is inescapable, and yet we wish to carry on talking? The answer, surely, is to exploit the fact that very often our unmade semantic decisions don't matter. Often, what you want to say will be true under all different ways of making the

[8] I do not think reference is entirely up to our choice. Some things are by their nature more eligible than others to be referents or objects of thought, and when we do nothing to settle the contest in favor of the less eligible, then the more eligible wins by default; see D. Lewis, "Putnam's Paradox", *Australasian Journal of Philosophy* 62 (1984): 221–236. That's no help here: nature is gradual, no handy joint in nature picks out one of the *c*'s from all the rest.

unmade decision. Then if you say it, even if by choice or by necessity you leave the decision forever unmade, you still speak truthfully. It makes no difference just what you meant, what you say is true regardless. And if it makes no difference just what you meant, likewise it makes no difference that you never made up your mind just what to mean. You say that a famous architect designed Fred's house; it never crossed your mind to think whether by "house" you meant something that did or that didn't include the attached garage; neither does some established convention or secret fact decide the issue; no matter, you knew that what you said was true either way.

This plan for coping with semantic indecision is van Fraassen's method of *supervaluations*.[9] Call a sentence *super-true* if and only if it is true under all ways of making the unmade semantic decisions; *super-false* if and only if it is false under all ways of making those decisions; and if it is true under some ways and false under others, then it suffers a super-truth-value gap. Super-truth, with respect to a language interpreted in an imperfectly decisive way, replaces truth *simpliciter* as the goal of a cooperative speaker attempting to impart information. We can put it another way: Whatever it is that we do to determine the "intended" interpretation of our language determines not one interpretation but a range of interpretations. (The range depends on context, and is itself somewhat indeterminate.) What we try for, in imparting information, is truth of what we say under all the intended interpretations.

Each intended interpretation of our language puts one of the cat candidates on the mat into the extension of the word "cat", and excludes all the rest. Likewise each intended interpretation picks out one cat candidate, the same one, as the referent of "Tibbles". Therefore it is super-true that there is just one cat, Tibbles, on the mat. Because it is super-true, you are entitled to affirm it. And so you may say what you want to say: there is one cat. That is how the method of supervaluations solves the paradox of 1001 cats.

Objection. Just one of the candidates is a cat, no more and no less. But don't try to say which one it is. Nothing you might say would be super-true. For it is exactly this semantic decision that remains unmade; it is exactly in this respect that the intended interpretations differ. Although it is super-true that something is a cat on the mat, there is nothing such that it is super-true of it that *it* is a cat on the mat. (It's like the old puzzle: I owe you a horse, but there's no horse such that I owe you that horse.) This is peculiar.

---

[9] B. van Fraassen, "Singular Terms, Truth-Value Gaps, and Free Logic", *Journal of Philosophy* 63 (1966): 481–495.

Reply. So it is. But once you know the reason why, you can learn to accept it.

Objection.[10] Supervaluationism works too well: it stops us from ever stating the problem in the first place. The problem supposedly was that all the many candidates had equal claim to cathood. But under the supervaluationist rule, that may not be said. For under any one way of making the unmade decision, one candidate is picked as a cat. So under any one way of making the decision, the candidates do *not* have equal claim. What's true under all ways of making the decision is super-true. So what's super-true, and what we should have said, is that the candidates do *not* have equal claim. Then what's the problem? And yet the problem was stated. So supervaluationism is mistaken.

Reply. What's mistaken is a fanatical supervaluationism, which automatically applies the supervaluationist rule to any statement whatever, never mind that the statement makes no sense that way. The rule should instead be taken as a defeasible presumption. What defeats it, sometimes, is the cardinal principle of pragmatics: The right way to take what is said, if at all possible, is the way that makes sense of the message. Since the supervaluationist rule would have made hash of our statement of the problem, straightway the rule was suspended. We are good at making these accommodations; we don't even notice when we do it. Under the supervaluationist rule, it's right to say that there's only one cat, and so the candidates have unequal claim. Suspending the rule, it's right to say that the candidates have equal claim, and that all of them alike are not definitely not cats. Suspending the rule, it's even right to say that they are all cats! Is this capitulation to the paradox?—No; it's no harm to admit that in *some* sense there are many cats. What's intolerable is to be without any good and natural sense in which there is only one cat.

Objection.[11] The supervaluationist's notion of indeterminate reference is conceptually derivative from the prior notion of reference *simpliciter*. But if the problem of the many is everywhere, and semantic indecision is inescapable, then reference *simpliciter* never happens. To the extent that we gain concepts by "fixing the reference" on actual examples, we are in no position to have the concept of reference. Then neither are we in a position to have the derivative concept of indeterminate reference due to semantic indecision.

---

[10] Here I'm indebted to remarks of Saul Kripke many years ago. At his request, I note that what I have written here may not correspond exactly to the whole of what he said on that occasion.

[11] Here I'm indebted to Andrew Strauss (personal communication, 1989).

Reply. We don't need actual examples to have the concept. We have plenty of imaginary examples of reference *simpliciter*, uncomplicated by semantic indecision. These examples are set in sharper worlds than ours: worlds where clouds have no outlying droplets, where cats shed their hairs instantaneously, and so on. When we picked up the concept of reference, in childhood, we probably took for granted that our own world was sharp in just that way. (When not puzzling over the problem of the many, maybe we half-believe it still.) We fixed the reference of "reference" on these imaginary examples in the sharp world we thought we lived in—and if any theory of reference says that cannot be done, so much the worse for it.

I conclude that the supervaluationist solution to the paradox of 1001 cats, and to the problem of the many generally, is successful. But is it the only successful solution?—I think not. I turn now to the other sort of solution: the kind which concedes that the many are cats, but seeks to deny that the cats are really many.

Relative Identity: The Many Are Not Different Cats

Geach himself favors one such solution. The paradox of 1001 cats serves as a showcase for his doctrine of relative identity.

> Everything falls into place if we realize that the number of cats on the mat is the number of *different* cats on the mat; and c_{13}, c_{279}, and c are not three different cats, they are one and the same cat. Though none of these 1001 lumps of feline tissue is the same lump of feline tissue as another, each is the same cat as any other: each of them, then, is a cat, but there is only one cat on the mat, and our original story stands.... The price to pay is that we must regard "_____ is the same cat as_____" as expressing only a certain equivalence relation, not an absolute identity restricted to cats; but this price, I have elsewhere argued, must be paid anyhow, for there is no such absolute identity as logicians have assumed.[12]

"Same cat" is a relation of partial indiscernibility, restricted to respects of comparison somehow associated with the term "cat", and discernibility by just a few hairs doesn't count. "Same lump of feline tissue" is a different relation of partial indiscernibility, and a more discerning one.

I agree that sometimes we say "same", and mean by it not "absolute identity" but just some relation of partial indiscernibility. I also agree that sometimes we count by relations of partial indiscernibility. As I

---

[12] Geach, *Reference and Generality*: 216.

once wrote:

> If an infirm man wishes to know how many roads he must cross to reach his destination, I will count by identity-along-his-path rather than by identity. By crossing the Chester A. Arthur Parkway and Route 137 at the brief stretch where they have merged, he can cross both by crossing only one road.[13]

I'll happily add that for that brief stretch, the two roads are the same. But though I don't object to this positive part of Geach's view, it doesn't ring true to apply it as he does to the case of the cats.

If you ask me to say whether c_{13}, c_{279}, and c are the same or different, I may indeed be of two minds about how to answer. I might say they're different—after all, I know how they differ! Or I might say they're the same, because the difference is negligible, so I duly ignore it. (Not easy to do while attending to the example as I now am; if I attend to my ignoring of something, *ipso facto* I no longer ignore it.) But if you add the noun phrase, either "same cat" or "same lump of feline tissue", it seems to me that I am no less hesitant than before. Just as I was of two minds about "same", so I am still of two minds about "same cat" and "same lump of feline tissue".

Other cases are different. If you ask me "same or different?" when you hold Monday's *Melbourne Age* in one hand and Tuesday's *Age* in the other, or when you hold one Monday *Age* in each hand, again I won't know how to answer. But if you ask me "same or different newspaper?" or "same or different issue?" or "same or different copy?" then I'll know just what to say. We can dispute his explanation of what happens, but at least the phenomenon happens exactly as Geach says it does. Not so, I think, for the case of "same cat" versus "same lump".

Something else is lacking in Geach's solution. In other cases where it comes natural to count by a relation other than identity, it seems that identity itself—"absolute identity"—is not far away. Local identity, as between the Arthur Parkway and Route 137 for the stretch where they have merged, is identity *simpliciter* of spatial parts. Likewise temporary identity, as between a thing and the matter that temporarily constitutes it, is identity *simpliciter* of temporal parts. Qualitative identity is identity *simpliciter* of qualitative character. The newspaper that Monday's *Age* is an issue of and the newspaper that Tuesday's *Age* is an issue of are identical *simpliciter*; likewise my copy and your copy of Monday's *Age* are copies of the identical issue. But Geach never tells us what the "same cat" relation has to do with identity *simpliciter*.

---

[13] D. Lewis, "Survival and Identity", in A. Rorty, ed., *The Identities of Persons* (Berkeley: University of California Press, 1976): 27.

He wouldn't, of course, because he thinks "there is no such absolute identity as logicians have assumed." (Nor would he accept all my examples above; certainly not the one about temporary identity and identity of temporal parts.) But Geach's case against absolute identity is unconvincing. It seems to come down to a challenge: If Geach is determined to construe all that I say in terms of relations of partial indiscernibility, is there any way I can stop him? Can I *force* him to understand? (What's more, can I do it with one hand tied behind my back? Can I do it, for instance, without ever using the second-order quantification that Geach also challenges?)[14] I suppose not. But I don't see why that should make me doubt that I know the difference between identity and indiscernibility.

We have the concept of identity, *pace* Geach; and if we are to justify denying that the cats are many, we need to show that they are interrelated by a relation closely akin to identity itself. Geach has not shown this, and wouldn't wish to show it. Nevertheless it can be shown, as we shall soon see. But at that point we shall have a solution that bypasses Geach's doctrine of relative identity altogether.

Partial Identity: The Many Are Almost One

What is the opposite of identity? *Non*-identity, we'd offhand say. Anything is identical to itself; otherwise we have two "different" things, two "distinct" things; that is, two non-identical things. Of course it's true that things are either identical or non-identical, and never both. But the real opposite of identity is distinctness: not distinctness in the sense of non-identity, but rather distinctness in the sense of non-overlap (what is called "disjointness" in the jargon of those who reserve "distinct" to mean "non-identical"). We have a spectrum of cases. At one end we find the complete identity of a thing with itself: It and itself are entirely identical, not at all distinct. At the opposite end we find the case of two things that are entirely distinct: They have no part in common. In between we find all the cases of partial overlap: things with parts in common and other parts not in common. (Sometimes one of the overlappers is part of the other, sometimes not.) The things are not entirely identical, not entirely distinct, but some of each. They are partially identical, partially distinct. There may be more overlap or less. Some cases are close to the distinctness end of the spectrum: Siamese twins who share only a finger are almost completely distinct, but not quite. Other cases are close to the identity end. For instance, any two of our cat-candidates overlap

---

[14] P. T. Geach, "Identity", *Review of Metaphysics* 21 (1967): 3–12. (Reprinted in *Logic Matters* [Oxford: Blackwell, 1972].)

almost completely. They differ by only a few hairs. They are not quite completely identical, but they are almost completely identical and very far from completely distinct.

It's strange how philosophers have fixed their attention on one end of the spectrum and forgotten how we ordinarily think of identity and distinctness. You'd think the philosophers of common sense and ordinary language would have set us right long ago, but in fact it was Armstrong who did the job.[15] Overshadowed though it is by Armstrong's still more noteworthy accomplishments, this service still deserves our attention and gratitude.

Assume our cat-candidates are genuine cats. (Set aside, for now, the supervaluationist solution.) Then, strictly speaking, the cats are many. No two of them are completely identical. But any two of them are almost completely identical; their differences are negligible, as I said before. We have many cats, each one almost identical to all the rest.

Remember how we translate statements of number into the language of identity and quantification. "There is one cat on the mat" becomes "For some x, x is a cat on the mat, and every cat on the mat is identical to x". That's false, if we take "identical" to express the complete and strict identity that lies at the end of the spectrum. But the very extensive overlap of the cats does approximate to complete identity. So what's true is that for some x, x is a cat on the mat, and every cat on the mat is almost identical to x. In this way, the statement that there is one cat on the mat is almost true. The cats are many, but almost one. By a blameless approximation, we may say simply that there is one cat on the mat. Is that true?—Sometimes we'll insist on stricter standards, sometimes we'll be ambivalent, but for most contexts it's true enough. Thus the idea of partial and approximate identity affords another solution to the paradox of 1001 cats.

The added noun phrase has nothing to do with it. Because of their extensive overlap, the many are almost the same cat; they are almost the same lump of feline tissue; and so on for any other noun phrase that applies to them all. Further, the relation of almost-identity, closely akin to the complete identity that we call identity *simpliciter*, is not a relation of partial indiscernibility. Of course we can expect almost-identical things to be very similar in a great many ways: size, shape, location, weight, purring, behavior, not to mention relational properties like location and ownership. But it is hard to think of any very salient respect in which almost-identical things are guaranteed to be entirely indiscernible. Finally, the relation of almost-identity, in other words extensive overlap, is

---

[15] D. M. Amstrong, *Universals and Scientific Realism* (Cambridge: Cambridge University Press, 1978), Vol. 2: 37–38.

not in general an equivalence relation. Many steps of almost-identity can take us from one thing to another thing that is entirely distinct from the first. We may hope that almost-identity, when restricted to the many cats as they actually are, will be an equivalence relation; but even that is not entirely guaranteed. It depends on the extent to which the cats differ, and on the threshold for almost-identity (and both of these are matters that we will, very sensibly, leave undecided). What this solution has in common with Geach's is just that we count the cats by a relation other than strict, "absolute" identity. Beyond that, the theories differ greatly.[16]

One Solution Too Many?

We find ourselves with two solutions, and that is one more than we needed. Shall we now choose between the way of supervaluation and the way of partial identity? I think not. We might better combine them. We shall see how each can assist the other.

Here is how to combine them. In the first place, there are two kinds of intended interpretations of our language. Given many almost-identical cat-candidates, some will put every (good enough) candidate into the extension of "cat"; others will put exactly one. Context will favor one sort of interpretation or the other, though not every context will settle the matter. Sometimes, especially in our offhand and unphilosophical moments, context will favor the second, one-cat sort of interpretation; and then the supervaluation rule, with nothing to defeat it, will entitle us to say that there is only one cat. But sometimes, for instance when we have been explicitly attending to the many candidates and noting that they are equally catlike, context will favor the first, many-cat sort of interpretation. (If we start with one-cat interpretations, and we say things that the supervaluation rule would make hash of, not only is the rule suspended but also the many-cat interpretations come into play.) But even then, we still want some good sense in which there is just one cat (though we may want a way to say the opposite as well). That is what almost-identity offers.

[16] There is another way we sometimes count by a relation other than strict identity. You draw two diagonals in a square; you ask me how many triangles; I say there are four; you deride me for ignoring the four large triangles and counting only the small ones. But the joke is on you. For I was within my rights as a speaker of ordinary language, and you couldn't see it because you insisted on counting by strict identity. I meant that, for some w, x, y, z, (1) $w, x, y,$ and z are triangles; (2) w and x are distinct, and...and so are y and z (six clauses); and (3) for any triangle t, either t and w are not distinct, or...or t and z are not distinct (four clauses). And by "distinct" I meant non-overlap rather than non-identity, so what I said was true.

This is one way that almost-identity helps a combined solution. It is still there even when we discuss the paradox of 1001 cats, and we explicitly choose to say that the many are all cats, and we thereby make the supervaluation solution go away.

Perhaps it helps in another way too. The supervaluation rule is more natural in some applications than in others. For instance it seems artificial to apply it to a case of unrelated homonyms. "You said you were going to the bank. Is that true? No worries, you bank at the ANZ, it's right down by the river, so what you said was true either way!"—I don't think such a response is utterly forbidden, but it's peculiar in a way that other applications of the supervaluation rule are not. The two interpretations of "bank" are so different that presumably you did make up your mind which one you meant. So the means for coping with semantic indecision are out of place. The supervaluation rule comes natural only when the alternative interpretations don't differ too much. If they are one-cat interpretations that differ only by picking almost-identical cats, that's one way for them not to differ much.

How, on the other hand, do supervaluations help the combined solution? Why not let almost-identity do the whole job?

For one thing, not every case of the problem of the many is like the paradox of 1001 cats. The almost-identity solution won't always work well.[17] We've touched on one atypical case already: if not a problem of the many, at least a problem of two. Fred's house taken as including the garage, and taken as not including the garage, have equal claim to be his house. The claim had better be good enough, else he has no house. So Fred has two houses. No! We've already seen how to solve this problem by the method of supervaluations. (If that seemed good to you, it shows that the difference between the interpretations was not yet enough to make the supervaluation rule artificial.) But although the two house-candidates overlap very substantially, having all but the garage in common, they do not overlap nearly as extensively as the cats do. Though they are closer to the identity end of the spectrum than the distinctness end, we cannot really say they're almost identical. So likewise we cannot say that the two houses are almost one.

For another thing, take a statement different from the statements of identity and number that have concerned us so far. Introduce a definite description: "The cat on the mat includes hair h_{17}." The obvious response to this statement, I suppose, is that it is gappy. It has no definite truth-value, or no definite super-truth-value, as the case may be. But how can we get that answer if we decide that all the cat-candidates are

---

[17] Here I'm indebted to Phillip Bricker (personal communication, 1990).

cats, forsake supervaluations, and ask almost-identity to do the whole job? We might subject the definite description to Russellian translation:

(R1) There is something that is identical to all and only cats on the mat, and that includes h_{17}.

Or equivalently:

(R2) Something is identical to all and only cats on the mat, and every cat on the mat includes h_{17}.

Both these translations come out false, because nothing is strictly identical to all and only cats on the mat. That's not the answer we wanted. So we might relax "identical" to "almost identical". When we do, the translations are no longer equivalent: (R1)-relaxed is true, (R2)-relaxed is false. Maybe we're in a state of semantic indecision between (R1)-relaxed and (R2)-relaxed; if so, we could apply the supervaluation rule to get the desired gappiness. Or we might apply the supervaluation rule more directly. Different one-cat interpretations pick out different things as the cat, some that include h_{17} and some that don't. Under any particular one-cat interpretation the Russellian translations are again equivalent, and different one-cat interpretations give them different truth-values; so the translations, and likewise the original sentence, suffer super-truth-value gaps. Or more simply, different one-cat interpretations differ in the referent of "the cat"; some of these referents satisfy "includes h_{17}" and some don't, so again we get a super-truth-value gap. Whichever way we go, supervaluations give us the gappiness we want. It's hard to see how else to get it.

3

Part and Whole in Quantum Mechanics

Tim Maudlin

It seems to be widely believed that there is a prevalent metaphysical doctrine that a whole is nothing but the sum of its parts. Or, at the least, it is remarked of some objects that they are more than the sum of their parts, as if this were a remarkable and unusual circumstance. It is also widely believed that the aim of modern physics, and perhaps of all modern science, is to understand wholes in terms of their parts. This is sometimes called "reductionism." Reductionism is to be opposed to "holism," according to which a whole is something more than the sum of its parts, or has properties that cannot be understood in terms of the properties of the parts. It is further often stated that contemporary physics, specifically quantum mechanics, surprisingly incorporates a form of holism absent from classical physics.

The most obvious attempts to make the reductionist credo clear, though, run into immediate difficulties. In what sense could my pocket watch be thought of as nothing but the sum of its parts? If the watch is just the sum of its parts, then it seems to follow that at any time the parts of the watch exist, the very same sum of those parts will exist, and hence the watch as a whole will exist. Furthermore, at any time the parts of the watch have all the same properties as they do now, the sum will have the same properties, and hence the watch as a whole. But, in at least one way of understanding these claims, they are plainly false. If I take the watch apart, I may be left with a heap consisting of its parts, the parts themselves being unchanged. It seems as if we now have the same parts but a different "sum": in one case a watch, in the other a pile of junk. Whether one wants to say that the watch exists at all after it has been disassembled is moot, but it is certain that the collection of parts, taken as a whole, no longer has all of the properties that the watch had. Yet whatever one means by a "sum" in this metaphorical sense, it ought to be the case that given the same summands, one gets the same sum. But can't I have all the parts of the watch yet not have the watch?

This collapse of reductionism seems rather too easy. Clockwork mechanisms are paradigms of complex objects whose workings are explained and understood by examination and analysis of their parts. If the sorts of explanations one has of mechanical clocks are not reductionist, then no scientific enterprise ever has been, or could be, reductionist. Indeed, holism is commonly opposed to the "mechanical" or "clockwork" picture of nature supposedly inherited from the Scientific Revolution. So we had best revive reductionism if its defeat at the hands of quantum mechanics is to mean anything at all.

One attempt to revive reductionism would insist that the summands are not in fact unchanged once the watch has been taken apart, since after disassembly they lose some of their *relational* properties. When the watch is put together, for example, the winding stem has the property of being attached to the mainspring, while the winding stem in the pile of junk fails to have this property. But the admission of such relational properties clearly casts the net too wide. One relational property the mainspring has is that of being a certain distance from the Eiffel Tower. So if in the "sum" of the parts we include all facts that can be deduced from even the relational properties of the parts, then not only the state of the watch but the state of the Eiffel Tower (and ultimately of the entire universe) is determined entirely by the state of the parts of the watch.

Perhaps we ought to try this: the watch is the sum of its parts iff all of the facts about the watch are determined by the intrinsic (nonrelational) state of the parts together with the relations between those parts. This formulation gets closer but fails due to the unclarity of the term "relation." A holist about watches presumably would say that although there are some facts determined by the intrinsic properties of the parts together with their relations, there are some further facts not so determined. The holist would say that there are properties of the watch *as a whole* that are not fixed by the state and relational disposition of its parts. But what is to prevent the reductionist from responding that these supposed holistic properties of the watch are nothing more than further relations of the parts? The holist says that, when put together, the watch acquires a heretofore unsuspected property, the reductionist that the parts instantiate a heretofore unsuspected relation. Any surprising behavior of the watch that one might ascribe to the holistic property could as easily be attributed to the new relation. The grammatical subjects of the two locutions differ, but it is not easy to see exactly what the supposed metaphysical distinction amounts to. It is further unclear what could possibly count as evidence favoring one locution over the other.

For reductionism to have any bite, then, some constraints must be put on the allowed relations between the parts of an object. And imme-

diately an obvious, plausible constraint presents itself: the only fundamental external relations between parts of an object are their spatial (or
spatiotemporal) relations. As a first approximation, this view asserts that
any whole can be divided into parts, each of which may be characterized by an intrinsic, nonrelational physical state, and further *all* physical
properties of the whole supervene on the nonrelational physical states
of the parts together with the spatial relations between the parts.

In the case of the watch, the commonly mentioned parts, the gears,
spring, and so forth, are all themselves spatially extended objects with
parts of their own. Proceeding in the same way, reductionism would con-
tend that the physical state of a gear is completely fixed by the physical
states of the parts of the gear together with the spatial relations between those parts. The analysis of parts into their parts will obviously
end only with the introduction of parts that themselves have no further
parts. Since we are analyzing objects into spatial parts, these partless
parts must be spatially unextended: they must be points. So we arrive finally at field theory: a classical field is specified by the attribution of a
physical quantity to every point in a region of space.

The clearest exposition of this form of reductionism may be found in
one of Einstein's letters to Max Born:

> If one asks what, irrespective of quantum mechanics, is characteris-
> tic of the world of ideas of physics, one is first of all struck by the
> following: the concepts of physics relate to a real outside world, that
> is, ideas are established relating to things such as bodies, fields, etc.,
> which claim "real existence" that is independent of the perceiving
> subject—ideas which, on the other hand, have been brought into as
> secure a relationship as possible with the sense-data. It is further
> characteristic of these physical objects that they are thought of as
> arranged in a space-time continuum. An essential aspect of this ar-
> rangement of things in physics is that they lay claim, at a certain
> time, to an existence independent of one another, provided these
> objects "are situated in different parts of space." Unless one makes
> this kind of assumption about the independence of the existence
> (the "being-thus") of objects which are far apart from one another
> in space—which stems in the first place from everyday thinking—
> physical thinking in the familiar sense would not be possible. It is
> also hard to see any way of formulating and testing the laws of
> physics unless one makes a clear distinction of this kind. This prin-
> ciple has been carried to extremes in the field theory by localizing
> the elementary objects on which it is based and which exist indepen-
> dently of each other, as well as the elementary laws which have been
> postulated for it, in the infinitely small (four-dimensional) elements
> of space.[1]

Einstein's presentation is notable both for the striking clarity of the fundamental ontological posit and for the distressing obscurity of his grounds for believing that the failure of this form of reductionism would spell the end of science. Presumably he was worried by the specter of a truly radical holism in which the physical state of any portion of space-time could not be specified without somehow taking account of every other portion of space-time, that is, of the whole universe. Be that as it may, quantum mechanics has been precisely formulated and rigorously tested, so if it indeed fails to display the structure Einstein describes, it also immediately refutes his worries.

From the viewpoint of general ontology, the central issue is not testability, but rather the question of whether the totality of all facts about the universe is determined by a salient subset of those facts. One thesis discussed in the philosophical literature, commonly associated with Hume, is that all the global facts about the universe supervene on the totality of local facts. Hume held that natural necessity could be reduced to nothing more than constant conjunction, so that specification of the propinquity and temporal order of events would automatically fix all facts of causation and natural law. By extension, any "Humean" theory maintains that some facts are local (i.e., some facts are determined completely by the intrinsic state of a limited space-time region) and that any two worlds that agree on all local matters also must agree on everything else. (Thus two Humean worlds cannot disagree on their laws without disagreeing on what happens in some smallish region of space-time.) Humean accounts of law and causation have been very popular, and presumably much of their appeal has been exactly the ontological reductionism implicit in them. All that there is is supposed to follow from the collection of local facts, together, of course, with the spatiotemporal relations between the local regions. We recover Einstein's view again, but without direct recourse to claims about wholes and parts.

Whether regarded as reductionism or Humeanism or the denial of holism, the attempt to account for the world by dividing it into spatial regions and "summing" has a long and illustrious history. If quantum mechanics refutes this program, then the quantum revolution goes much deeper than merely the rejection of determinism, or of the determinateness of all physical properties. Both indeterminism and Bohr's philosophy of complementarity are still compatible with reductionism. David Bohm has long contended that what is radically new about the quantum theory is the "undivided wholeness" that it posits, and if Bohm is right, philosophical commentaries on the quantum theory have long been preoccupied with the wrong features of the theory.

Quantum Threats to Reductionism

We now have a reasonably clear question: according to the quantum theory, can the physical state of a system be completely specified by the attribution of physical states to the spatial parts of the system, together with facts about how those parts are spatiotemporally related? The quantum formalism poses an immediate problem here, which arises from its treatment of the spatial location of particles. Classical particles are always definitely located somewhere or other, so the division of a complex classical object into its spatial parts automatically entails a corresponding division of it into sets of particles, namely the particles present in each spatial part. But the quantum wave function from which one extracts predictions about particle locations is typically widely spread out. On the standard interpretations of quantum theory, this means that the particles themselves have no definite location at all, so the notion of dividing a complex of many particles into its spatial parts becomes problematic. Worse, the wave function for a collection of particles isn't even defined on physical space: it is defined on the configuration space of the system. In the case of a single particle, this configuration space is isomorphic to physical space, so one can be lulled into the mistaken notion that at least the wave function can be analyzed into spatial parts. But as soon as more than one particle is involved, even this becomes impossible: one cannot ask for the state of the wave function "here" or "there" (pointing to regions of physical space) since, for a collection of N particles, the wave function lives in a $3N$ dimensional space which cannot be pointed at.

I want to leave these problems aside. In part, this is because some of the interpretation involved is controversial. In Bohm's theory, for example, particles always have a determinate location, so picking out a region of space does pick out a definite set of particles that inhabit it. In part it is because these problems are tied in a deep way to the particle ontology. In a field ontology, the fundamental physical properties are exactly the physical states associated with space-time points, so problems of indefinite location cannot arise (or cannot arise in the same way). But, mostly, I want to leave questions of spatial location aside since they don't seem to lie at the heart of the quantum holism. The problems with reductionism appear even when the spatial location of objects is not at issue. Too much attention to the location of particles might mislead us into misidentifying the real root of the holism of quantum theory.

Let us turn, instead, to spin. The property of spin is associated with a Hilbert space. For a single spin-1/2 particle, such as an electron, the Hilbert space is a two-dimensional complex space. What sorts of spin properties might such a single electron have?

The answer to this question is already in dispute. One account holds that the spin of a single particle is represented by a ray in the associated Hilbert space. Since each such ray is the eigenvector for some spin operator, this account would entail that any single electron must always have a definite spin in some direction. Let us call this the *ray* view. Opposed to the ray view is the *statistical operator* view, which maintains that spin states are to be associated with statistical operators in the Hilbert space. Statistical operators are akin to convex sums of rays, but cannot be identified with them since the mapping from convex sums to statistical operator is many-one. So, for example, consider first the eigenstates of z-spin and of x-spin: $|z\text{-up}\rangle$, $|z\text{-down}\rangle$, $|x\text{-up}\rangle$ and $|x\text{-down}\rangle$. If one knew that a particle definitely was either $|z\text{-up}\rangle$ or $|z\text{-down}\rangle$, but was entirely uncertain which, one would naturally represent the *state of one's knowledge* by the convex sum $(.5)|z\text{-up}\rangle + (.5)|z\text{-down}\rangle$, or perhaps more perspicuously by $\{(.5)|z\text{-up}\rangle, (.5)|z\text{-down}\rangle\}$. If one had to calculate expected values for any operator given this state of knowledge, one would calculate the expected value for $|z\text{-up}\rangle$, the expected value for $|z\text{-down}\rangle$, and average the two (with obvious adjustments if one's state of uncertainty were instead $(.9)|z\text{-up}\rangle$ and $(.1)|z\text{-down}\rangle$). Thus any convex sum $\{\alpha|A\rangle + \beta|B\rangle + \gamma|C\rangle \ldots\}$ naturally leads to an expectation value for every Hermitean operator on the Hilbert space. By Gleason's theorem, every such set of expectation values is associated with some statistical operator on the space (if the space has more than two dimensions). The converse, however, does not hold. If I knew instead that the electron is either in the state $|x\text{-up}\rangle$ or $|x\text{-down}\rangle$, with a 50 percent chance of each, I would represent the state of my knowledge by $\{(.5)|x\text{-up}\rangle, (.5)|x\text{-down}\rangle\}$. This state yields identical expectation values for all Hermitean operators as $\{(.5)|z\text{-up}\rangle, (.5)|z\text{-down}\rangle\}$, and so is associated with the same statistical operator. Statistical operators are therefore useful but blunt instruments for representing purely epistemic uncertainty: if I am unsure about the state of a system, some statistical operator will tell me how I should place bets about the outcomes of measurements, but my uncertainty will have a finer structure than is captured in the statistical operator.

The issue before us, though, is not subjective uncertainty but physical states. Quite apart from what anyone knows, can a single electron be in a physical state properly represented by a statistical operator (a mixed state) or only by a ray in the Hilbert space (a pure state)? To put the question in its most stark and dramatic terms, if there were only a single particle in the whole universe, would its spin state have to be a ray in its associated Hilbert space or not?

The ray view and the statistical operator view simply disagree here. For example, according to the ray view, such an isolated particle would have to be in an eigenstate of spin in some direction, while according to the statistical operator view it would not. I do not know of any way to establish decisively either view. There are interesting consequences of each view, when combined with other elements of an interpretation of quantum mechanics. For example, if one accepts a collapse postulate according to which measurements eventuate in a collapse of the wave function to an eigenstate of the measured observable, then the statistical operator view entails that the collapses can be physically irreversible. For in a world with a single electron, there may be a physical process (collapse upon measurement of spin) that takes the spin state of the particle from a mixed state to a pure one, but no possible physical process (either Schrödinger evolution or collapse) that can take the pure state back to the mixed. Similarly, on the statistical operator view, a particle can be in a physical state even though there is no possible way to take a single particle and prepare it in such a state (since state preparation is achieved by collapses). It is perhaps also of interest to note that if no compelling reason can be given for preferring the ray view to the statistical operator view, then we have found a new source of radical indeterminacy in physical theory: no experiments that we can do could ever assure us about the physically possible spin states of an isolated electron. However, our purpose here is not to adjudicate between the ray and statistical operator views, but only to note them, and their consequences in the sequel.

Let us move on to pairs of particles. Let us suppose we have two particles, and that the particles are easily distinguishable (e.g., an electron and a proton), and that they unproblematically occupy distinct regions of space (one is in New York, the other in San Francisco). By Einstein's criterion, they should each have an independent "being-thus"—that is, an independent physical state—and, according to reductionism, the physical state of the pair should be nothing but the sum of the physical states of the parts. How does quantum theory represent the spin states of this pair? The pair of particles is associated with a four-dimensional complex Hilbert space. This space contains some states that can be easily interpreted. So, for example, if particle 1 is definitely z-spin up and particle 2 is definitely x-spin down, the Hilbert space contains a ray we can write as $|z\text{-up}\rangle_1|x\text{-down}\rangle_2$ which represents just this state of affairs. Indeed, for any ray in the Hilbert space of particle 1 and any other ray in the Hilbert space of particle 2, the joint Hilbert space contains a ray that uniquely represents their product. In these cases, the state of the whole is reducible to the states of the parts.

But the joint Hilbert space contains other states as well. Consider the four unproblematic product states $|z\text{-up}\rangle_1|z\text{-up}\rangle_2$, $|z\text{-down}\rangle_1|z\text{-up}\rangle_2$, $|z\text{-up}\rangle_1|z\text{-down}\rangle_2$, and $|z\text{-down}\rangle_1|z\text{-down}\rangle_2$. These four vectors span the joint Hilbert space, which means that the space contains not only them, but all (normalized) linear combinations of them. In particular, the joint space contains states like the singlet state

$$S: 1/\sqrt{2}|z\text{-up}\rangle_1|z\text{-down}\rangle_2 - 1/\sqrt{2}|z\text{-down}\rangle_1|z\text{-up}\rangle_2.$$

Both the ray view and the statistical operator view agree that such a joint state exists. The question then is: if the pair of particles is in spin state S, and the particles occupy distinct and distant spatial locations, what can be said of the spin states of the particles taken individually? Does each individual particle have a spin state and, if so, what is it (or what might it be), and further is the state of the pair nothing over and above the states of the parts taken individually?

Here the analyses of the two views may partially diverge. Let us begin with the ray view. The state S is obviously not an eigenstate of the z-spin of either particle. If the z-spin of particle 1 is measured, the result might be up or might be down. And, although it is not obvious, the same holds for spin in any other direction. Indeed, the state S can alternatively be written as

$$S: 1/\sqrt{2}|x\text{-up}\rangle_1|x\text{-down}\rangle_2 - 1/\sqrt{2}|x\text{-down}\rangle_1|x\text{-up}\rangle_2,$$

or

$$S: 1/\sqrt{2}|y\text{-up}\rangle_1|y\text{-down}\rangle_2 - 1/\sqrt{2}|y\text{-down}\rangle_1|y\text{-up}\rangle_2,$$

and so forth, being expressed in terms of the spin in any direction. For any direction one chooses, the state S predicts that there is a 50 percent chance of spin up and 50 percent chance of spin down for each particle.

Since on the ray view the spin states of isolated particles are always eigenstates of spin in some direction, that view must insist that the particles in state S are not, taken individually, in states that an isolated particle could possibly be in. The proponent of the ray view can adopt two modes of speech here, although they amount to the same thing. According to the *no-state* mode, each particle in the state S has no spin state at all. Only the pair taken together has a state, which is, appropriately, represented by a ray in the Hilbert space of the joint system. According to the *joint-state* mode, each particle has a spin state, but one that can only be specified by referring to the other particle. Thus one could say that the spin state of particle 1 is just that given by being particle 1 in the state S. In this case the spin state of particle 1 simply cannot be specified without mentioning particle 2, or at least without mentioning the existence of some other particle.

Whichever mode of speech the ray view theorist adopts, reductionism is dead. For the total physical state of the joint system cannot be regarded as a consequence of the states of its (spatially separated) parts, where the states of the parts can be specified without reference to the whole. According to the no-state mode of speech, neither of the parts has a spin state, so the joint state can clearly not be recovered from the information that neither part has a state. And according to the joint-state mode, neither part has a state that can be specified without reference to the other part. The ray view implies that no isolated particle (i.e., no particle that is the sole particle in a world) can exist with the same spin state as either of the members of S. So S simply cannot be seen as being built up from single-particle states, where single-particle states are possible states of a sole existing particle.

The radical failure of reductionism on the ray view may seem to militate in favor of the statistical operator view. And indeed, on the statistical operator view, things are not quite so bad. Given the state S, one can calculate expectation values for all of the Hermitean operators in the spin space of each single particle. The calculation is quite simple: for any given direction, there is a 50 percent chance that the spin will be found up and a 50 percent chance that it will be found down. Again, there is a statistical operator on the spin space for each individual particle that yields just these expectation values. So the statistical operator view can maintain that each individual particle has a spin state even when the joint state is S. Indeed, the spin state of one particle is just that which arises from "tracing over" the degrees of freedom associated with the other particle in the familiar way. And the statistical operator view can also maintain that the resulting spin state is one the particle could have had even if it had been the only particle existing in the world, even though no state preparation could have produced a particle in such a state.

But even the statistical operator view does not recover reductionism. For although each particle in S has a state, and even though each could continue to have that state in the absence of the other, still S does not supervene on the states of the parts.

The easiest way to see this is to note that, on any view, the state S is distinct from

$$S': 1/\sqrt{2}|z\text{-up}\rangle_1|z\text{-up}\rangle_2 - 1/\sqrt{2}|z\text{-down}\rangle_1|z\text{-down}\rangle_2.$$

S and S' make different predictions for some observables on the joint system. In particular, if one measures the z-spin of both particles, then S entails that those measurements will have different outcomes (one up and the other down), while S' predicts that they will have the same outcome. But the statistical operators one gets from S' for particle 1 and particle 2 are identical to those one gets from S. So even on the statistical operator view, the state of the joint system does not reduce to, or supervene on, the states of the parts. Given the states of particle 1 and particle 2 individually, the joint state could equally well be either S or S'. And this holds even though the two particles might inhabit completely disjoint spatial regions, indefinitely far apart. Einstein's picture cannot be maintained by any interpretation that takes the quantum formalism seriously, even though such interpretations may themselves differ in important ways.

Consequences of Holism

In quantum theory, then, the physical state of a complex whole cannot always be reduced to those of its parts, or to those of its parts together with their spatiotemporal relations, even when the parts inhabit distinct regions of space. Modern science, and modern physics in particular, can hardly be accused of holding reductionism as a central premise, given that the result of the most intensive scientific investigations in history is a theory that contains an ineliminable holism. What lessons can be learned from this?

For the sake of completeness, we should note that this form of holism has no general moral or methodological consequences. The defenders of the so-called holistic health moment, for example, can draw no comfort from quantum theory. The "holism" advocated there has to do with the idea that humans are very complex organisms, and that the biological state of any part may be influenced by a myriad of factors, including diet, exercise, and mental state. These are possibilities that can easily be accommodated by a "mechanical" ontology, and that ought not to be ruled out a priori by any decent methodological principle.

Nor does the holism of quantum mechanics imply an ethically charged view of the universe, according to which all things are one, or are morally interconnected. If quantum mechanics inspires people to become vegetarians, or strengthens their resolve to promote world peace, then the effect is merely via a vague metaphor, which could as easily be suggested by Newton's theory of universal gravitation.

It is tempting to think that the holism indicates a fundamental failing in our very notions of space and time. After all, if the spin states of particles 1 and 2 are so fundamentally interconnected, perhaps we are mistaken to think that they are really distinct particles, or occupy different regions of space-time. And perhaps our notions of space and time could do with a fundamental revision. But the failure of reductionism in quantum mechanics does not force this option on us. Einstein's picture seems to fail not in the assumption that there are "objects situated in different parts of space," but rather because the "being-thus" (i.e., the physical states) of those objects are not independently specifiable. One can try to promote Einstein's intuition into an infallible principle by always readjusting one's notion of "being situated in different parts of space" to fit, but it is hard to see this as a hopeful project. As far as quantum mechanics tells us, the particles in state S can have any spatial relation at all. If we have to readjust the "true" spatial structure constantly so they don't really occupy different spaces, then no independent account of spatial structure would seem to be possible.

If the pair of particles forms an indivisible whole, though, part of the modern account of space-time structure seems to be at risk. The problem is that we have been rather blithely assuming that the division of the world into parts and wholes is unproblematic. In the classical regime, this is true in at least one respect: the relevant parts of a whole are parts that all exist *at the same time*. Since the notion of simultaneity is absolute in classical physics, this posit comes at no cost. But in the relativistic regime, there is no unique way of carving up extended spatiotemporal objects into sets of interrelated parts. This can lead to very puzzling situations, as will be depicted here.

Consider particle 1 and particle 2 which are initially in the singlet state S. Suppose that at some point the z-spin of particle 1 is measured and found to be up. Further suppose that particle 2 is sufficiently far away that there exists a point P on its worldline that, *in one reference frame*, occurs before particle 1 is measured and *in another reference frame* occurs after particle 1 has been measured (see figure 3.1). We have already found that we cannot, on any view, capture all there is to say about the physical state of particle 2 at point P without regarding it as a part of a larger whole. But which whole? Should we take particle 2 at P together with particle 1 before the measurement or with particle 1 after the measurement? On any view, the choices seem to be incompatible.

If we take particle 2 at point P together with particle 1 before the measurement, then their joint state will be state S. As we have seen, on the ray view, this means that either particle 2 has no state at point P, or that its state must be specified via its joint state with particle 1. On the statistical operator view, particle 2 will be in a mixed state at point P, with further reference to the joint state needed to complete the physical description. But if we take particle 2 at P together with particle 1 after the measurement, we get a completely different story. The joint state is now $|z\text{-up}\rangle_1|z\text{-down}\rangle_2$, and on either account, particle 2's state is unproblematically $|z\text{-down}\rangle$! Furthermore, since the joint state now factorizes, no reference to particle 1 is needed. So on any view, the state

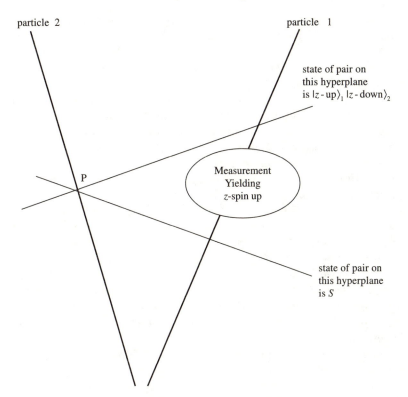

particle 2

particle 1

state of pair on
this hyperplane
is $|z\text{-up}\rangle_1 \, |z\text{-down}\rangle_2$

P

Measurement
Yielding
z-spin up

state of pair on
this hyperplane
is S

Fig. 3.1. Particle pair in space-time.

one gets for particle 2 at P is quite different depending on which temporal stage of particle 1 is chosen to complete the whole. And according to relativity, none of the spacelike separated stages of particle 1 bears any special relationship to P, due to the relativity of simultaneity.

Relativity transforms holism from a mere surprise to a real theoretical problem. It now seems not just that a particle may not have an intrinsic state, not just that one must supplement the physical description of the parts with a further specification of the whole, but that the parts may be attributed seemingly conflicting states depending on the whole that is considered. Taken as part of one whole, particle 2 at P has no z-spin: there is no fact about how it would behave in a spin measurement. Yet taken as part of another whole, it has a definite spin state, and would react deterministically to any z-spin measurement. This is not the harmless relativity of relational states, familiar from Socrates being tall relative to Cebes and short relative to Theatetus. In that case, Socrates has an unproblematic intrinsic state (his height) and all that changes are relational properties. But now the only possible intrinsic property (the spin state) itself changes in two wholes. How are we to make sense of this?

There seem to be but two choices. One is simply to reject the basic posit of relativity. There is, in fact, a preferred reference frame in space-time, and the wholes governed by physics are wholes all of whose parts exist simultaneously in that frame. This is the route taken (at least provisionally) by Bohm. Having accepted the holism of quantum theory perhaps more enthusiastically than anyone else, Bohm simply bites the bullet when it comes to relativity: at the most fundamental level, the relativistic account of space-time structure must be incomplete or incorrect.

The other choice is to accept somehow that particle 2 has one spin state (or lack of a spin state) as part of one whole and another as part of another. Ultimately, this means that we won't be able to say anything at all about particle 2 at P (not even if it has or fails to have a spin state!) until we further specify the remaining parts of some whole. Since the candidate wholes all have parts that exist simultaneously in some reference frame, this means that we must specify a simultaneity slice through P. This idea, that all the properties of an object are relative to a hyperplane, has been championed by Gordon Fleming in his hyperplane-dependent formulation of quantum mechanics.[2] Despite Fleming's occasional protests to the contrary, this radical hyperplane dependence is truly new. In relativistic physics, lots of quantities are hyperplane dependent *locally*: how you describe the physical quantities at some point may depend on how you chop up space-time into slices *around that point*. But this new hyperplane dependence posits something else: how you describe the physical state at a point depends on picking a hyperplane through it and *on what goes on arbitrarily far away on that hyperplane*. Indeed no amount of information about how things are described on one set of hyperplanes (e.g., that the pair of particle is in a singlet state before, during, and immediately after P) can give a clue about how they will be described on another set of hyperplanes through P (e.g., that particle 2 has z-spin up). The familiar examples of hyperplane dependence in relativity don't have this feature at all. In any case, the result is much more radical than the whole being something more than the sum of the parts: the parts have no specifiable physical state at all (or even a lack of a physical state) until they have been specified as parts of some larger whole, with a multiplicity of such specifications equally acceptable.

Aside from the problems with relativity, quantum holism ought to give some metaphysicians pause. As has already been noted, one popular "Humean" thesis holds that all global matters of fact supervene on local matters of fact, thus allowing a certain ontological parsimony. Once the local facts have been determined, all one needs to do is distribute them throughout all of space-time to generate a complete physical universe. Quantum holism suggests that our world just doesn't work like that. The whole has physical states that are not determined by, or derivable from, the states of the parts. Indeed, in many cases (according to the ray view), the parts fail to have physical states at all.

As an example of a current metaphysical doctrine that must take this sort of holism into account, consider David Lewis's theory of possible worlds. In an attempt to guarantee the existence of a large number of worlds, Lewis seeks a principle by which the existence of some possible objects ensures the existence of others:

> To which end, I suggest that we look to the Humean denial of necessary connections between distinct existences. To express the plenitude of possible worlds, I require a *principle of recombination* according to which patching together parts of different possible worlds yields another possible world. Roughly speaking, the principle is that anything can coexist with anything else, at least provided they occupy distinct spatiotemporal positions. Likewise, anything can fail to exist with anything else.[3]

Lewis's basic idea is clear enough to anyone who has used a computer graphics program: one should be able to "cut and paste" parts of different worlds together, where the pieces being cut are given by a spatiotemporal boundary. And if Einstein's vision were correct, such a thing would be possible. But the failure of Einstein's principle casts doubt on the cogency of Lewis's principle. If I cut out one of the particles in state S and paste it into an otherwise empty world, what do I get? On the ray view, the isolated particle must have a spin in some direction, even though the member of the original pair did not. This would violate Lewis's stricture that the copies be physically identical. On the statistical operator view, I can get a single particle in a mixed state. But what if I cut and paste both particles? Do I get just a pair of particles each in a mixed state, or a pair of particles in S? What if I cut and paste two particles that come from different pairs, each original pair being in S? The principle of recombination seems at least incomplete, and perhaps just false.

On Fleming's hyperplane dependent theory, things get even worse. Let us grant that the statistical operator view is correct. If I try to copy the particle at P into an otherwise empty world, do I get a particle with a mixed spin state or one that is z-spin down? Relative to different hyperplanes, the particle at P has these different states, but in the new world to which it is copied, no partner exists on any hyperplane. Perhaps all I can recombine on Fleming's view are complete hyperplanes, but this will restrict the principle of recombination so it can't do the work Lewis requires of it.

Of course, our modal intuitions support the principle of recombination, but that is because we have come to think of the world just as Einstein did. If we are wrong about how this world is, perhaps we are also wrong about how it might be. In any case, quantum mechanics gives us good reasons to reconsider with some suspicion Humean principles such as that cited by Lewis. The world is not just a set of separately existing localized objects, externally related only by space and time. Something deeper, and more mysterious, knits together the fabric of the world. We have only just come to the moment in the development of physics that we can begin to contemplate what that might be.

Notes

1. M. Born, *The Born-Einstein Letters* (New York: Walker, 1971), 170–171.
2. G. Fleming, "Lorentz Invariant State Reduction and Localization," in A. Fine and J. Leplin, eds., *PSA 1988*, vol. 2 (East Lansing, Mich.: Philosophy of Science Association, 1989).
3. D. Lewis, *On the Plurality of Worlds* (Oxford: Blackwell, 1986), 87–88.

4

The Genidentity of Quantum Particles

Hans Reichenbach

The relation of genidentity, which connects different states of the same thing at different times, was studied in §5 [of *The Direction of Time*].[1] A thing is a series of events succeeding one another in time; any two events of this series are genidentical. The conception of physical identity, of an individual thing that remains the same throughout a stretch of time, is based on the properties of this relation.

Speaking of things and speaking of events represent merely different modes of speech. When we wish to translate one of these modes into the other, we must be careful not to use a mixed language. The sentence of the *thing language*, "This tree is old," must be translated into *event language* not in the form, "All events constituting this tree are old," but in the form, "The first events of the series constituting this tree are separated by a long time stretch from the present event." Similarly, the sentence, "Peter is taller by two inches than he was a year ago," is translated as "The present event of the series constituting Peter occupies a spatial volume which is two inches longer than a corresponding volume was a year ago." If the translation is carefully done, there will be no difficulties, and the many pseudoproblems are avoided which philosophers have taken pains to discover in the existence of things that change and yet persist in time.[2]

However, we must investigate the problem of defining the kind of genidentity which allows us to regard a series of events as states of a thing. We saw that genidentity is restricted to events which are members of a causal chain. This condition, obviously, is not sufficient to define genidentity. An event may be situated on several causal chains, which intersect at the event; and the question arises how to select the chain of physical identity.

For macroscopic objects, we define *material genidentity* in terms of several characteristics, which can be divided into three groups. First, we associate material genidentity with a certain *continuity* of change. When a billiard ball moves, there is a continuous transition from position to position. If it hits another billiard ball, we would not accept an interpretation according to which the two balls exchange their physical identity, because such exchange would require the jumping of one ball into the position of the other. This jumping would be discontinuous; the center of one ball would jump into the center of the other. Obviously, such a discrimination is possible only because our spatial observation is exact enough to distinguish between positions of the ball the distance between which is small compared with the diameter of the ball.

Second, it is a characteristic of material objects that the space occupied by one cannot be occupied by another. Two billiard balls occupy different volumes of space. There may arise ambiguities; a nail driven into a piece of wood may be said to be inside the space occupied by the wood. But we can redefine the space filled by the wood in such a way as to exclude the cavity filled by the nail. Thus *spatial exclusion* can always be regarded as a necessary characteristic of solid material objects.

Third, we find that, whenever two material objects exchange their spatial positions, this fact is noticeable. We usually recognize this change of position by the use of specific marks on the objects. Billiard balls are distinguishable by their color, or by little scratches on their surfaces, or by a very precise measurement of their weights, and so on. These marks remain on the object in accordance with the continuity criterion, and permit an identification of the objects even when no observation during the change of spatial positions was made and the continuity criterion cannot be applied. In other words, an *interchange of spatial positions is a verifiable change* even though no records of the act of interchanging are available.

These three characteristics of material genidentity are necessary criteria, but they are not sufficient. When a house is torn down, the resultant heap of bricks and beams represents the same accumulation of matter as the house, but the heap is not called the same thing as the house. The definition of a house requires certain interrelationships between its parts which are no longer satisfied by the heap.[3] But the following considerations require only the necessary criteria just given.

Some investigation shows that not all marks are transferred in accordance with continuity. If a rolling billiard ball hits another ball which is at rest, it can impart all its speed to the other ball and come to a full stop. Here the mark consisting in the motion is transferred, contrary to material genidentity. This allows us to construct another chain of genidentity: we say that the kinetic energy travels from one ball to the other, and by this usage of words we single out another physical entity, the energy, whose genidentity follows different rules. This is not a material genidentity; it might be called a *functional genidentity*, a genidentity in a wider sense. Also of this functional kind is the genidentity of water waves, which we distinguish from the material genidentity of the water particles; these water particles move up and down, as indicated by a floating piece of wood, while the wave moves horizontally.

Things of a mere functional genidentity may readily violate the second and third characteristics, whereas the first is usually adhered to. Two quantities of energy can exist in the same place; energy satisfies the principle of superposition. And energy cannot be earmarked. When two billiard balls of equal speeds enter into a head-on collision, they return the same way with the same speeds. Were their energies exchanged, or does each ball continue to possess its original energy, while merely its velocity is reversed? This is a meaningless question, because there is no way of identifying energy. It should be noted that here the continuity criterion, though it is satisfied, does not supply an answer either, because both interpretations lead to a continuous genidentity. We have here a situation in which two world lines intersect and we cannot tell which line is which after the intersection. We might try to find a way out of this ambiguity by considering two balls of different speeds. It is true that if one ball traveled originally at a higher speed, this higher speed will be found, after the collision, in the other ball. But we need not conclude that the energies were completely exchanged; perhaps only the surplus energy traveled into the other ball. Such ambiguities are typical for mere functional genidentity. They create no difficulties if it is recognized that genidentity can here be defined in various ways.

Is there some such thing as a "true" genidentity? This question appears nonsensical. We can define "genidentity" to suit our purposes. However, we can distinguish between different kinds of genidentity; and it makes sense to distinguish the material genidentity of the billiard balls from the functional genidentity of the traveling kinetic energy, or the material genidentity of the water particles from the functional genidentity of the waves. The reason is that material genidentity is definable in terms of the three characteristics mentioned—continuity of change, spatial exclusion, and noticeable difference when macrocosmic objects exchange positions.

On closer inspection, we discover that what we call material genidentity is often, in fact, a functional genidentity. The human body exchanges its material substance continually; after seven years, it is said, no part of it is materially the same as before. As a consequence, the human body satisfies the continuity criterion only approximately; a precise examination would show that parts of the body separate from it. But the amount of matter that is exchanged at any moment is small as compared with the quantity of matter which is not affected by change during that moment. When a sailboat is repaired repeatedly and finally every one of its boards and supports has been exchanged, one can still quite meaningfully call it the same boat, though again the continuity criterion is only approximately satisfied. A river replaces its water all the time and thus is functionally rather than materially genidentical. When Heraclitus realized this fact, he concluded that change contradicts the conditions of a "true" genidentity and that one cannot step twice into the same river. He should have concluded instead that functional genidentity is a legitimate concept and that one can step twice into the same river.

Even when we accept the latter conclusion, however, we should like to insist that a material genidentity can always be carried through, if we select the genidentity suitably. If a living body exchanges its substance all the time, the parts of matter which enter into it, stay inside it, and then leave it, seem to define lines of material genidentity. And if these parts of matter disintegrate chemically and divide into other substances, which in turn may be divided by corrosion, by chemical processes, or in other ways, it appears to be a legitimate statement that the elementary particles of which the parts of matter are composed define strict lines of material genidentity. From this viewpoint the theory of atoms was constructed. If macrocosmic genidentity is functional throughout, then there is a material genidentity behind it in the microcosm; the atoms are those last units which in their immutable sameness draw lines of material genidentity through the physical world. This conception stood in the background of the atomic theories of the ancients; and although modern atoms are no stable lumps of matter any more, they are supposed to have elementary constituents ready to play the part of ultimate physical entities attributed to atoms by Democritus.

In particular, atoms help us understand the material nature of such things as gases and liquids. Two gases mixed and filling the same container violate the second characteristic of material genidentity, since they occupy the same space. By breaking down the gases into very small parts, into atoms, or molecules, their genidentity is reduced to the material genidentity of the atoms. The genidentity characteristics of the solid macroscopic body are thus construed as properties of the ultimate particles of matter, and the liquid and gaseous states of matter are interpreted as reducible to the solid state of atoms.

The question arises in what sense one can speak of the material genidentity of elementary particles. The usual methods of identification break down in the atomic domain. We cannot observationally test the continuity of motion, though there remained the hope for classical physics that some day we might be able to do so, a supposition which is made physically meaningless, for quantum physics, by the indeterminacy relation. Reference to the second characteristic, spatial exclusiveness, cannot be made, because there exists no crucial experiment deciding whether the interphenomena consist of waves or particles; if they are waves, they satisfy the principle of superposition. And the presence of the third characteristic cannot be observed either, at least not in the way in which it is observed for macroscopic objects. We cannot make marks on atoms which they keep during their travels, as the zoologist labels birds to trace them during their migration. It is true, we can observe traces of individual particles in Wilson's cloud chamber; and the physiologist uses radioactive atoms as tracers for the spreading of chemical substances within a living organism. But the question is whether we thus trace a material or merely a functional genidentity. How can we distinguish between the two, if the macroscopic methods of identification are inapplicable?

A good answer to this question was given in classical thermodynamic statistics. It suggests that we have to replace an individual examination of particles by inferences based on statistical properties of an assemblage of particles. This program can be made clear by the use of the statistical methods discussed in §8,[4] where we explained the computation of probabilities concerning various distributions of molecules. There we considered arrangements of specific, individual particles situated in different cells, or compartments, either of physical space or of some parameter space. We recall that a distribution is a class of arrangements and that counting arrangements depends on what we understand by *different* arrangements. It was emphasized in §8 that, in classical physics, two arrangements are regarded as different when they result from one another by an interchange of molecules between two given compartments. For instance, if we have two particles, which we name a_1 and a_2, and we wish to arrange them in three compartments, this can be done in nine different ways (see table 4.1). The third characteristic of material genidentity, the requirement that exchange of spatial position should lead to an observable difference, is thus translated into a statistical property; it supplies the definition of "arrangement."

Table 4.1
MAXWELL-BOLTZMANN STATISTICS
(particles are distinguishable)

Compartment	Arrangement								
	1	*2*	*3*	*4*	*5*	*6*	*7*	*8*	*9*
1	a_1 a_2	—	—	a_1	a_2	—	—	a_1	a_2
2	—	a_1 a_2	—	a_2	a_1	a_1	a_2	—	—
3	—	—	a_1 a_2	—	—	a_2	a_1	a_2	a_1

This way of counting is characteristic for Maxwell-Boltzmann statistics. It was shown in (8, 3)[5] that, if we have n particles and m compartments, or states, the number of possible arrangements is given by

$$\text{Maxwell-Boltzmann:} \quad N(n, m) = m^n. \tag{1}$$

In a similar way, other statistical results are found; for instance, formula (8, 5)[6] is based on the given definition of arrangement. Since this formula, in turn, enters into the value of the entropy, and the laws of entropy are accessible to experimental test, experiments of this kind constitute a test for the adequacy of the given definition of "arrangement" and thus for the assumption of material genidentity of the molecules. The same applies to other physical consequences derived from the definition of "arrangement" in terms of computed probabilities.

Whereas these tests have been positive in the realm of occurrences studied by classical physics, they have turned out negative whenever the quantum nature of elementary occurrences can no longer be neglected—for example, when gases are subjected to very low temperatures. Let us first remark that, for quantum phenomena, the compartments of particle statistics have a peculiar physical significance. They are given by the various quantum states in which a particle can be and which are represented by the *eigenvalues* resulting from solutions of Schrödinger's equation.[7] For instance, the possible orbits of the electrons in Bohr's atom model represent such states. The energy of gas particles, likewise, is divided into discrete states. For gas statistics, quantum states find a natural interpretation as follows. In §10[8] we discussed the phase space built up by the $2un$ coordinates resulting from the position and momentum parameters $q_1 \ldots q_u, p_1 \ldots p_u$ of n particles. The product $q_i p_i$ of corresponding parameters has here the dimensions ml^2t^{-1} (m = mass, l = length, t = time), which is the dimension of action, that is, the product of energy and time. This is precisely the dimension of Planck's quantum of action, h; thus the phase space is divided naturally into cells of the size h^{un}, which represent the compartments of quantum statistics. These quantum cells take over the function of the cells into which Boltzmann divides the phase space; in contrast to the cells of the phase space, they do not result from an arbitrary division of space, but are given, so to speak, in an absolute way. They are defined by the fundamental rules of quantum physics.

When the statistics of quantum processes are computed in terms of quantum states, it turns out that the usual Maxwell-Boltzmann methods do not lead to correspondence with observational results. It was shown by Bose and Einstein that the deviation springs from the way of counting arrangements, and that in order to arrive at statistical results conforming to experimental data, we must give up regarding two arrangements as different when the particles in different compartments are interchanged. This means that we must regard the particles as indistinguishable, and hence as violating the third characteristic of material genidentity. The computation in table 4.1 is then to be replaced by the one given in table 4.2, in which we have omitted the subscript at the letter "*a*," and thus have indicated that the particles cannot be distinguished. Here we have only six different arrangements for two particles in three compartments.

The extension of this way of counting to *n* particles and *m* states, or compartments, is given as follows. Let us arrange the *m* compartments all in a row; they then have $m - 1$ inner dividing walls, or partitions. When we now arrange the *n* particles, in some chosen fashion, in the *m* compartments, we will put the particles within the same compartment in a row. The *n* particles and the $m - 1$ partitions constitute a row of $n + m - 1$ things. The number of their permutations is $(n + m - 1)!$. Since interchange of particles, or interchange of partitions, leaves the arrangement identical, the number must be divided by $n!(m - 1)!$. For instance, when we consider the first arrangement, that is, the first column of table 4.2, we can permute the two particles and the two inner partitions. We then obtain, not only all the other arrangements, but each in four replicas, the number 4 resulting as the product 2!2!. We thus find:

$$\text{Bose-Einstein:} \quad N(n, m) = \binom{n + m - 1}{n}. \qquad (2)$$

This way of counting enters into all kinds of computations of thermodynamic quantities. The results have been well confirmed by experiments. Among such observational confirmations, the computation of the chemical constant of a gas may be mentioned; furthermore, liquid helium

Table 4.2
BOSE-EINSTEIN STATISTICS
(particles are indistinguishable)

Compartment	1	2	3	4	5	6
			Arrangement			
1	*a* *a*	—	—	*a*	—	*a*
2	—	*a* *a*	—	*a*	*a*	—
3	—	—	*a* *a*	—	*a*	*a*

obeys Bose statistics, as is seen from the strange behavior of its specific heat, the mathematical theory of which was given by F. London.

There are other physical processes which require a third form of statistics. Here, too, the particles are indistinguishable; but, in addition, they are subject to an exclusion principle first discovered by W. Pauli and extended to gas statistics by E. Fermi and P.A.M. Dirac. There cannot be more than one particle in a given state. This kind of statistics is presented in table 4.3. We have here only three different arrangements for the two particles in three compartments.

For n particles and m states, or compartments, we arrive at the result:

$$\text{Fermi-Dirac:} \quad N(n, m) = \binom{m}{n}. \tag{3}$$

This formula is found as follows. For one particle, we have m possible arrangements. When we add another particle, there are only $m - 1$ compartments free for it, because of the exclusion principle; thus we have $m(m - 1)$ arrangements. This must be divided by 2! because interchanging the particles leaves the arrangement unchanged. The method is easily extended to n particles.

Again, the values of $N(n, m)$ enter into various kinds of statistics. It applies to certain groups of physical phenomena—for example, to the "gas" of conduction electrons within a wire through which an electric current flows.

That the nonclassical statistical formulas (2) and (3) yield results that correspond to observational facts cannot be questioned. The question is what these facts prove with respect to the problem of genidentity. It is usually agreed that they prove that the particles are indistinguishable, and we have thus far followed this conception. We must now investigate whether this conception is correct, or subject to qualifications.

We said above that the material genidentity of particles is not accessible to individual examination and that instead of making such an

Table 4.3

FERMI-DIRAC STATISTICS

(particles are indistinguishable and
subject to an exclusion principle)

Compartment	Arrangement		
	1	2	3
1	a	—	a
2	a	a	—
3	—	a	a

examination we have to draw inferences based on statistical properties of an assemblage of particles. What kind of inferences are we making here? Obviously, we are drawing inferences concerning interphenomena, since genidentity refers to physical identity extending from one observable phenomenon to another. These inferences, therefore, presuppose certain extension rules. Now we cannot ask whether an extension rule is true; we can only ask at what consequences we arrive if we use a certain extension rule. The extension rule assumed for the given interpretation, which is the rule usually employed in physics, is that all distinguishable arrangements must be equally probable; it then follows that, for instance for Bose statistics, only the arrangements presented in table 4.2 are distinguishable from one another. The question is, whether this is the only possible extension rule, or whether some other rule would reintroduce a physical identity of individual particles.

This question can be answered by the following method. We use as the extension rule not the equiprobability of distinguishable arrangements, but the rule that the particles be considered as individually distinguishable. Since the probability metric at which we must arrive is given, say, by Bose statistics, this rule leads to unequal probabilities for distinguishable arrangements. Let us see what this means.

When we toss two coins with one hand, we have four possible arrangements. If we denote heads by "H" and tails by "T" and indicate the individual coin by a subscript, these possible arrangements can be written:

$$H_1H_2, \ H_1T_2, \ T_1H_2, \ T_1T_2. \tag{4}$$

The assumption that these four arrangements are equiprobable is made in classical statistics; and actual observation confirms this assumption when the coins are thrown repeatedly. The frequency of the result "heads only" amounts to 1/4; the frequency of the result "tails only," likewise to 1/4; but that of the result "1 head and 1 tail" amounts to 1/2, because this result includes two possible arrangements.

Now let us suppose that statistics lead to different results and that each of the three cases named occurs in 1/3 of all throws. This is the statistics of *nonindividualized combinations*, that is, Bose statistics. It would be symbolized by the three cases

$$H \ H, \ H \ T, \ T \ T, \tag{5}$$

in which we have omitted the subscripts. If frequent observation confirmed these statistics, for which the three arrangements named in (5) would be equiprobable, would we be compelled to say that the coins are indistinguishable?

Such a conclusion would be unjustified. Since we can easly distinguish the coins—by a visible mark, for example—we would say that the four possible arrangements (4) still occur, but that they are not equally probable. The first and the last arrangement of (4) would each have the probability 1/3, and the two middle ones would each have the probability 1/6. On this assumption, the three arrangements of (5) would be equally probable.

This interpretation would mean that the throws of the coins are not independent events. If one coin shows heads, the other would have a tendency to show the same result; and likewise for tails. Denoting the events heads on the first and the second coin by "B_1" and "B_2" respectively, we have here:

$$P(B_1) = \tfrac{1}{2}, \qquad P(B_2) = \tfrac{1}{2},$$
$$P(B_1, B_2) = \tfrac{2}{3} > P(B_2), \quad P(B_2, B_1) = \tfrac{2}{3} > P(B_1), \tag{6}$$

and correspondingly:

$$P(\bar{B}_1, B_2) = \tfrac{1}{3} < P(B_2), \quad P(\bar{B}_2, B_1) = \tfrac{1}{3} < P(B_1). \tag{7}$$

This means that the events would be causally dependent on one another. The sequence of events given by the throws of one coin would be statistically, and thus causally, dependent on the other sequence.

Corresponding relations are easily constructed for a disjunction of n events and two sequences.[9] If three or more sequences are involved, the dependence relations are more complicated. For instance, for three sequences the attributes of which are denoted by "B," "C," and "D," we have:

$$P(B, C) \neq P(C), \quad P(C, B) \neq P(B),$$
$$P(B.C, D) \neq P(D), \quad P(C.D, B) \neq P(B), \tag{8}$$

and similar relations resulting when the letters are interchanged. The quantities with two letters before the comma are here unequal to those with one letter before the comma. This means that one sequence is dependent, not only on each other one, but also on the two others together.[10]

We arrive at the following result. Assume that we could assign material genidentity to each particle of a Bose ensemble; then we would find that the particles are mutually dependent in their motions. If one particle is in a certain state, then there exists a tendency for the others to go into the same state; and even more complicated dependence relations exist for combinations of particles, similar to (8). These causal relationships would represent action at a distance, since the particles can be far apart; that is, the dependence relations would constitute causal anomalies. In other words: any assignment of physical identity to Bose particles leads to causal anomalies.

We see now how the thesis concerning indistinguishable particles is to be qualified. In precise language we cannot simply say: the particles are indistinguishable. We must say: either the particles are indistinguishable, or their behavior displays causal anomalies. We are left the choice of selecting the one or the other interpretation. Neither interpretation is "more true" than the other; two are equivalent descriptions.

However, only one of the two descriptions supplies a normal system, that is, a system free from causal anomalies; this is the description according to which the particles are indistinguishable. When we follow the usual rule of employing a normal system whenever it is possible, we may therefore say, without hesitation, that the particles are indistinguishable.

For Fermi statistics, the situation is somewhat different. Fermi statistics does not offer any difficulties to an interpretation in which the particles are distinguishable. For this interpretation, each arrangement of Fermi statistics would be multiplied by the constant factor $n!$, because this is the number of permutations of n particles. The number of all possible arrangements given in (3) is thus multiplied by $n!$; but the equiprobability relations would not be changed. This means that any two Fermi arrangements, which are now classes of individual arrangements, would still be equiprobable.

The causal anomaly of Fermi statistics consists, rather, in the exclusion principle, if that principle is interpreted as a force acting between individual particles. This means that if we assign physical identity to the particles, the exclusion represents a strong mutual dependence of a negative kind. If one particle is in a certain state, then no other particle is in that state.

These considerations supply an important contribution to the problem of determinism. This problem was discussed in §25[11] as a question of inductive evidence. We derived certain consequences which would follow if the hypothesis of determinism were true, and then turned the unlikely nature of these consequences into inductive evidence against the hypothesis. We can construct a similar argument from the result of our analysis of Bose statistics.

If the synoptic principle of quantum mechanics were false and we could somehow observe position and momentum of a particle simultaneously, we could follow the particle in its path and thus verify its physical identity by means of the continuity characteristic. Since particles are supposed to satisfy the characteristic of spatial exclusiveness, this identification would lead to a material genidentity. Now we saw that for materially genidentical particles Bose statistics leads to causal anomalies. We thus arrive once more at a situation in which a causal anomaly has become a property of observables. This conclusion supplies inductive evidence against the hypothesis of determinism and for the synoptic principle.

Conversely, if we accept the synoptic principle, its combination with Bose statistics leads to the result that there is no material genidentity for elementary particles, provided we describe the physical world in terms of a normal system, that is, a system without causal anomalies. It follows that, in the sense of the term as defined above, there is no material genidentity at all in the physical world; there is only functional genidentity. The conception of material genidentity turns out to be an idealization of the behavior of certain macroscopic objects, the solid bodies, which behavior, however, corresponds only approximately to the idealization. Extending the idealization to atoms in the hope of finding it satisfied in this domain—the program of the classical theory of atoms— does not help. In the atomic domain, material genidentity is completely replaced by functional genidentity.

Notes

1. H. Reichenbach, *The Direction of Time*, ed. M. Reichenbach (Berkeley: University of California Press, 1991; originally published in 1956), 32–42. [Note of the editor.]
2. For a discussion of the two languages, see H. Reichenbach, *Elements of Symbolic Logic* (New York: Macmillan, 1947), §48.
3. See H. Reichenbach, *Experience and Prediction* (Chicago: University of Chicago Press, 1938), 105.
4. Reichenbach, *The Direction of Time*, 56–64. [Note of the editor.]
5. Ibid., 58. [Note of the editor.]
6. That is, the formula giving the number of arrangements belonging to a distribution of n molecules in m cells. See ibid., 58. [Note of the editor.]
7. See, for instance, H. Reichenbach, *Philosophical Foundations of Quantum Mechanics* (Berkeley: University of California Press, 1944), 73.
8. Reichenbach, *The Direction of Time*, 72–81. [Note of the editor.]
9. See H. Reichenbach, *The Theory of Probability* (Berkeley: University of California Press, 1949), 156.
10. Ibid., 105.
11. Reichenbach, *The Direction of Time*, 211–224. [Note of the editor.]

5

The Problem of Indistinguishable Particles

Bas C. van Fraassen

In the quantum-mechanical description of nature, an elementary particle is first of all characterized by some constant features, such as mass and charge. These features serve to classify them into basic kinds or types; physicists sometimes refer to particles characterized by the same constants as "identical particles." (In deference to philosophical usage I shall use "identical" only in the strict sense in which no two distinct entities are identical.) In addition to these constant features, each particle is capable of various states of motion (represented by a Hilbert space). And that is all.

So if two particles are of the same kind, and have the same state of motion, nothing in the quantum-mechanical description distinguishes them. Yet this is possible. We have a dilemma: either this possibility violates the principle of identity of indiscernibles, or the quantum-mechanical description of nature is not complete. The dilemma could also be undercut: perhaps to conceive of such a particle as an individual, to which such a principle even could apply, is one of those many conceptual mistakes fostered by an upbringing in classical physics. A closer look very quickly reveals a whole cluster of problems, of which this dilemma is the center. Some sorts of particles obey the exclusion principle, and cannot have two in the same state of motion—but they too have been cited as a violation of identity of indiscernibles. Both sorts of particles exhibit (for reasons that appear to be related) statistical correlations in their behavior which seem to defy causal explanation. And so forth.

In this paper I shall only try to identify, relate, and clarify the problems in this problem cluster. Though I will describe attempts at solution, including my own, I advocate none at this point. The very attempt to describe the problems systematically may reveal presuppositions whose denial could open up the way to a more satisfactory overall view.

1. Some Notes on the Literature

Elementary particles described by quantum mechanics fall into two classes distinguished by whether or not the exclusion principle applies. This is a principle concerning their aggregate rather than their individual behavior; the principle is roughly formulated as ruling out occupancy of the same state by more than one particle of the given sort. There is a second division by aggregate behavior: *fermions* are particles whose assemblies obey Fermi-Dirac statistics and *bosons*, those which obey Bose-Einstein statistics. The two divisions coincide: fermions are the particles to which the exclusion principle applies. The two dividing principles are not logically independent, but neither do they logically coincide. The first division is logically exhaustive, but other types of statistics exist (including the classical Maxwell-Boltzmann statistics, and various nonclassical "parastatistics"). There is a third division, again factually, but not logically coincident with the first: fermions are the particles with half-integral spin, bosons the ones with integral spin.

Bose statistics and Pauli's exclusion principle were both introduced before the definitive formulation of either matrix or wave mechanics.[1] The quantum-mechanical treatment, properly speaking, of assemblies of identical particles was developed by Dirac (1926) and Heisenberg (1926);[2] the general case of N particles and its relation to group theory was first presented by Wigner (1927).[3] Central to this progress was the recognition of the permutation-invariance requirement for aggregate states of identical particles (see section 2).

The conceptual situation has continued to lead to debate and research in many areas; I list recent examples: in the foundations of physics (Aerts and Piron; Aerts),[4] foundations of statistics (Sudarshan and Mehra; Costantini, Galavotti, and Rosa),[5] philosophy of physics (Margenau; Reichenbach),[6] quantum logic (Mittelstaedt),[7] and general epistemology and metaphysics as related to science (e.g., the connected series by Cortes, Barnette, Ginsberg, and Teller).[8] In a paper presented and commented on by Wesley Salmon in 1969 (but published in 1972),[9] I addressed some of these issues, each of which has seen new contributions since then. There was a crucial aspect of Margenau's argument that I did not appreciate until after I had studied his and his students' writings on measurement, the Einstein-Podolski-Rosen paradox, and mixed states more closely (see section 3). We must be careful, I think, not to become overly preoccupied with the principle of identity of indiscernibles in the discussions; yet Margenau was right to give it a central place, for it appears to relate to some aspect of each problem we shall encounter.

2. The Basic Invariance Requirement

Pauli's exclusion principle is generally presented in popular writings and textbooks as: no two electrons (more generally, fermions of the same sort) can be in the same state at once. The more basic invariance requirement, to which it was related by Fermi, Heisenberg, and Wigner, also entails that this is an inaccurate and sometimes misleading formulation.[10]

Given two systems X and Y with states in Hilbert spaces H_1 and H_2, we represent the states of the composite system $(X + Y)$ in the tensor product space $H_{12} = H_1 \otimes H_2$. When X and Y are "identical" (of the same sort), and $H_1 = H_2 = H$ then $H_{12} = H \otimes H$. If X, Y are mutually isolated and respectively in the pure states ϕ, ψ, we assign $(X + Y)$ the product state $\phi \otimes \psi$. In the identical case, $\psi \otimes \phi$ would be another possible state, a permutation of the first one. More generally, the composite states take the form

$$\Phi = \sum c_{ij}\phi_i \otimes \psi_j,$$

a permutation of which would be

$$\Phi' = \sum c_{ij}\psi_i \otimes \phi_j.$$

States are in general not identical with, or multiples of, their permutations. The question is whether the mathematical distinction reflects a real physical difference. That question appears to be answered in the negative by the basic postulate of

Permutation Invariance. *If Φ is the state of a composite system of identical particles, the expectation value of any observable A is the same for all permutations (i.e., $(\Phi, A\Phi) = (\Phi', A\Phi')$ for each observable A).*

Which vectors in $H \otimes H$ have this property? It can be deduced that they form two classes (henceforth omitting the symbol $\otimes$ when convenient):

Symmetric states: $\quad \Phi = \sum c_{ij}(\phi_i\psi_j + \psi_j\phi_i).$

Antisymmetric states: $\quad \Psi = \sum c_{ij}(\phi_i\psi_j - \psi_j\phi_i).$

If you substitute the symbols 'ϕ' and 'ψ' for each other, the symmetric state is left the same, while the antisymmetric state is multiplied by -1 (which also does not affect expectation values). It is easy to see that in the antisymmetric case, for each pair of indices i, j for which $\phi_i = \psi_j$ we have a factor of *zero* $(c_{ij}(\phi_i\phi_i - \phi_i\phi_i) = 0)$. This gives us the correct

QM formulation of the exclusion principle. Call Φ in $H \otimes H$ an *exclusive product state* exactly if $\Phi = \phi \otimes \psi$ with $\phi \neq \psi$. Then we deduce:

QM Exclusion Principle. *Any antisymmetric state is a superposition of exclusive product states.*

Returning now to the division of elementary particles, we note that they are classified by constants of motion (mass, charge, ...). The quantum mechanical evolution operators described by Schrödinger's equation also do not turn symmetric states antisymmetric, or conversely (the *symmetry-type* is also a constant of motion). So we have here in principle a further classificatory distinction, but in fact all known particles already fell into one class (all states symmetric, e.g., photons) or the other (all states antisymmetric, e.g., electrons).

The permutations in question are permutations of the particles: intuitively the attribution of $\phi \otimes \psi$ to $(X + Y)$ is that of $\psi \otimes \phi$ to $(Y + X)$. Permutation invariance is therefore limited to empirical indistinguishability of the individuals. But this raises the *completeness question*: is there no real difference, or a real difference which quantum mechanics does not represent? This is just the sort of question raised by Einstein, Podolski, and Rosen, and (see below) is similarly related to nonclassical correlations.[11]

3. The Exclusion Principle and the Completeness Question

The idea of a connection with Leibniz's principle of the identity of indiscernibles (henceforth, PII) was perhaps first suggested by Herman Weyl, who referred to the exclusion principle as "the Pauli-Leibniz principle of exclusion." The nomenclature suggests the connection (which I proposed in my 1969 paper, unaware of Weyl's earlier suggestion). Consider two distinct orbital electrons in an atom. Each is characterized by certain constants, definitive of electrons, plus a state of motion. The quantum-mechanical description admits nothing further, so if we assume that description to be complete, then the identity of indiscernibles requires their states to be different.

The assumption of completeness here may have a "metaphysical" air, but it is involved in the very application to atomic structure for which the exclusion principle was introduced. To show this, let us look at the application of this principle in the theory of atomic structure, and its reconstruction of the periodic table of chemical elements. For the structure of the hydrogen atom we introduce already three quantum numbers n, l, m, which together determine the hydrogen-atom wave functions. The *principal* number n determines the total energy E_n, and the number of nodes (radial and angular), which is $n - 1$. This number n can take any positive integral value. The *azimuthal* number l is the number of angular nodes; it is thus less than or equal to $n - 1$, and can otherwise take any nonnegative integral value. It determines the square of the angular momentum; the *magnetic* quantum number m determines one component of angular momentum, which equals $m\hbar$. This quantum number takes the values $0, \pm 1, \pm 2, \ldots, \pm l$. Considering only one electron, when $n = 1$ (electron in the lowest orbit), l and m are obviously constrained to be zero. Hence, if these three numbers told us all, and the exclusion principle applied, there could only be one electron in the lowest orbit. But this is not so. In 1925 Goudsmith and Uhlenbeck introduced a fourth property to characterize the atomic electron, an intrinsic magnetic moment independent of its orbital motion, the *spin*, quantum number s being associated with total spin (1/2 for all electrons) and the quantum number m_s associated with one component thereof ($m_s = \pm 1/2$). Having thus a new parameter with two possible values, we have at least two possible states for the case $n = 1$. If we now assume the description to be complete, there are exactly two possible states available, and if the exclusion principle is then applied, a maximum number of two electrons in the first orbit. This gets us as far as the model of the helium atom, with a nucleus of charge $+2e$ and two orbital electrons. Application to the three-atom lithium atom entails that the third electron cannot also have the lowest orbital state. For the second energy level ($n = 2$), there are four orbital states: as we have already seen l can then have value 0 and m value 0, or l have 1 and $m = 0, \pm 1$. Adding that each of these states can be further distinguished by m_s, it follows that there are at least eight possible states available. Assuming completeness of this description suggests there are exactly eight states, and applying the exclusion principle, we conclude that there can be at most eight electrons in the second orbit. And so forth. So the completeness assumption, used earlier in the suggested deduction of the exclusion principle, is also essential to its primary application in atomic theory, and not extraneous to the scientific context.

In the light of the foregoing, it is with some surprise that we find Margenau (in 1944 and again in 1950) citing the electron as violating Leibniz's principle, in writings specifically devoted to the exclusion principle. But looking more closely at the permutation invariance principle, I believe that we can reconstruct an argument that leads to Margenau's conclusion (which I did not appreciate when I wrote my 1969 paper). Consider a two-particle system in the simple antisymmetric state $\phi_{12} = \phi_1 \otimes \phi_2 - \phi_2 \otimes \phi_1$. What states, if any, can we attribute to the individual components?

Well, suppose I make a measurement of observable A on one component. This is equivalent to measuring $A \otimes I$ (with I the identity operator) on the composite system. Calculating the expectation value, we find exactly the same answer as for the case of measuring A on a single system in a mixed state, which is a half-and-half mixture of ϕ_1 and ϕ_2. From other discussions by Margenau and his students (of the measurement problem and the quantum mechanical paradoxes) we know that in these circumstances they conclude that the attribution of a pure state to the component system is not only unwarranted, but *false*. Instead they assign each particle that same mixed state. There are strong, though not entirely uncontroversial, consistency arguments for their conclusion.

The view at issue here is the so-called ignorance interpretation of mixed states. It is true of course that if I am sure that a given particle was prepared either in state ϕ or in state ψ, but I have no idea which, I can adequately represent the situation by a mixed state, a half-and-half mixture of these two pure states. In that case my ascription of a mixed state reflects my ignorance. But mixed states are encountered also in a different context. Sometimes a complex system has a pure state, and it is different from any pure state we would ascribe to it, in view say of the spatial separation of its components, on the basis of any supposition of pure states for those components. This happens typically after past interactions; Schrödinger called it *the* distinguishing feature of quantum mechanics (in a paper related to the Einstein-Podolski-Rosen paradox). In that case however, predictions about observables relating to *one* component can be based on a "reduction of the density matrix," which ascribes a mixed state to the component part. This ascription is ex hypothesi incompatible with the idea that the mixed state is a mixture of pure states that the component may have for all we know. One view is that the component has no state at all (of its own). Another view, which I attribute here to Margenau, is that the mixture is the state of the component; and, correlatively, that mixtures *are* the possible states of motion, with pure states representing only an unprivileged special case.

But on this view, the two particles discussed in the second last paragraph (which are as an aggregate in a superposition of exclusive product states) are not themselves in different states at all. Hence they are literally indiscernible—though not identical.

It is clear that the crucial step in Margenau's argument about the exclusion principle must consist in a stronger completeness claim for quantum-mechanical description—one that rules out the universal applicability of an "ignorance" interpretation of mixed states. Could the stronger claim be independently disputed? A careful inquiry into the nuances of the completeness question is needed. Or could the very individuality of the component systems be denied in such a fashion that PII becomes inapplicable?

One possible reaction (I won't put it more strongly) is to consider the weakened completeness claim to be found in modal interpretations of quantum mechanics.[12] These interpretations distinguish between *states* and *events*—an event happening exactly when a (nonconstant) observable has some definite value. The state is incomplete in that it gives only probabilistic information about events, but complete in that no other information has any predictive value for future events. In the most restrictive version (Copenhagen version) of the modal representation, a pure state is actually complete in both respects, but mixed states are not. (Thus at the end of the measurement, the apparatus is in a mixture of pointer-reading "states," but the pointer is *actually* at one specific number on the dial.) At least formally, we have here a resolution of the tensions between the identity of indiscernibles and the exclusion principle. If it is held that the fermions have no features left undescribed by the quantum mechanical formalism, this interpretation implies that a pair of fermions cannot be in a product of two identical pure states, but each could be in the same mixed state (while *actually* subject to different ones of the possible events allowed by that state).

4. Considerations of Genidentity for Bosons

Bosons are particles capable only of symmetric aggregate states; they too have been suggested as examples of numerically distinct entities that may be indiscernible. There has appeared a connected series of articles dealing with this: Reichenbach (1956, sect. 26), van Fraassen (1969), Cortes, Barnette, Ginsberg, and Teller—all but the first referring to the preceding one.

Bose's introduction of his statistics was the last step in a historical development directly concerned with electromagnetic radiation and statistical mechanical analogies. If a certain amount of light, say, is introduced into an evacuated enclosure with perfectly reflecting walls, we have a situation in some ways similar to an enclosed body of gas. Specifically, the "radiation gas" exerts a pressure on the walls and work must be expended to decrease the volume. If now a piece of matter is introduced, capable of emitting radiation in every frequency, then emission and absorption will happen until their two rates are equal, and remain so: an equilibrium is reached. Experiments suggested that in this equilibrium situation, the intensity of light of a given frequency in the enclosure is a function solely of that frequency and the temperature of that enclosure. The description of that function is exactly the subject of Stefan's, Wien's, Rayleigh's, and finally Planck's laws of radiation. While Stefan's law is based on experimental results, and was accepted as a partial constraint on the required function, Wien's and Rayleigh's were based respectively on a thermodynamical argument and a deduction from the classical laws of electromagnetism (both using additional assumptions). These latter two turned out to be erroneous on the whole though approximately correct in certain limits. It was exactly at this point that Planck introduced his "quantum theory," and was able to deduce his empirically satisfactory radiation law. But the deduction was based partly on classical assumptions and partly on assumptions incompatible with classical physics, not a theoretically satisfactory situation.

Einstein's treatment of the photoelectric effect, in which corpuscular properties were attributed to energy quanta (radiation of frequency v consisting of photons having energy hv and momentum hv/c) made the statistical mechanical view more than a mere analogy. The pressure that the radiation exerts on the walls can now be attributed to the impact of the photons, exactly the same mechanism as for an ordinary gas. If we now apply Boltzmann's classical statistical mechanics to the distribution of numbers of photons over the various energy levels (corresponding to intensities of radiation over various frequencies) for an equilibrium situation, we obtain Wien's law. But Planck's law should result. Hence, Bose introduced a nonclassical assumption of equiprobability. The classical assumption would be that each arrangement of individual particles, classed together when they have the same energy level, is equiprobable. Bose's assumption was that the identity of the particles is to be ignored, and each possible assignment of occupation numbers to the different energy levels is equiprobable. This was ad hoc: it led to Planck's law. But in retrospect, the quantum-mechanical treatment justified the relevant dismissal of individuality: the correct division into equiprobable cases is obtained if we assume all possible aggregate states of the "photon-gas" to be equiprobable, but suppose that these are symmetrical composite states. (The correct linkage between the Bose-Einstein treatment of radiation and symmetric wave functions is indicated by Wigner [1927, p. 495].) One main suggestion recurring in the literature is that the lack of individuality or identifiability of the bosons thus appealed to in the usual explanation of Bose's statistics has to do with identity across time.

The traditional questions concerning identity through time were reformulated precisely by Reichenbach, first in connection with relativity (1928, section 43), then for quantum mechanics (1956, sect. 26). He uses the classical particle/wave distinction as illustration. A floating cork bobs up and down when a wave reaches it; thus we see that no water moves laterally, although the wave moves across the surface. If the individual water droplets or, better, its molecules are entities persisting in time, the wave is merely a changing configuration of these entities. In Reichenbach, a particle has *material identity* (its temporal stages, or the events involving it, are *genidentical* with each other) and the wave does not.

Hence questions of individuation or identification of waves (which may form superpositions) are either misplaced, or settleable by convention.

Even in the context of the classical world picture, this distinction may be challenged. Are groupings of events into individual histories, by contiguity and succession, anything more than just that—more than a conventional if practically important classification? Reichenbach's answer was, in part, that one grouping may give us a world subject to, and another a world devoid of, causal anomalies. This distinction, he thought, could bestow objectivity on the genidentity relation. In the case of the quantum-mechanical world, it was not clear (and now seems entirely unlikely) that any grouping of events into individual histories will eliminate all causal anomalies in his sense. (His precise form of the problem was that neither a classical particle nor a classical wave picture will by itself fit all the phenomena.) But at least it was possible to conclude, according to him, that to regard bosons as entities persisting through time (having material identity) entails causal anomalies.

The causal anomaly to which he points is the statistical correlation in boson behavior, even in the absence of perturbing forces. This correlation I shall discuss in the next section. Let us here just address the suggestion that bosons are "not genidentical." This means that where intuitively we have, say, an assembly of n photons each persisting in time, we really have only at each moment n photon stages (temporal slices), and there is no objectivity of any sort to the classification of one of these photon stages at time t belonging to the same photon as one or other of the stages at time $t + d$. A photon stage at a certain time is really no more than an event—the *being-occupied* of a certain photon state. But now we recall that the boson aggregate states are symmetric, which entails that several or even all of them may be in the same pure state at once. ($\phi \otimes \phi$ is a symmetric two-particle state.) Hence all n events may have exactly the same character—there are n being-occupieds of the same photon state. Not being individuated by historical connections to prior stages, and the quantum mechanical state being assumed to give a complete description, we conclude that we have here to do with true numerically distinct indiscernibles. Hence PII is violated (a conclusion that Reichenbach does not draw). Were PII to operate, nongenidentity would entail that there is no multiple occupancy of boson states.

In the next two sections I shall discuss the statistical correlations and take issue with Reichenbach's views on causality. For now let us note that if we accept PII, then for the case even of instantaneous particle stages we have a dilemma: either there cannot be more than one of exactly the same sort (as characterized by the quantum-mechanical description) or else they are really distinguished by some (non-quantum-mechanical) "hidden factor" (which may be genidentity, or something else—e.g., perhaps the particle stage is a configuration of some underlying medium whose parts are separately individuated, as for classical waves). If we ignore other possible hidden variables, then it is exactly the nongenidentical entities that must obey the exclusion principle, according to PII, and the genidentical ones that need not. Presence and absence of material identity through time could therefore "explain" what makes a particle a fermion or a boson.

This inversion of Reichenbach's classification, which I described in my 1969 paper, was challenged by Cortes, who argued that it is better to reject PII than to accept as real an empirically vacuous hidden factor. Barnette accused Cortes of confusing metaphysics and epistemology; Ginsberg showed that in quantum field theory Barnette's reasoning looked much less plausible; and Teller argued cogently that these discussions left a number of unsolved problems. (Aerts and Piron, I should add, take exactly the view that bosons are distinguished by some feature ignored in the physical description—see Aerts (1981), p. 402.) I note here that, at a crucial point in the preceding paragraph, a completeness claim occurred again; resolution here probably depends on the issues of the preceding section. Again we must also ask: could we perhaps deny the individuality of component systems in a much more radical way (beyond denial of genidentity) so as to dissolve the problem? (But for bosons, this denial of individuality would have to apply even to an assembly each of whose members is in a *pure* state.)

5. Theoretical Unification of the Different Statistics

The classical Maxwell-Boltzmann and the nonclassical Bose-Einstein and Fermi-Dirac statistics can (and are) also studied abstractly, independent of their physical basis. These studies look disconcertingly classical, but Sudarshan and Mehra show that Bose and Fermi statistics, with their major physical consequences, can be formulated for a classical phase space with a preferred cell size (e.g., h^3) as the only nonclassical feature. To give a feeling for what they are like, consider the case of the two individuals distributed over two cells (which I shall call H and T for "heads" and "tails," but these names imply nothing about them). (See table 5.1.)

Here p^{mb} is the familiar equiprobability assignment to all *logically* possible cases. The others may be understood thus: p^{fd} obeys the exclusion principle (zero probability for multiple occupancy) and p^{be} treats as equiprobable all numerical distributions (two in the first cell; one in each cell; two in the second cell). The symbols a_1, a_2 may of course be no more than indices of the sort we used in description of quantum-mechanical states.

Table 5.1

Case	H	T	p^{be}	p^{fd}	p^{mb}
1	a_1, a_2		1/3	0	1/4
2	a_1	a_2	1/6	1/2	1/4
3	a_2	a_1	1/6	1/2	1/4
4		a_1, a_2	1/3	0	1/4

All three obey the condition that permutation of these indices does not change probabilities: isomorphic cases 2 and 3 are always assigned the same probability. (Carnap called this "*symmetry.*")[13] We also see that on the case of *one* individual all three agree: the proposition that a_1 is in H is the disjunction of cases 1 and 2, to which each of them gives the sum 1/2. Thus the differences concern aggregate behavior. Can we give a unified account of them—an account that places each in a systematic classification, and may throw some light on the basic physical differences they model?

The most far-reaching recent studies in this abstract vein are undoubtedly those by Costantini and his colleagues (1979, 1982, 1983). For the case k cells and n individuals, they define the characteristic *relevance quotient*

$$e(p) = \frac{p \,(a \text{ is in cell } V_i \mid b \text{ is in cell } V_j)}{p \,(a \text{ is in cell } V_i)} \quad \text{for } i \neq j; a \neq b,$$

which they prove to be well-defined (i.e., the same for all a, b, i, j) on the basis of certain general conditions satisfied by the three statistics. They then show that the three statistics are uniquely differentiated by the value of e:

$$e(p^{mb}) = 1,$$
$$e(p^{be}) = k/(k+1),$$
$$e(p^{fd}) = k/(k-1).$$

This is a precise measure of the correlation of individual behavior modeled by the three statistics. In table 5.1 we can quickly verify: $p^{mb}(Ha_1 \mid Ta_2) = p^{mb}(Ha_1)$ so $e(p^{mb}) = 1$; $p^{be}(Ha_1 \mid Ta_2) = (1/3)$ so $e(p^{be}) = (1/3) \div (1/2) = (2/3)$; $p^{fd}(Ha_1 \mid Ta_2) = 1$ so $e(p^{fd}) = 1 \div (1/2) = (2/1)$. The case $e = 1$ is that of total statistical independence between the individuals. We shall go further into this topic of correlation in the next section. For now, note, however, that their systematic classification has room for other statistics ("parastatistics") characterized by other relevance quotients. Hence it does not illuminate why only those three cases should appear in physical situations.

The preceding results were presented in Carnap's framework for probability theory, in one of their publications (1982). In my 1969 paper I also characterized the three statistics in the terms of that framework, though proposing a unification of a rather different sort. To explain it I must say something about Carnap's program. He took it that just about any assignment of probabilities could be correct in some possible situation, but also thought that this assignment would be the result of conditionalizing a certain basic "ur-probability," a logically determined prior probability function, on a set of propositions characterizing that situation. (Since his interest was in confirmation theory, the situations were epistemic, and the characteristic set of propositions exactly the known data or information given.) The properties that single out the ur-function, such as *symmetry* (see above) and *regularity* (assignment of *zero* only to logically impossible propositions) are not preserved under conditionalization. (For example, the information that a coin has landed heads up destroys symmetry [since this coin is now distinguished from other coins] and regularity [because the contingent proposition that it landed tails up is now assigned *zero*].) Hence these properties do not generally characterize the correct probability assignment for a particular sort of situation.

Could we find an ur-probability such that p^{mb}, p^{be}, and p^{fd} are all conditionalizations of it? In fact, Carnap himself proposed p^{be} (his m^*) as *the* candidate (relative to his program) for the ur-probability (though later he revised this opinion). Let us try it as a candidate for the ur-probability for assemblies of physical particles.

For Carnap, the cells are characterized by families of predicates— thus in the study of an urn-problem, the predicates might be 'cubical' and 'red', the cells being cubical-red, cubical-nonred, noncubical-red, and noncubical-nonred. The four complex predicates representing the cells are called *Q-predicates* (logically strongest consistent predicates in the language). Let us say that a family of predicates *individuates* a set of individuals if no two of them can be alike with respect to all these predicates (i.e., no Q-predicate formed from this family can characterize more than one such individual). We now have three possible situations:

1. The family does not individuate the individuals (every logically possible state description can be true).
2. The family as a whole individuates (only those state descriptions in which each individual satisfies a different Q-predicate can be true).
3. A proper subfamily individuates (each individual satisfies a different Q-predicate *of that subfamily*, in each state description that can be true).

Cases 2 and 3 can hold only relative to some postulates, on which the ur-probability is to be conditionalized, for *logically* speaking of course every state description could be true. In fact, it is easy to formulate the relevant postulates. Let $a_1, \ldots, a_n$ be individual constants and let $\{F_1, \ldots, F_k, G_1, \ldots, G_m\}$ be the total family of predicates. Let $\{Q_1, \ldots, Q_q\}$ with $q = 2^{k+m}$ be the set of Q-predicates for the whole family and $\{Q'_1, \ldots, Q'_r\}$ with $r = 2^m$ the set of Q-predicates for $\{G_1, \ldots, G_m\}$ alone. Then those postulates are

Situation 2: $\quad Q_h(a_i) \supset \sim Q_h(a_j) \quad$ for $h = 1, \ldots, q,$

Situation 3: $\quad Q'_h(a_i) \supset \sim Q'_h(a_j) \quad$ for $h = 1, \ldots, r,$

in both cases for each $i \neq j$ from 1 to n.

In table 5.1 it is already illustrated that p^{fd} is p^{be} conditionalized on the situation 2 postulates. That is, with H, T as the two Q-predicates, we have

$$p^{fd}(--) = p^{be}(-- Ha_1 \supset Ta_2 \ . \ \& \ . \ Ha_2 \supset Ta_1).$$

This is easily checked by noting that the odds between the *remaining* state descriptions are the same in both cases $(1/6 \div 1/6, 1/2 \div 1/2)$. This is a trivial case but the argument is general: the situation 2 postulates rule out all *structure* descriptions exhibiting multiple occupancy of cells. The remaining structure descriptions each contain the same number of state descriptions, so all remaining state descriptions (as well as, separately considered, all remaining structure descriptions) are treated as equiprobable by p^{be}—just as p^{fd} does.

To illustrate the effect of the situation 3 postulates we need a bigger table (see table 5.2). Let $k = m = 1$, so we have only four cells (and $G_1, \sim G_1$ are the Q-predicates of the relevant subfamily), and let $n = 2$.

Table 5.2

Case	F_1	$\sim F_1$	G_1	$\sim G_1$	
1	a_1, a_2	—	a_1	a_2	1/8
2	a_1	a_2	a_1	a_2	1/8
3	a_2	a_1	a_1	a_2	1/8
4	—	a_1, a_2	a_1	a_2	1/8
5	a_1, a_2	—	a_2	a_1	1/8
6	a_1	a_2	a_2	a_1	1/8
7	a_2	a_1	a_2	a_1	1/8
8	—	a_1, a_2	a_2	a_1	1/8

Here cases 1–4 are mutually nonisomorphic; case i is isomorphic to case $4 + i$ (for $i = 1, 2, 3, 4$). All other cases are nonisomorphic to these and ruled out by the situation 3 postulates. So we have here four of the original structure descriptions, in fact. By p^{be} all structure descriptions were equiprobable, and conditionalization leaves the "internal odds" the same, so the remaining four are now still equiprobable (1/4 each). Now we notice that the *state descriptions* in the *other* subfamily $\{F_1\}$ correspond to those remaining structure descriptions in the whole family $\{F_1, G_1\}$, and hence are equiprobable. But that means that p^{be}, so conditionalized, *coincides with p^{mb} on the other subfamily* (on the remainder of the overall family). An example is: we have two coins (a_1, a_2) one of which is scratched (G_1) and the other not; each can be heads (F_1) or tails $(\sim F_1)$ independently of the other. If we now look at heads versus tails *alone*, p^{be} conditionalized on the relevant postulate $(G_1 a_1 \supset \sim G_1 a_2)$ gives us the effect of p^{mb}.

To sum this up then, we can see the three statistics as special cases of the same principle (prior equiprobability for structure descriptions) for situations of different extent of individuation by the predicates considered. Again we see here the relevance of completeness and the PII. A claim of individuation is a completeness claim for a family of predicates; the PII entails that there must always be some family of predicates that individuate. Thus in this perspective, situation 1, the boson case, provides the challenge to PII.

6. Causality and Correlation

Viewed from a classical perspective, the Bose and Fermi statistics entail correlations that cry out for causal explanation. This is made clear by Costantini's relevance quotient classification, but let us begin with the corresponding physical considerations. The nonclassical correlations are easily illustrated by the quantum mechanical model (briefly considered by Margenau) for two identical particles moving in parallel (say, along the x-axis) with sharp velocities u and v.

In this simple illustration I will not normalize the state vectors; hence the value $f(x, x')$ derived is *proportional* to the probability that the two particles are found at positions x and x'. We are given two particles in momentum eigenstates, with motion along the x-axis. Treated individually, they would be assigned states $\psi_1 = e^{iux}$ and $\psi_2 = e^{iux'}$; setting $p = (ux + vx')$ and $q = (ux' + vx)$ and choosing first the antisymmetric case, we assign the composite system the state

$$\psi_{12} = e^{ip} - e^{iq}.$$

Applying the Born rule for probabilities of position, we must evaluate

$$\psi_{12}^*\psi_{12} = (e^{ip} - e^{iq})^*(e^{ip} - e^{iq})$$
$$= (e^{-ip} - e^{-iq})(e^{ip} - e^{iq})$$
$$= 2 - e^{i(q-p)} - e^{-i(q-p)}$$
$$= 2 - [\cos(q - p) + i\sin(q - p)] - [\cos(q - p) - i\sin(q - p)]$$
$$= 2[1 - \cos(q - p)] = 2[1 - \cos(ux' + vx - vx' - ux)]$$
$$= 2[1 - \cos((v - u)(x - x'))].$$

Hence the probability $f(x, x')$ is proportional to $[1 - \cos((v - u) \cdot (x - x'))]$. Taking secondly the symmetric case, we assign state $\psi'_{12} = e^{ip} + e^{iq}$ and evaluate

$$\psi_{12}^*\psi'_{12} = (e^{ip} + e^{iq})^*(e^{ip} + e^{iq})$$
$$= (e^{-ip} + e^{-iq})(e^{ip} + e^{iq})$$
$$= 2 + e^{i(q-p)} + e^{-i(q-p)}$$
$$= 2[1 + \cos(q - p)]$$
$$= 2[1 + \cos((v - u)(x - x'))],$$

and hence the probability $f'(x, x')$ is in this case proportional to $[1 + \cos((u - v)(x - x'))]$.

Obviously we get *zero* for $a_1 = a_2$ and positive probability values increasing with d (up to a point) for $a_1 = a_2 + d$ in the antisymmetric case. Thus it looks as if the two particles repel each other, and do so the more if their velocities are more nearly the same. In the symmetric case, the probability is always positive, but increases as we take a_1, a_2 closer together. Thus it looks as if they attract each other. The absence of forces of attraction and repulsion (though there were speculations about "Pauli forces," and the conscious metaphor of "exchange forces") makes one want to say that the particles seem to know each other's state, and either shy away from it (fermions) or try to follow suit (bosons).

The abstract formulation of the statistics shows how they model these correlations in their basic equiprobability divisions. In table 5.3, each structure description is represented by a single comprised state description, with the isomorphic ones indicated by names ('S_i') only. The numbers are assignments of probability to the *structure* descriptions.

Table 5.3

p^{fd}	p^{be}	p^{mb}	F_1	F_2	F_3	F_4	Isomorphic State Description	
1/6	1/10	2/16	a	b			S_1	S_2
1/6	1/10	2/16	a		b		S_3	S_4
1/6	1/10	2/16	a			b	S_5	S_6
1/6	1/10	2/16		a	b		S_7	S_8
1/6	1/10	2/16		a		b	S_9	S_{10}
1/6	1/10	2/16			a	b	S_{11}	S_{12}
0	1/10	1/16	a, b				S_{13}	
0	1/10	1/16		a, b			S_{14}	
0	1/10	1/16			a, b		S_{15}	
0	1/10	1/16				a, b	S_{16}	

p^{fd}, p^{be}, p^{mb} each assign $1/4$ to the event F_1a (a in cell F_1), that is, to the class of state descriptions $\{S_1, S_3, S_5, S_{13}\}$. The conditional probabilities are not the same:

$$p^{fd}(F_1a|F_3b) = 1/3 \quad \text{positive correlations (``repulsion''),}$$

$$p^{be}(F_1a|F_3b) = 1/5 \quad \text{negative correlation (``attraction''),}$$

$$p^{be}(F_1a|F_1b) = 2/5 \quad \text{positive correlation,}$$

$$p^{mb}(F_1a|F_3b) = 1/4 \quad \text{no correlation.}$$

Reichenbach had proposed that every genuine (persistent, resilient) positive correlation must have a causal explanation. The quantum mechanical description does not bear this out. The challenge quantum mechanics presents here (correlations that fit no causal model) is therefore fundamentally the same as in the EPR and Bell Inequality cases.[14] If the three statistics can be unified (with p^{be} as basic) in the way suggested earlier, we may have a way of reconciling our intuitions. The intuitive idea would be that correlations "built into" the basic statistics require no explanation, but only divergences therefrom. To make this precise we need some measure of such divergence (Jaynes-Kullback relative information is a prima facie measure, as are some functions described by I. J. Good), and I think also a more liberal criterion of explanation than Reichenbach's (something like redistribution of odds that "tells for" the feature to be explained). But these are at the moment only tentative suggestions.

7. Naming and Describing in Quantum Logic

The assertion that at least two photons were emitted during a certain interval by a certain atom is at first blush easily formulable. The formula will have the form '$(Ex)(Ey)(Fx \,\&\, Fy \,\&\, x \neq y)$'. Semantic analysis of the usual sort entails that this formula is true exactly if there are entities such that if one is (momentarily) the referent of 'x' and the other of 'y', then '$Fx \,\&\, Fy \,\&\, x \neq y$' is true. But what exactly is required for an entity to be the referent of 'x'? At one extreme we have what Putnam contemptuously calls the *magical theory of reference*. It says in effect that really nothing is required: if there are two individuals, there also exist functions mapping the set 'x', 'y' into this couple of individuals. The semantic analysis can be taken as saying simply that there exists some function such that if we regard 'x' as denoting $f('x')$ and 'y' as denoting $f('y')$, then '$Fx \,\&\, Fy \,\&\, x \neq y$' is true.

This attitude may or may not suffice as long as we look only at such quantificationally closed sentences. It becomes vastly more precarious if we try to extend it from momentary, arbitrary naming by free variables to reference by full-fledged individual terms such as names or descriptions. At the other extreme we find the *causal theory of reference*, which requires that some causal chain connects the (production or use of the) term and its referent. However obscure the notion of causal chain may be, this view of reference would seem to preclude differential naming of two photons in the same state—since entering into distinct causal chains would surely distinguish them in a way that quantum mechanics does not recognize. Unless of course the notion of causality be sufficiently metaphysical, or the quantum mechanical description of nature sufficiently incomplete!

It is not surprising therefore that the problem of indistinguishable particles has recently begun to fascinate writers in the area of quantum logic (see especially Mittelstaedt; Dalla Chiara and Toraldo di Francia).[15] This is the youngest of our problem areas. I can do no better at this point than to refer to the fact that some recent approaches to quantification and singular terms are much less sensitive than others to application problems for standard semantic concepts. There is first the theory of quantification, names, and identity for complete lattices in terms of abstractors (van Fraassen 1982).[16] Second, and perhaps less abstruse, there is the general notion of quantifiers suggested by Kit Fine and elaborated by Charles Daniels.[17] On this second view, a quantifier is based on a function that takes propositions into numbers. Let 'A' be a wff, 'v' a variable, 'X' a proposition; then $V('vA')$ is a function f such that $f(X) = n$ just if there are exactly n valuations V' differing from V at most at v such that $V'('A') = X$. Then the formula 'NvA' is true at world w exactly when 'N' denotes the number $\sum \{V('vA')(X): w \in X\}$. I know this looks rather complicated at first; but less so if you assume the lattice of propositions to be finite and take most of the summed numbers to be zero. In any case, the maneuver is suggestive of the idea that although in a permutation-invariant state we do not treat the particles as inherently distinguishable, *number* remains a well-defined observable.

8. Instead of a Conclusion

Although I have not been entirely neutral in my exposition, and have suggested some approaches to specific problems in the cluster, I was quite sincere in my initial announcement that I have no general or over-all solution to offer. Unfortunately, there are too many loose ends to pretend otherwise, even if we suppose that all my favorite approaches will turn out to work.

When it comes to the interpretation of quantum mechanics I do think, despite the many problems and disagreements, that there has been a great deal of progress in recent years (or, at least, recent decades). Foundational studies of many sorts have provided us with a much deeper understanding of the structure of the theory. As a result alternatives have been sharpened and their consequences clarified. If we now disagree, for example, on the ignorance interpretation of mixtures, or the (relative) completeness of quantum mechanics, we can all quickly locate exact problem areas and consistency questions for our views. Most important perhaps has been the progressive shift of inquiry into facets of aggregate behavior and the structure of composite systems. Revolutionary as the introduction of indeterminism and discreteness and absence of joint distributions were, the most radical features of the quantum-mechanical world are undoubtedly those pertaining to wholeness—the challenge to the ingrained idea that what is distinguishable is separable or separate. (This is also the theme of the conclusion of the recent paper by Dalla Chiara and Toraldo di Francia—a theme that goes back of course to Bohr's replies to Einstein, but which returns with devastating new impact at every new turn in our *problematique*.) Addressing issues in the foundations of quantum statistics, which by definition goes beyond the theory of the individual system, we take this shift one step further in the same direction.

Notes

1. S. N. Bose, "Plancks Gesetz und Lichtquanten-hypothese," *Zeitschrift für Physik* 26 (1924): 178–181; A. Einstein, "Quantentheorie des einatomigen idealen Gases," *Preussische Akad. der Wissenschaften (Phys.-math. Klasse) Sitzungsberichte* 1924: 261–267, and 1925: 3–14; W. Pauli, "Ueber die Zusammenhang des Abschlusses der Elektronengruppen im Atom mit der Komplexstruktur der Spektren," *Zeitschrift für Physik* 31 (1925): 765–783.
2. P.A.M. Dirac, "On the Theory of Quantum Mechanics," *Proceedings of the Royal Society of London*, ser. A 112 (1926): 661–677; W. Heisenberg,

"Schwankungserscheinungen und Quantenmechanik," *Zeitschrift für Physik* 40 (1926): 501–506.

3. E. Wigner, "Über nicht kombinierende Terme in der neueren Quantentheorie," *Zeitschrift für Physik* 40 (1927): 492–500; E. Wigner, "Über nicht kombinierende Terme in der neueren Quantentheorie. II Teil," *Zeitschrift für Physik* 40 (1927): 883–892.

4. D. Aerts and C. Piron, "Physical Justification for Using the Anti-symmetric Tensor Product" (preprint TENA, Free University of Brussels, 1979); D. Aerts, "Description of Compound Physical Systems," in E. Beltrametti and B. van Fraassen, eds., *Current Issues in Quantum Logic* (New York: Plenum, 1981).

5. E. Sudarshan and J. Mehra, "Classical Statistical Mechanics of Identical Particles and Quantum Effects," *International Journal of Theoretical Physics* 3 (1970): 245–251; D. Costantini, "The Relevance Quotient," *Erkenntnis* 14 (1979): 149–157; D. Costantini, M. C. Galavotti, and R. Rosa, "A Rational Reconstruction of Elementary Particle Statistics," *Scientia* 117 (1982): 151–159; D. Costantini, M. C. Galavotti, and R. Rosa, "A Set of 'Ground Hypotheses' for Elementary Particle Statistics," *Il Nuovo Cimento* 74B (1983): 151–158.

6. H. Margenau, "The Exclusion Principle and Its Philosophical Importance," *Philosophy of Science* 11 (1944): 187–208; H. Margenau, *The Nature of Physical Reality* (New York: McGraw-Hill, 1950), chap. 20; H. Reichenbach, *The Philosophy of Space and Time*, M. Reichenbach and J. Freund (New York: Dover, 1957), sect. 43; H. Reichenbach, *The Direction of Time*, ed. M. Reichenbach (Berkeley: University of California Press, 1956), sect. 26, reprinted here as chapter 4.

7. P. Mittelstaedt, "Naming and Identity in Quantum Logic," presented at the 7th International Congress, Logic, Methodology and Philosophy of Science, Salzburg, 1983.

8. A. Cortes, "Leibniz's Principle of the Identity of Indiscernibles: A False Principle," *Philosophy of Science* 43 (1976): 491–505; R. L. Barnette, "Does Quantum Mechanics Disprove the Principle of the Identity of Indiscernibles?" *Philosophy of Science* 45 (1978): 466–470; A. Ginsberg, "Quantum Theory and Identity of Indiscernibles Revisited," *Philosophy of Science* 48 (1981): 487–491; P. Teller, "Quantum Physics, the Identity of Indiscernibles, and Some Unanswered Questions," *Philosophy of Science* 50 (1983): 309–319.

9. B. van Fraassen, "Probabilities and the Problem of Individuation," presented at the American Philosophical Association, 1969, in S. Luckenbach, ed., *Probabilities, Problems and Paradoxes* (Encino, Calif.: Dickenson, 1972); W. Salmon, "Commentary on van Fraassen" (1969), summary in Luckenbach, *Probabilities, Problems and Paradoxes*.

10. E. Fermi, "Zur Quantelung des idealen einatomigen Gases," *Zeitschrift für Physik* 36 (1926): 902–912; Heisenberg, "Schwankungserscheinungen und Quantenmechanics"; Wigner, "Über nicht kombinierende Terme"; R. F. Streater and A. S. Wightman, *PCT, Spin and Statistics, and All That* (New York: W. A. Benjamin, 1964).

11. J. Bub, "On the Completeness of Quantum Mechanics," in C. A. Hooker, ed., *Contemporary Research in the Foundations and Philosophy of Quantum Theory* (Dordrecht: Reidel, 1973); B. van Fraassen, "The Einstein-Podolski-Rosen

Paradox," *Synthese* 29 (1974): 291–309; A. Fine, "Antinomies of Entanglement: The Puzzling Case of the Tangled Statistics," *Journal of Philosophy* 79 (1982): 733–747.

12. B. van Fraassen, "A Modal Interpretation of Quantum Mechanics," in E. Beltrametti and B. van Fraassen, eds., *Current Issues in Quantum Logic* (New York: Plenum, 1981).

13. R. Carnap, *Logical Foundations of Probabitility* (Chicago: University of Chicago Press, 1950).

14. B. van Fraassen, "The Charybdis of Realism: Epistemological Implications of Bell's Inequality," *Synthese* 5 (1982): 25–38.

15. Mittelstaedt, "Naming and Identity"; M. L. Dalla Chiara and G. Toraldo di Francia, "Individuals, Kinds and Names in Physics," in E. Agazzi and M. Mondadori, eds., *Logica e Filosofia della Scienza, oggi* (Proceedings Soc. Italiana di Logica e Filos. delle Scienze, San Gimignano, 1983) (Bologna: Clueb, 1986).

16. B. van Fraassen, "Quantification as an Act of Mind," *Journal of Philosophical Logic* 11 (1982): 343–369; B. van Fraassen, "Semantic Analysis of Quantum Logic," in C. A. Hooker, ed., *Contemporary Research in the Foundations and Philosophy of Quantum Theory* (Dordrecht: Reidel, 1973).

17. C. Daniels, "Towards an Ontology of Numbers" (unpublished manuscript, 1980).

6

On the Withering Away of Physical Objects

Steven French

1. Introduction

In his 1988 Presidential Address to the Philosophy of Science Association, Arthur Fine urged the assembled cohorts to "actively engage philosophy with on-going science" and reminded us of "the potential in science itself for addressing virtually all the sorts of interpretative questions and issues that philosophy traditionally pursues."[1]

Someone who has taken this latter naturalistic claim very seriously is Dudley Shapere, who has argued that science can in fact resolve traditional philosophical questions concerning, for example, identity and existence.[2] At the same 1988 meeting, he suggested that "by being internalized into the scientific process, even such concepts as explanation and existence can be subject to alteration in the light of what we learn."[3]

I want to point out a problem for this program of "reading metaphysics off current physics," to put it crudely, which arises from what might be called the "underdetermination" of metaphysics by physics. The piece of metaphysics that I want to use as an example concerns, of course, individuality.

2. The Evaporation of Physical Objects

In his contribution to the Boston Studies Memorial Volume for Lakatos, Quine presented a short piece entitled "Whither Physical Objects?"[4] There he argued that developments in physics this century support the evaporation of physical objects—at foundation, elementary particles—into nothing more than regions of space-time bearing certain properties. More recently, Resnick has sought support for his "structuralist" view of mathematics in the same developments.[5] In particular he claims that epistemic differences between mathematical and physical objects are blurred by recent physics.[6]

Quine's approach is explicitly based on a well-known presentation by Heinz Post where it is argued that elementary particles cannot be regarded as individuals in the classical (philosophical) sense, but must be seen as "non-individuals" in some way.[7] This conclusion is reached by noting that quantum statistics treats the permutation of indistinguishable elementary particles very differently than does classical statistical mechanics. Put simply, in the latter a particle permutation is counted as a new complexion whereas in the former it is not, leading to a difference in the "weights" assigned to the arrangements of particles over states—hence the conclusion that quantum statistics does not treat elementary particles as distinct individuals. Let's consider this result a little more formally.

3. Indistinguishability and Individuality

In his beautifully clear book, *Individuality*, Gracia distinguishes six different problems concerning individuality: its intension, its extension, its ontological status, the principle of individuation, the discernibility of individuals, and their reference.[8] Although, as Gracia notes, an overall meshing condition can be imposed on this set, in that the solution of any one of these problems may constrain those of the others, it is important to appreciate, at least, the differences in the issues involved: the first is logical, the second, third, and fourth are metaphysical, the fifth is epistemological, and the sixth is semantic.

It is not my intention to delve into all these issues and their differences here but what I would like to emphasize is the conceptual distinction that can (and should!) be drawn between distinguishability and individuality: that which distinguishes an individual from others may not be that which makes an individual an *individual*.[9] The former involves consideration of more than one entity of the kind concerned, whereas the latter relates to the entity taken on its own; it relates, as Suarez put it, to the "fundamental unity" of the individual.[10]

This distinction underlies the classical view (in more than one sense) of elementary particles as indistinguishable individuals, whose individuality is conferred by something that "transcends" the properties of the particles and is designated by a particle label.[11] However, a tension arises with regard to this view in the quantum context.

Consider two indistinguishable particles, labeled 1 and 2, distributed over two states. There would seem to be four possibilities:

1. particles 1 and 2 in state *a*;
2. particles 1 and 2 in state *b*;
3. particle 1 in state *a*, particle 2 in state *b*;
4. particle 1 in state *b*, particle 2 in state *a*.

In classical statistical mechanics, possibilities 3 and 4 are counted as distinct and given equal weight in the assignment of probabilities. However, to get the correct results in quantum statistics possibilities 3 and 4 must be counted as one and the same. But this seems to run counter to the whole point of regarding the particles as individuals and labeling them. From the point of view of the statistics, the labels are otiose, which suggests that the particles should be regarded as nonindividuals, in some sense.

More formally, the above result can be expressed in the form of the indistinguishability postulate:

$$\langle P\Psi | Q | P\Psi \rangle = \langle \Psi | Q | \Psi \rangle, \forall | \Psi \rangle, \forall Q, \forall \Psi,$$

where the $|\Psi\rangle$ represent physically realizable states, the Ps are particle (label) permutation operators, and the Qs are operators representing physical observables.

Now we have to be careful how we interpret this principle. The foregoing argument depends on understanding it as imposing restrictions on the set of possible observables, such that particle permutation operators cannot be so regarded. However, if we interpret it as a restriction on the set of states, then it says that nonsymmetric states, such as possibilities 3 and 4, are rendered inaccessible to the particles. This interpretation is consistent with the metaphysical view of particles as individuals; quantum statistics is recovered by regarding such states as possible but never actually realized.[12]

Thus the formalism can be taken to support two very different metaphysical packages, one in which the particles are regarded as "nonindividuals" in some sense and another in which they are regarded as (philosophically) classical individuals for which certain sets of states are rendered inaccessible. (As we shall shortly see, this latter view needs to be supplemented with a particular view of the relations holding between the particles when quantum entanglement is considered.)

4. The Underdetermination of Metaphysics by Physics

There is, then, a kind of "metaphysical underdetermination" of metaphysics by the physics and thus good reason to be skeptical of claims that physics *forces* us to drop the view of elementary particles as named individuals (having said that, a fundamental conceptual problem with the naming side of things does exist, as we shall see).[13] Quine is certainly on shaky ground in drawing firm ontological conclusions from Post's particular interpretation of the indistinguishability postulate and Resnick, while not engaging the formal details of quantum physics, explicitly appeals to the supposed loss of individuality of quantum particles in support of his approach. Even John Lucas can be set among these odd bedfellows as he argues that the disappearance of transcendental individuality from the "categories of ultimate reality"[14] has contributed to the opening up of "new vistas of rationality" and supports, in particular, an antireductionist theory autonomy.[15]

And the more extreme kind of naturalist program espoused by Shapere grinds to a halt before it really begins to pick up speed: the weight of metaphysics—*some* metaphysics—proves to be too much.

Is there any way of breaking this underdetermination, or, put slightly differently, are there any reasons for preferring one metaphysical package over another?

Let's look more closely at the two packages on offer.

5. Individual Particles + Inaccessible States

This has the obvious advantage of allowing us to maintain a classical ontology, at least where it concerns the individuals themselves. Perhaps that is its only advantage but it is a significant one. Indeed, it is sometimes claimed that such a view draws support from the very practice of experimental physics itself, with its individual tracks in a bubble chamber, distinct clicks from a counter, and individual flashes on a scintillation screen.[16] There may be problems with regard to the issues set out by Gracia above but whatever the outcome of that particular philosophical discussion, quantum particles can be considered as individuals, just like classical particles, chairs, tables, and people. This meshes very nicely with that very general approach to QM, which seeks to interpret the theory in terms that, ontologically speaking, differ as little as possible from classical mechanics (or, better, classical statistical mechanics). Thus it could serve as the underlying ontology of some sort of hidden-variables approach in general or, more particularly, of the Bohm-Hiley interpretation, where you have individual particles chugging along well-defined spatiotemporal trajectories.[17]

But what of Bell? The determined Bohmian, with her acceptance of a form of nonlocality that sends shivers down the spines of ordinary mortals, has no fear of Bell's theorem, of course. Nevertheless, it has been argued that the metaphysical implications of the latter run counter to the kind of ontology being proposed here.

As is, by now, very well known, Jarrett demonstrated that the locality principle used in proofs of this theorem (or theorems) could be decomposed into what he called "locality," which states that the measurement result in one wing of a Bell-type experiment is stochastically independent of the *setting of the apparatus* in the other, and "completeness," which states that the measurement result in one wing of a Bell-type experiment is stochastically independent of the *result* in the other.[18]

The connection with issues of individuality has been made by Howard, who has urged that Jarrett's "completeness" condition, what Shimony calls "outcome independence," be interpreted as a "separability" condition.[19] With spatiotemporal separability regarded as a principle of individuation, the violation of this condition by QM, which, according to Howard, is what Einstein really objected to, implies a nonseparable, and hence nonindividuatable, ontology.[20]

However, nonindividuatable, understood in the sense of not being able to individuate or distinguish, does not mean that the entity concerned is a nonindividual. Nonseparability may have implications for discernibility but not necessarily for individuality. Recalling what we've learned from Gracia and the Scholastics about the distinction between distinguishability and individuality, we can still maintain that quantum particles are individuals, provided we come up with a suitable reinterpretation of the infamous entanglement that lies behind nonseparability. Teller has done this in terms of nonsupervenient relations—that is, relations that are neither determined by nor dependent on nonrelational properties of the relata.[21]

The argument goes like this:

1. The peculiar nature of so-called entangled states only arises when one considers correlation measurements between observables relating to the particles.
2. Such statistical correlations express a relational property of the particles.
3. State functions representing such correlations cannot be reduced to simple products of the functions for each particle separately.
4. A system possesses a certain property if and only if it is in the corresponding eigenstate.

Therefore:

5. A system in the eigenstate represented by a superposition possesses a relational property which cannot be reduced to a monadic property of the particles. These relational properties are said to be strongly nonsupervenient in the sense that they are not dependent on any (intrinsic or extrinsic) monadic property of the particle.[22]

Thus we might push a view of individual particles + nonsupervenient relations holding between them.[23] The latter aspect might seem pretty strange but this may only be because we have been wedded to what Teller calls the "particularist" position for so long—and the quantum strangeness has to surface somewhere. The ontological status of relations is an interesting topic in itself[24] and the connection with a possible formal representation of "relational holism" might be explored through Stairs's realist quantum logic, which attempts to formalize the notion that "there can be facts about a pair of quantum systems which are, in a clearly specifiable sense, about the whole system but are neither reducible to nor implied by facts about the parts."[25]

Nevertheless, there are further philosophical problems that this package must face. Consider the particle labels which tag them as individuals, which *name* them. How are these names to be regarded—as rigid designators which pick out the same individual across possible worlds, or as disguised definite descriptions?

Maidens has recently argued that the Kripkean view of names as rigid designators cannot, in fact, accommodate quantum statistics.[26] Her very persuasive argument is based on a passage in the preface to *Naming and Necessity* where Kripke introduces his view of possible worlds by way of an elementary example from probability theory. There Kripke asks us to consider the possible states of a pair of dice and presses his view that these possibilities are just abstract states and not "complex physical entities" composed of "phantom" counterparts by noting the absurdity in asking for criteria of "transstate" identity to identify which of the dice in an abstract state are identical with one of the actual die. Such states—possible worlds—are stipulated, rather than "viewed from afar" and "the state (die *A*, 6; die *B*, 5) is *given* as such (and distinguished from the state (die *B*, 6; die *A*, 5)). . . . The 'possibilities' simply are not given purely qualitatively (as in: one die, 6, the other, 5). If they had been, there would have been just twenty-one distinct possibilities, not thirty-six."[27]

As Maidens points out, Kripke has just stipulated his way into a commitment to classical Maxwell-Boltzmann statistics!

A possible way out might be articulated along the following lines. It has been suggested on a number of occasions that there are constraints on what we can conceive when it comes to possible worlds. An early example of this is given in the Leibniz-Clarke correspondence, when Leibniz asserts that, although we can *imagine* a situation in which there are two indistinguishable individuals, such a situation is not a *genuine* possibility being contrary to God's will. (Poor old Clarke just doesn't get the point and accuses Leibniz of blatant inconsistency here.) A similar tack might be taken on stipulation; indeed, Hacking has appealed to just such an approach in an attempt to save Leibniz's principle of identity of indiscernibles from the standard Blackian two-globe counterexamples: "In arguing that in a certain possible world there exist two distinct but indistinguishable objects, bland assertion is not enough. There must be argument."[28] Although I have argued that Hacking's attempt to save PII in these terms by paying due regard to what physics tells us about the space-time background actually fails (on the grounds that there exists a kind of metaphysical underdetermination similar to the one discussed here),[29] I am broadly sympathetic to *this* kind of naturalism. Returning to Kripke's example, we might say, à la Hacking, that "cold naked stipulation" is not enough and that we must apply a form of stipulation dressed up with what we have discovered about the actual world in the form of quantum statistics; that is, the answer in the probability example is *not* given a priori but only on the basis of our understanding of what the actual world is like.

In response it could be argued that invocation of the actual world in this manner smacks of Lewis's view in which assessments are made of the "closeness" to the actual world of possible worlds containing counterparts to individuals in the actual world. But giving an account of which possible worlds are physically possible in this way still isn't *discovery*.[30] It seems to me that what is essential and attractive about Kripke's approach is the idea that possible worlds are stipulated, rather than "viewed from afar." Where he went wrong is in choosing the analogy of a pair of dice and their possible states and then claiming that the latter are just "given as such." In a sense they are, for objects that obey Maxwell-Boltzmann statistics; for other kinds of objects the set of permissible states will be different and fully clothed stipulation is required to ensure we end up with the "right" set.[31] What is important is to avoid having to specify criteria of counterparthood and I see nothing here that compels us to do that. Kripke himself invoked an aspect of the actual world in drawing on the dice example; his mistake lay in failing to note the other possibilities that exist in this context.[32]

What about the descriptivist view of names, or particle labels, according to which they are regarded as disguised definite descriptions? Of course, if this is to account for the apparent rigidity of most names, some form of essentialism must be embraced. Unfortunately, this view won't work either, but the descriptivist can at least explain why: entities with nonsupervenient relations holding between them just don't admit of uniquely distinguishing descriptions.[33] Not surprisingly this has unfortunate consequences for the identity of indiscernibles.

6. The Principle of the Identity of Indiscernibles

Leibniz himself was a kind of "superessentialist" in that he regarded *every* property (excluding existence) possessed by an individual as essential; the set of all such properties forms the individual's "complete notion," which effectively acts as a principle of individuality, making that individual the individual that it is. The principle of the identity of indiscernibles then drops out of this as a principle of individuation: if a and b possess *all* properties in common, then they must, in fact, be identical in the sense of being the same individual, since their complete notions will be the same. PII effectively guarantees unproblematic individuation.[34]

But PII itself is, of course, problematic. Different versions of it arise depending on what properties we take to be within its scope (i.e., what properties are taken to be included in the complete notion). The strongest form results if we take this set to include only monadic, nonrelational properties; a weaker form can be obtained by excluding only those properties that can be described as spatiotemporal; and the weakest form of all states that it is not possible for there to be two individuals possessing all properties of whatever kind (monadic, relational, spatiotemporal, and otherwise) in common.[35]

Given an impenetrability assumption, to the effect that no two particles can exist at the same spatiotemporal location, it would seem that the weakest form is acceptable within classical physics, since particles may be individuated by their distinct spatiotemporal trajectories. The other two, however, are ruled out by the indistinguishability of the particles in terms of their nonspatiotemporal, intrinsic, properties.[36]

In the quantum domain, PII is in even deeper trouble, as is well known. Identifying the state-dependent properties of the particles with all the monadic and relational properties that can be expressed in terms of physical magnitudes associated with self-adjoint operators that can be defined for the individual particles, it can be shown that two bosons *or* two fermions in the appropriate superposition state possess the same properties, one to another. The weakest form of PII is therefore violated and the others obviously fall as well.[37]

Attempts to rescue PII by introducing some "extradynamical unused structure" to which the particle labels refer reveal the extreme lengths to which people are willing to go in order to save this principle.[38] Thus, in a work explicitly based on Mirman's attempt to analyze the "experimental meaning" of individuality,[39] de Muynck has pressed the view that the particle labels themselves should be regarded as intrinsic properties of the particles.[40] This suggests a rather bizarre metaphysics of property, as particle labels are not the subject of any theory, nor are they invoked to account for the behavior of the particles. Regarding them as "extradynamical" suggests an analogy with, say, the color of a billiard ball, at least as far as collision phenomena are concerned.[41] But this is a misleading analogy: the color of a billiard ball can be regarded as a secondary property related, by some body of theory, to the primary properties of the fundamental particles of which the ball is composed. These latter are not taken to possess such secondary properties and it is difficult to see what meaning could be given to a primary property that was both intrinsic and yet unconnected to the dynamical behavior of the particles. The claim that such labels can generate the qualitative difference necessary to preserve PII is simply not plausible.

The upshot for the descriptivist view is less than encouraging. Without PII, there is nothing to guarantee that we will never be faced with a situation where we have two individuals that are descriptively equivalent in terms of possessing the same set of properties.[42] We could never guarantee individuation.

7. Individuality and Distinguishability Revisited

Are the Kripkean and descriptivist positions the only shows in town, however? Gracia has elaborated an alternative, "threefold" view of (proper) names, based on his distinction between individuality and discernibility.[43] Thus, he argues there are two aspects to the process of "tagging" an individual: the first concerns our epistemic access by means of which we become aware of the individual; the second has to do with the act of denoting, or giving the individual a name. The former involves description, whereas the latter "has no descriptive mediation."[44] According to Gracia's view, the primary function of (proper) names is to refer and they are established through an act of baptism, in accordance with the Kripkean view. On the other hand, the descriptivist's theory nicely accounts for how we *learn to use* these names effectively. Thus, a description may play an epistemic role in distinguishing an individual from others, in terms of some set of properties, and in fixing reference through baptism, but may not be necessarily tied to the name.[45]

This approach seems to make sense in the context of the standard philosophical examples, such as "Socrates," but can it be extended to cover the quantum domain? I think that it can, although the conceptual gap between ontology and epistemology is perhaps so wide as to give even the most hardened scholastic metaphysician pause. The essential message Gracia's approach carries is that we should be very careful about making claims to the effect that what names do is enable us to isolate and maintain contact with individuals.[46] The latter concerns distinguishability, rather than individuality itself, and here, as Gracia points out, a description may be invoked but is not necessarily tied to the individual. We may continue to regard quantum particles as named individuals, once we recognize what the proper function of the names is.

Of course, quantum particles are indistinguishable, not merely in terms of sharing the same set of intrinsic properties, but also in terms of relations that do not supervene on such properties, and here the gap between ontology and epistemology yawns. At this point one may feel that the former should be more closely tied to the latter; this is a legitimate feeling, my point is simply that we are not driven by the physics to give up this particular package.

But now, what of the alternative? This asserts that particle labels are the result of the whole enterprise getting off on the wrong foot to begin with and that quantum particles should not be regarded as individuals at all.

8. Nonindividuality and QFT

According to this view, quantum particles are nonindividuals, in some sense. But in what sense? If the idea of nonsupervenient relations is hard enough to swallow, that of entities which cannot be regarded as individuals must surely stick in the metaphysician's throat. How could such an idea even be expressed in logicomathematical terms? The sense of unease one might have in thinking of things along these lines might have a good deal to do with the fact that we don't have, or rather haven't had up until now, a logical formalism capable of accommodating such a notion (classical set theory, for example, is explicitly based on collections of distinct individuals). The development of quasiset theory by da Costa and Krause, might provide just such a formalism,[47] where a quasiset remains the same under the interchange of its (indistinguishable) elements and is regarded as a collection of nonindividual entities.[48]

Such a formalism would provide the underpinning to this particular package, thereby supplying the necessary formal structure of the categorial framework, as it were. The questions touched on earlier, as to the nature and role of proper names, may then be simply obviated, as this view suggests a wholesale abandonment of extensional semantics and the development of an intensional alternative.[49]

According to this metaphysics, broadly understood, tagging the particles with labels was a metaphysical mistake by the founding fathers of quantum theory who were unable to break free of the classical world view in this respect. Looking back from the vantage point of quantum field theory, we can smile indulgently at Schrödinger's faster-than-the-eye-can-see shuffling of particle labels and assert confidently that this shuffling subverts the very classical picture in which it was introduced.

Thus it is often claimed that we should prefer this package of nonindividuals because it "meshes" nicely with quantum field theory, where talk of individuals is avoided right from the word go. This was suggested by Post himself and, more recently, by Redhead and Teller.[50] Three broadly "methodological" considerations (although they might be more properly regarded as metaphysical) can be given in support of this move.

1. In the context of the quantum individuals package, and under fairly reasonable assumptions, the principle of identity of indiscernibles can be ruled out, as we have just seen. Abandoning this package then restores hope to those who might wish to maintain this principle as being at least contingently true.

However, its an empty sort of hope, as the principle remains true only for "compound" objects and is simply obviated in the quantum domain (where on this view we no longer have individuals at all). Anyone who pays more than lip service to some notion of ontological reduction faces difficulties here as well. Given the exclusion of the quantum world as a legitimate domain of applicability, why should we wish to maintain PII anyway? Of course, Leibniz himself would have been utterly unfazed by the results sketched out here since quantum particles would be regarded as nothing but well-founded phenomena. The complete notion of an individual, from which PII follows, is applicable to "individual substances"—monads—only.[51] It is only with regard to monads that one can have a version of PII which is both true and nontrivial and according to which monads adopt the role of basic particulars.

As a methodological principle *about* possible worlds, to the effect that two indiscernible worlds are actually *the same* world, PII might have some epistemic plausibility, but taken out of its monadological context the grounds for plausibility disappear when applied to situations *within* a possible world.

2. We recall that in order to account for the appropriate statistics on the particles-as-individuals view, we had to introduce the notion of state accessibility restrictions, imposed on the system as a kind of initial condition. Once in a certain set of states, bosonic or fermionic, the dynamics will ensure that the particle never gets out of it.

However, there is a lack of ontological economy in positing individual particles together with states loitering out there, as it were, but utterly inaccessible to the particles concerned. Redhead and Teller express this point in terms of Redhead's notion of "surplus structure": "A theory may describe states of affairs which, initially, do not seem to correspond to anything found in nature. Sometimes this is because one does not yet understand how the theoretical description should be interpreted or because the means of detection are not available."[52] The inaccessible, nonsymmetric states, however, represent real surplus structure in the formalism, since if they were to be found in nature we would observe classical statistics in the quantum domain.

3. The individuals package sucks one down into the morass of philosophical difficulties concerning the principle of individuality.

Both of these last two problems can be avoided by shifting to the Fock space description of quantum field theory, which is, as I mentioned, free of attributions of particle individuality from the start. Just as the under-determination of theory by the data might be broken through the introduction of a new set of data, the evidential connections to which favor one theory over another, so, it might be claimed, metaphysical underdetermination can be resolved by considering the relationship between our packages and some other theory.

However, with regard to number 2, one must be careful not to be too dismissive of these accessibility restrictions as merely "ad hoc," in some way. First of all, this notion makes better sense in the context of a broader account of quantum statistics that includes the possibility of parastatistics—that is, the statistics of particles with "mixed symmetry" states. The indistinguishability postulate can then be thought of as a kind of "superselection" rule which divides up the appropriate Hilbert space into a number of irreducible subspaces,[53] with transitions between states corresponding to certain of these subspaces forbidden by the dynamics.[54] In this way the indistinguishability postulate can be regarded as an extra postulate of QM, or as an additional initial condition in the specification of the situation. I say "additional" because we are accustomed, in considering classical statistical mechanics, to think of the energy integral as imposing constraints on the region of phase space that may be occupied by a system. The fact that in QM further constraints are imposed relating to the symmetry properties of the states merely highlights the greater role of symmetry principles in quantum as compared with classical statistics.[55]

Furthermore, the ultimate ground of these symmetry constraints will presumably lie in some future "theory of everything" since what they do is express the difference between bosons, fermions, parabosons, and parafermions of order n and so on. The expression of this difference may involve real surplus structure in this context but one wonders how it can be avoided.[56]

Even the Fock space description must incorporate it in some form and, as Redhead and Teller acknowledge, the distinction between bosons, unable to get into antisymmetric states, and fermions, unable to get into symmetric ones, gets replaced by a distinction between field quantization in terms of commutators and anticommutators, respectively (with suitable generalizations for parafields). Its not so much the existence of surplus structure that's bothersome as accounting for this distinction, and here the Fock space formulation fares no better than the labeled particles one.[57]

With regard to point 3, van Fraassen has recently claimed that QFT cannot resolve the underdetermination, since "It is equivalent to a somewhat enriched and elegantly stated theory of [individual] particles. That we can take it as a description of a world that is particle-less only masquerades as an incompatible alternative."[58] The basis of this claim is the relationship of representation that is purported to hold between models of QFT and "concrete" constructions of Fock space carried out within de Muynck's "labeled particle" approach.[59]

Such a claim has been roundly, and soundly, criticized by Butterfield, who has cast doubt on the empirical equivalence that is supposed to exist between QFT and many-particle quantum theory.[60] More relevantly in this context, he has suggested that the existence, within QFT, of states that are superpositions of particle number render this claim false. The idea here is that on any reasonable account of individuality, the number of individuals must be definite. To deny the latter leaves one open to the objection that the entities concerned are not individuals at all.

Nevertheless, difficulties regarding individuality emerge in this context also, albeit suitably shifted.

There is a general question as to how fields should be regarded from this perspective. One possibility is to claim that they are themselves individuals, in some sense, more or less on the grounds that the concept of an individual captures what we mean by a particular, and fields are particulars. It might further be pressed that they are substantial in that they carry energy, the role of substance in physics having changed from that which is impenetrable to that which is the recipient and carrier of energy.

The alternative is to regard the field quantities, and thus the fields themselves, as simply properties of the points of space-time. But then this merely shifts all the problems associated with individuality, the nature of the principle of individuality, and the like back on to the space-time continuum, where they disport themselves with renewed vigor.

9. Metaphysical Underdetermination and Realism

Van Fraassen completes his own discussion of these matters by waving good-bye to metaphysics and concluding in good (constructive) empiricist fashion that "There cannot be in principle, but only as a historical accident, convergence to a single story about our world."[61] Can we be realists in the face of the underdetermination of metaphysics by physics? Perhaps, but it will have to be a very different kind of realism from the traditional varieties. What are we going to be realists about if the very metaphysical nature of particles—whether they can be regarded as individuals or not—is so radically underdetermined?[62]

It is very tempting, in these circumstances, to fall back on something like Putnam's "internal realism," and deny that there is a "fact of the matter" as to which of these metaphysical schemes is "really" true. Adopting a Carnapian principle of tolerance, the choice between them can be regarded as merely a choice of language, determined by local convenience. Thus Putnam compares the conceptual scheme of "commonsense" objects and the "scientific-philosophical" scheme of fundamental particles and comments: "Each of these schemes contains, in its present form, bits that are 'true' (or 'right') and bits that will turn out to be 'wrong' in one way or another—bits that are right and wrong *by the standards appropriate to the scheme itself*—but the question 'which kind of "true" is really Truth' is one that internal realism rejects."[63]

However, in the case of our metaphysical underdetermination, the appropriate standards *are the same for each scheme*. Thus, there is no way of separating "true" into different kinds: there simply is no ground for asserting that the view that quantum particles are individuals is true relative to the epistemic standards of one package and the view that they are nonindividuals is true relative to standards of the other.

Things are further complicated by the fact that these metaphysical packages both lie within Putnam's "scientific-philosophical" scheme. If the notion of a "conceptual scheme" is taken so broadly as to embrace both nonindividuals and individuals-plus-nonsupervenient-relations, it can legitimately be doubted whether there is any one coherent "scientific-philosophical" scheme to speak of. Davidson, of course, has questioned the very idea of a conceptual scheme, using his principle of charity as a blunt instrument to drive all translation into determinate form. Putnam's response is that the argument assumes that the translator possesses only one language and that it is this that renders conceptual relativism unintelligible, whereas the "radical translator" may have more than one "home" conceptual scheme. Shifting focus from Putnam's vacuously incoherent conceptual schemes to our more humble metaphysical packages, the logical languages of each have barely begun to be articulated. In the case of quantum nonindividuals the language of quasisets looks to be the best prospect so far, whereas in that of individuals the semantics of nonsupervenient relations is still up for grabs.

This point has also been made by Zahar in his elaboration of Poincaré's version of structural realism,[64] some form of which might be regarded as the most natural view for a realist to take up in this context.[65] Defenders of this approach typically point to historical changes in ontology, while simultaneously emphasizing structural commonalities preserved by correspondence. Metaphysical underdetermination could be thought of as another arrow in the structural realist's quiver.

There may be as many structural realisms as there are structural realists but the most recent form of this view suggests that what it is about a theory that corresponds to reality are certain structural relations, while the relata themselves are regarded as ontologically eliminable.[66] Metaphysical underdetermination might be used to support this claim, in that it indicates that we cannot draw determinate metaphysical conclusions about the most basic, intrinsic nature of the entities denoted by elements of a theory from the theory itself. Pulling back from this level of metaphysics, although not as far back as the constructive empiricist, the structural realist argues that what is real is the fundamental structure itself of which the foregoing metaphysical packages can be regarded as offering differing interpretations or representations.[67] Spelling out the semantics of structure is then the major challenge for this view.

Whither physical objects? Left in a curious state of limbo by their multiple describability while, metaphysically agnostic, we turn our epistemic attention to the net cast over them. Whither that?

Notes

A version of this paper was presented to the Σ-Club, Department of History and Philosophy of Science, University of Cambridge. I would like to thank Michael Redhead, Jeremy Butterfield, Simon Saunders, Alberto Cordero, and the other participants in the meeting for useful comments and criticisms. I am also deeply grateful to Anna Maidens and James Ladyman for going over an early draft, numerous enlightening comments, and general all round encouragement; the responsibility for what results below is entirely mine, of course.

1. A. Fine, "Interpreting Science," in A. Fine and J. Leplin, eds., *PSA 1988*, vol. 2 (East Lansing, Mich.: Philosophy of Science Association, 1989).

2. D. Shapere, "The Origin and Nature of Metaphysics," *Philosophical Topics* 18 (1990): 163–174.

3. D. Shapere, "Modern Physics and the Philosophy of Science," in Fine and Leplin, *PSA 1988*.

4. W.V.O. Quine, "Whither Physical Objects?," in R. S. Cohen, P. K. Feyerabend, and M. W. Wartofsky, eds., *Essays in Memory of Jmre Lakatos* (Dordrecht: Reidel, 1976).

5. M. Resnick, "Between Mathematics and Physics," in A. Fine, M. Forbes, and L. Wessels, eds., *PSA 1990*, vol. 2. (East Lansing, Mich.: Philosophy of Science Association, 1991).

6. Hale, in an accompanying piece, has pushed this line even further, drawing on the discovery of antimatter and Heisenberg's "antifoundationalist" view of elementary particles. (S. Hale, "Elementarity and Anti-Matter in Contemporary Physics," in Fine, Forbes, and Wessels, *PSA 1990*.) However, her invocation

of the history of modern physics on this point fails to take account of the decline of S-matrix theory to which Heisenberg's view was tied and the rise of the gauged field theoretic formulation.

7. H. Post, "Individuality and Physics," *Listener* 70 (1963): 534–537.

8. J. J. Gracia, *Individuality* (Albany, N.Y.: State University of New York Press, 1988).

9. It is this which lies behind Redhead and Teller's disfinction between "Property Transcendental Individuality" and "Label Transcendental Individuality." M. Redhead and P. Teller, "Particle Labels and the Theory of Indistinguishable Particles in Quantum Mechanics," *British Journal for the Philosophy of Science* 43 (1992): 203.

10. It is worth noting that Gracia gets into more trouble than he realizes when he presses the distinction by asking us to imagine a possible world in which there is only one entity; this entity, he claims, cannot be regarded as distinguishable from others—since there are none—yet it may surely be considered to be an individual. The problem, of course, concerns the spatiotemporal background and how that is regarded—drawing this conceptual distinction in clear terms may be harder than we think!

11. Two (or more) particles are said to be indistinguishable if they share the same set of intrinsic, state-independent properties, such as rest-mass, charge, spin, and so on. (For a nice discussion of the difference between intrinsic and extrinsic properties in this context, see J. Butterfield, "Interpretation and Identity in Quantum Theory," *Studies in History and Philosophy of Science* 24 [1993]: 443–476.)

12. For further details see S. French, "Identity and Individuality in Classical and Quantum Physics," *Australasian Journal of Philosophy* 67 (1989): 432–446; S. French and M. Redhead, "Quantum Physics and the Identity of Indiscernibles," *British Journal for the Philosophy of Science* 39 (1988): 233–246; Redhead and Teller, "Particle Labels"; N. Huggett, "What Are Quanta, and Why Does It Matter?," in D. Hull, M. Forbes, and R. M. Burian, eds., *PSA 1994*, vol. 2 (East Lansing, Mich.: Philosophy of Science Association, 1995); N. Huggett, "Identity, Quantum Mechanics, and Common Sense," *The Monist* 80 (1997): 118–130.

13. For an example of the preceding claim, see M. L. Dalla Chiara and G. Toraldo di Francia, "Individuals, Kinds and Names in Physics," in G. Corsi, M. L. Dalla Chiara, and G. C. Ghirardi, eds., *Bridging the Gap: Philosophy, Mathematics, and Physics* (Dordrecht: Kluwer, 1993).

14. J. R. Lucas, "The Nature of Things," *Presidential Address, British Society for the Philosophy of Science*, June 7, 1993, p. 2.

15. "Explanations are inherently universalisable, and if the physical universe is one of qualitatively identical features that cannot, even in principle, be numerically distinguished, then the explanations offered by other disciplines are ones that cannot, even in principle, be improved upon by a fuller physical explanation. Indistinguishability and indeterminism imply a looseness of fit on the part of physical explanation which take away its Procrustean character. The new worldview makes room for there being different sciences which are

autonomous without invoking any mysterious causal powers beyond the reach of physical investigation." (Lucas, "The Nature of Things," 8.)

16. Falkenburg suggests that this claim can be construed as forming the basis of a kind of "transcendental argument" for particle individuality. See B. Falkenburg, "The Analysis of Particle Tracks: A Case against Incommensurability," *Studies in History and Philosophy of Modern Physics* 27 (1996): 337–371.

17. It is precisely this which motivates so much of the interest in such approaches. However, as Harvey Brown has recently pointed out, Bohmian metaphysics may be rather more peculiar than people realize, as even the intrinsic, state-independent properties such as mass can be regarded as affected by the quantum potential. (H. Brown, "Bohm Trajectories and Their Detection in the Light of Neutron Interferometry," 3rd Annual Conference on the Foundations of Quantum Theory and Relativity, Cambridge, September 13–16, 1994.) If such properties are rendered as "nonlocal" as the state-dependent ones, it is difficult to see how they can be regarded as possessed by individual particles.

18. J. Jarrett, "On the Physical Significance of the Locality Conditions in the Bell Arguments," *Nous* 18 (1984): 569–589; for a very clear exposition of these results, see J. T. Cushing, "A Background Essay," in J. T. Cushing and E. McMullin, eds., *Philosophical Consequences of Quantum Theory: Reflections on Bell's Theorem* (Notre Dame, Ind.: University of Notre Dame Press, 1989).

19. D. Howard, "Einstein on Locality and Separability," *Studies in History and Philosophy of Science* 16 (1985): 171–201; D. Howard, "Holism, Separability and the Metaphysical Implications of the Bell Experiments," in Cushing and McMullin, *Philosophical Consequences of Quantum Theory*.

20. In a recent work Howard has claimed that this idea of separability as a principle of individuation can be traced to Schopenhauer, locating the latter in Einstein's pantheon of personal heroes. However, despite the discussion of Suarez, one can detect throughout the paper a conflation of distinguishability and individuality: thus, Locke, for example, is understood as emphasizing the role of space as the "ground of individuation," although one could argue that this role is that of a principle of distinguishability, whereas the "principium individuationis" is "existence itself." See D. Howard, "A Peek behind the Veil of Maya: Einstein, Schopenhauer, and the Historical Background of the Conception of Space as a Ground for the Individuation of Physical Systems," in J. Earman and J. D. Norton, eds., *The Cosmos of Science: Essays of Exploration* (Pittsburgh: University of Pittsburgh Press; Konstanz: Universitätsverlag, 1997), 120–121.

21. P. Teller, "Relational Holism and Quantum Mechanics," *British Journal for the Philosophy of Science* 37 (1986): 71–81; P. Teller, "Relativity, Relational Holism and the Bell Inequalities," in Cushing and McMullin, *Philosophical Consequences of Quantum Theory*. See especially n. 8 of the latter work, where a comparison with Howard's view is given that is diametrically opposed to that set out here. Teller suggests that his conception of particularism refines Einstein's idea of separability, so that a similar position results if either is given up. However, it seems distinctly odd to talk of (nonsupervenient) relations holding between nonindividuatable relata. And certainly there is nothing to

prevent the relata from being regarded as individuals in the manner outlined here.

22. Cleland has argued that spatiotemporal relations between bodies are weakly nonsupervenient in the sense that they are dependent on but not determined by monadic properties of the bodies concerned. (C. Cleland, "Space: An Abstract System of Non-Supervenient Relations," *Philosophical Studies* 46 (1984): 19–40.) For further discussion see S. French, "Individuality, Supervenience and Bell's Theorem," *Philosophical Studies* 55 (1989): 1–22.

23. See ibid.

24. In his discussion of Poincaré's structural realism, Zahar has noted that we do not yet possess an appropriate semantics of relations independently of their representation in terms of monadic properties of the relata. What is required, at minimum, is release from a Quinean insistence on quantifying only over first-order variables. See E. G. Zahar, "Poincaré's Structural Realism and His Logic of Discovery," (forthcoming), 14.

25. A. Stairs, "Sailing into the Charybdis: van Fraassen on Bell's Theorem," *Synthese* 61 (1984): 351–360. Indeed it was Stairs's thoughts along these lines that prompted Teller to develop his "Antiparticularist" message; see Teller, "Relativity, Relational Holism and the Bell Inequalities," 222, n. 12.

26. A. Maidens, "Particles and the Perversely Philosophical Schoolchild: Rigid Designation, Haecceitism, and Statistics," *Teorema* 17 (1998): 75–87.

27. S. Kripke, *Naming and Necessity* (Oxford: Blackwell, 1980), 17.

28. I. Hacking, "The Identity of Indiscernibles," *Journal of Philosophy* 72 (1975): 249–256.

29. S. French, "Hacking Away at the Identity of Indiscernibles: Possible Worlds and Einstein's Principle of Equivalence," *Journal of Philosophy* 92 (1995): 455–466.

30. Of course some account must also be given as to the relationship between physical predicates and "logical" ones and Hacking has taken some interesting steps in this direction.

31. Cf. Dorling's remark: "You can't get out the statistics without putting in the dynamics." J. Dorling, "Probability, Information and Physics," preprint, Department of History and Philosophy of Science, Chelsea College, University of London, 1978.

32. Something similar is hinted at in the analogy Schrödinger, Hesse, Teller, and, most recently, Maidens all draw between quantum particles and pounds in a bank account.

33. These brief considerations were prompted by an earlier unpublished essay by A. Maidens, entitled "Trans-World Identity and Entities in Physics," parts of which feature in "Particles and the Perversely Philosophical Schoolchild."

34. Since the complete notion comprises all properties, past, present, and future, the issue of counterfactual situations becomes problematic: shift to a possible world by imagining a change in any of the individual's properties and you have a different individual. Mondadori has managed to come up with a Leibnizian account of counterfactuals that is consistent with superessentialism and which construes the counterpart relation as holding, not between individuals,

as Lewis has it, but between complete notions. (F. Mondadori, "Reference, Essentialism and Modality in Leibniz's Metaphysics," *Studia Leibnitiana* 5 (1973): 74–101.) This involves a treatment of proper names that Mondadori calls "deferred naming": we begin with an actual individual and its complete notion; we then change elements of the complete notion as required by the counterfactual being considered and arrive at a new complete notion; we end up with whatever individual would have possessed the complete notion in question had the possible world to which this notion belongs been actualized. Can such a superessentialist view help us in the quantum situation? Perhaps: there are interesting connections with discussions of the role of histories in supplying the necessary individuation and whether such histories make sense in the quantum context.

35. This is not necessarily the superessentialist form: Leibniz appears to have held that all the different kinds of properties could be reduced to monadic ones, although this interpretation has been famously disputed by Ishiguro and Hintikka.

36. See S. French, "Why the Identity of Indiscernibles Is Not Contingently True Either," *Synthese* 78 (1989): 141–166.

37. Ibid.

38. See, for example, B. van Fraassen, *Quantum Mechanics: An Empiricist View* (Oxford: Oxford University Press, 1991), 432–433 and 479–480. The reason for the empiricist's desperation to maintain PII is not hard to find—the collapsing of the distinction between distinguishability and individuality which the principle's effects eliminates the need for recourse to the substantial substratum or the spatiotemporal plenum as principles of individuality.

39. R. Mirman, "Experimental Meaning of the Concept of Identical Particles," *Nuovo Cimento* 18 B (1973): 110–122.

40. W. de Muynck, "Distinguishable and Indistinguishable-Particle Descriptions of Systems of Identical Particles," *International Journal of Theoretical Physics* 14 (1975): 327.

41. E. Sudarshan and J. Mehra, "Classical Statistical Mechanics of Identical Particles and Quantum Effects," *International Journal of Theoretical Physics* 3 (1970): 245.

42. That is, as noted earlier, entities related nonsuperveniently do not admit of uniquely distinguishing descriptions.

43. Gracia, *Individuality*, chap. 6.

44. Ibid., 219.

45. Ibid., 220.

46. See G. McCulloch, *The Game of the Name* (Oxford: Oxford University Press, 1989).

47. See D. Krause, "On a Quasi-Set Theory," *Notre Dame Journal of Formal Logic* 33 (1992): 402–411, and D. Krause and S. French, "A Formal Framework for Quantum Non-Individuality," *Synthese* 102 (1995): 195–214.

48. Related developments have been explored by Dalla Chiara and Toraldo di Francia, although their quaset theory differs from the da Costa and Krause approach in certain interesting ways; see M. L. Dalla Chisra, D. Krause, and

R. Giuntini, "Quasiset Theories for Microobjects: A Comparison," chapter 8 in this volume.

49. D. Krause, "Non-Reflexive Logics and the Foundations of Physics," in C. Cellucci, M. C. Di Maio, and G. Roncaglia, eds., *Logica e filosofia della scienza: problemi e prospettive* (Proceedings Soc. Italiana di Logica e Filos. delle Scienze, Lucca 1993) (Pisa: Edizioni ETS, 1994); Dalla Chiara, Krause, and Giuntini, "Quasiset Theories for Microobjects: A Comparison."

50. Redhead and Teller, "Particle Labels." See also P. Teller, *An Interpretive Introduction to Quantum Field Theory* (Princeton: Princeton University Press, 1995), and S. French and D. Krause, "The Logic of Quanta," in T. Y. Cao, ed., *Proceedings of the Boston Colloquium in the Philosophy of Science: A Historical Examination and Philosophical Reflections on the Foundations of Quantum Field Theory* (Cambridge: Cambridge University Press, forthcoming).

51. C. Brown, *Leibniz and Strawson: A New Essay in Descriptive Metaphysics* (Munich: Philosophia Verlag, 1990).

52. Redhead and Teller, "Particle Labels," 217.

53. O. W. Greenberg and A. M. L. Messiah, "Symmetrization Postulate and Its Experimental Foundation," *Physical Review* 136 B (1964): 251; R. H. Stolt and J. R. Taylor, "Correspondence between the First- and Second-Quantized Theories of Paraparticles," *Nuclear Physics* 19 B (1970): 10.

54. For a discussion of this in the case of paraparticles, see S. French, "First-Quantised Para-Particle Theory," *International Journal of Theoretical Physics* 26 (1987): 1141–1163.

55. The symmetry type of any suitably specified set of states is an absolute constant of the motion, equivalent to an exact uniform integral in classical terms. See P.A.M. Dirac, *The Principles of Quantum Mechanics*, 4th ed. (Oxford: Oxford University Press, 1978), 213.

56. Thus, as already noted, de Muynck goes so far as to regard the statistical behavior of quantal particles as an "independent and fundamental property of the particles." See de Muynck, "Distinguishable and Indistinguishable-Particle Descriptions," 334.

57. Again, as Redhead and Teller note, an analysis of the spin-statistics theorem may shed some light on this issue. See Redhead and Teller, "Particle Labels," 217, n. 7.

58. Van Fraassen, *Quantum Mechanics*, 436.

59. De Muynck, "Distinguishable and Indistinguishable-Particle Descriptions."

60. Butterfield, "Interpretation and Identity"; de Muynck himself argued that the standard theory and his own formalism were, in fact, inequivalent.

61. Van Fraassen, *Quantum Mechanics*, 482.

62. Of course, it is not the case that "anything goes," metaphysically speaking, as some packages just don't fit; thus, as I have indicated, under fairly plausible assumptions the principle of the identity of indiscernibles can be ruled out of court.

63. H. Putnam, "Truth and Convention: On Davidson's Refutation of Conceptual Relativism," *Dialectica* 41 (1987): 73.

64. Zahar, "Poincaré's Structural Realism."

65. Interestingly enough, Zahar also invokes the supposed implications of quantum theory concerning individuality—or the lack of it—in supporting his view of Poincaré's philosophy.
66. See J. Ladyman, "Structural Realism and the Model-Theoretic Approach to Scientific Theories," *Studies in History and Philosophy of Modern Physics* (forthcoming).
67. Cf. Butterfield: "the individual described by the theory is the underlying quantum system, with its Hilbert space of states. Particle and field are now both matters of a representation of a selected set of states. The glory of quantum field theory is that it allows and uses all these representations." Butterfield, "Interpretation and Identity."

7

Quantum Mechanics and Haecceities

Paul Teller

1. The Problem, and the Problem with the Problem, of Identical Particles and Quantum Statistics

Suppose we have a box with two qualitatively identical particles, particle 1 and particle 2, bouncing around inside. We think of the box as having a left (*l*) and a right (*r*) side of equal volumes. We suppose the particles to be moving around at random without interacting, so that their motions are independent; in particular we suppose the particles to be small enough so that we may neglect collisions. What are the chances for finding one or both on one side or the other?

Many find the following reasoning persuasive. There are four possible cases: both particles in *l*, both in *r*, particle 1 in *l* and 2 in *r*, and 2 in *l* and 1 in *r*. These should be equally likely, so that each has a probability of 1/4, or a probability of 1/4 for two in *l*, 1/4 for two in *r*, and 1/2 for one on each side.

This stylized example is a simple mock-up for a kind of situation that can occur with quantum entities and properties. For many of these situations the probabilities are in fact found to be 1/3 for each of the three cases: two in *l*, two in *r*, and one on each side. Many interpreters have found this fact utterly astonishing.

But on the face of it, there is a very simple resolution of the puzzle: give up supposing that there are two qualitatively identical but numerically distinct particles. Instead say that there are two *quanta*, as I'll put it, to which the notion of being numerically distinct does not apply. I can clarify my intent with an analogy.

Suppose that on Monday I put a silver dollar into my piggy bank, and on Tuesday I put in a second, qualitatively identical one. On Wednesday I pull one of the sliver dollars out and ask, is this the first or the second that I deposited yesterday? Even supposing these two coins to be qualitatively identical, we take there to be a fact of the matter. One of

"This coin was deposited first" and "This coin was deposited second" is true and one false, even though I may not be in a position to determine which coin is which.

Contrast the last case with one in which on Monday I make a deposit of one dollar in my checking account, and on Tuesday I make a second one dollar deposit. On Wednesday I appear at the bank wanting to withdraw one dollar. But the teller will not be able to make anything of it if I insist: "And be sure the dollar you give me is the one I deposited on Monday, not the one I deposited on Tuesday!"

If we think of the "contents" of the box as quanta, like dollars in a checking account, instead of particles, like silver dollars in a piggy bank, there is no 1 and 2. There are only two undifferentiated units and so only three basic cases: two in l, two in r, and one on each side. Now probabilities of 1/3, 1/3, 1/3 no longer seem astonishing.

Anyone who has had a little experience with this kind of issue knows that there is a big problem with the argument I have outlined: it appears to appeal to the suspect principle of indifference. This principle assigns probabilities in a situation by breaking the situation down into basic cases, the most specific outcomes that could arise. The principle then assigns these basic cases equal probabilities. Even if we wave the worry of why basic cases should all be treated alike, there is a problem with what counts as a basic case. This is clearest when probabilities are distributed over values for a continuous parameter. Suppose I am randomly casting darts at a target. Which are basic cases: equal areas, or equal distances from the center? The principle of indifference applied to different basic cases gives different probability distributions. Once we recognize that we need discretion in choosing "basic cases," it looks irrelevant whether we think of the things in the box as particles, with numerical identities, or as quanta, without. If I lump together the cases, 1 in l and 2 in r with 2 in l and 1 in r, and call them one case, one on each side, I get the 1/3, 1/3, 1/3 distribution with no fancy metaphysics.

If the imprecise argument I sketched at the beginning of this section is supposed to show that a 1/3, 1/3, 1/3 distribution is to be expected for quanta and not for particles, the argument will have to appeal to the problematic principle of indifference. But we can extract an argument for a weaker conclusion which does not need this appeal. My objective here is to defuse the surprise that we don't get probabilities of 1/4, 1/2, 1/4 (classical statistics) and get the 1/3, 1/3, 1/3 (quantum statistics) instead. We are driven to expect classical statistics as soon as we take what is in the box to be particles, suppose them not to interact with each other, and suppose each of them to be randomly distributed within the box. We can—and do—have indirect but very good evidence for the second and third assumptions, thus leading to the inference from particles to classical statistics without assuming the indifference principle. But if we suppose that we are dealing with quanta, with no applicability of the idea of a "this one and that one," then three cases, two in *l*, two in *r* and one on each side is as fine as our descriptions can get. Perhaps we have no reason to think that these finest descriptions can be expected a priori to be equally likely, but that they should be equally likely is no less plausible an alternative, so that when we find the quantum 1/3, 1/3, 1/3 statistics we have no occasion to be surprised.

This argument is far from conclusive: in particular I claimed that one has indirect evidence for the premise that the contents of the box, thought of as particles, do not interact. One could dispute this claim, and indeed there was a period in the history of quantum mechanics where practitioners talked about "exchange forces." But no good theory of these exchange forces was ever developed. The quantum model suggested by the money analogy seems entirely more straightforward.

If it's all so simple, where did the big problem about quantum statistics come from? I propose to examine two plausible and closely related sources.

2. Haecceities

First some terminology. Some philosophers carp about physicists' use of the word 'identical' to describe things like electrons. These philosophers insist that no matter how similar, two electrons are different things, and so are not identical (viz., not the same thing). But 'identical' is ambiguous and has just the physicists' intended usage in common speech, as when someone remarks, "Gosh, our neckties are identical!" Two neckties can be identical in the relevant sense even though they are not in the same place, even though one may be tied differently than the other, and the like. 'Identical' can also mean 'the very same object'. To disambiguate between the physicists' sense and the philosopher's "strict" sense of 'identical', I will distinguish as I did in the preceding section between being qualitatively and being strictly (or numerically) identical.[1] Clearly the words, 'the same', suffer the same ambiguity. When physicists say that two electrons are identical they mean what we do when we say that two neckties are identical, but they make what they say more precise. Quantum objects have state-dependent properties, like position and momentum, which vary from state to state; and state-independent properties, such as mass and charge, which are fixed. Two electrons are qualitatively identical inasmuch as they share all of their state-independent properties.

Now let's get to haecceities. The idea involves a contrast between any distinct objects, but the idea is easier to bring out if we consider two qualitatively identical objects. If your necktie and mine are qualitatively identical, many people nonetheless feel that there is a "this one" and a "that one" about them, a contrast between them independent not only of their relatively stable properties, which in this case they share, but also of their exact position, history, indeed independent of all their distinguishing properties and relations. The world would have been in that small way different if I had put on your tie and you had put on mine, indeed if your tie had gone through my tie's entire history and my tie had gone through yours.

Traditionally, philosophy has talked about an object's "haecceity" to mark the idea that an object is distinct from all others in some manner that transcends all properties in any usual sense of the word 'property'.[2] But in discussing haecceities with others I have sensed a great deal of confusion about what talk of haecceities is supposed to come to. Let me try to lay out the distinct ideas that seem to me often not to be clearly enough distinguished.

To begin with, let us take for granted the idea that strict identity applies to objects. That is, if we have two putatively distinct objects, a and b, then it makes sense to say that $a = b$, it also makes sense to say that $a \neq b$, and exactly one of these statements is correct. And let us take for granted some things that presuppose the applicability of strict identity: that names can refer "directly," that is without operating as definite descriptions; that repeated use of the same name picks out the same referent; that repeated use of the same variable bound by the same quantifier picks out the same referent; and that sets are defined extensionally, that is by their membership. Now, a metaphysician might ask: in virtue of what does strict identity apply to an object? Haecceities, in what I will call the metaphysically robust sense, are supposed to be some metaphysical feature, principle, characteristic, or "non-qualitative property" which answers this question. A thing's haecceity is not a property or quality in any ordinary sense. Instead it is some inscrutable aspect, unlike anything we would ordinarily think of as a property, which a thing must be supposed to have to make sense of its identity and its distinctness from other things.[3]

The metaphysically parsimonious will now protest: why on earth do we need any extra metaphysical ingredient to account for identity? To say that something exists is already to say that it has its own identity. Or, in the material mode: for a thing to be is already for it to be identical with itself and distinct from all other things. There is no extra metaphysical work that needs doing here.

The discussion of haecceities in the metaphysically robust sense proceeded on the presupposition that the subject matter concerns *things*, understood as a subject matter to which strict identity applies. Talk about haecceities in what I will call the minimalist sense is just a way of talking about this presupposition. What makes it so hard to discuss haecceities in this minimalist sense is the attitude of most that this presupposition is beyond questioning. Why bother giving a name to something that can't be questioned! Much worse, people seem to make no sense of my wanting to question these minimalist haecceities, that is, my presupposing a subject matter free from minimalist haecceities. Or, in a charitable effort to make sense of me, people insist that I must mean haecceities in the metaphysically robust sense. But I don't.

I get much more consternation and claims of unintelligibility than concrete argument for the inviolable status of the applicability of strict identity. But let us see why minimalist haecceities seem so secure. (Henceforth when I use 'haecceity' without qualification I will mean 'minimalist haecceity'.) To predicate a property of an object would seem to require that the object of predication have a definite identity, and be determinately distinct from other objects. Otherwise what sense would there to be made of the idea that it is the selected object of predication, as opposed to something else, which has the property? Again, as soon as one starts to talk about some*thing* we are poised to use the devices of repeated reference to the *same* thing. I say that something is on my table and that *it* weighs at least five pounds. There is no sense to be made of this statement unless there is some one thing, identical with itself and distinct from all others, that is the object of repeated reference. What goes for the word 'thing' goes equally for 'object' and 'entity', and, as illustrated in the last example, for the use of pronouns in natural languages and the devices of variables and quantification in first order logic.

Nonetheless we do have models where all this does not apply, where we use count nouns that clearly do not have the benefit of minimalist haecceities. Waves provide an example.[4] I like best the preceding section's example of dollars in my checking account. I have, let us say, an even 1,000 dollars in my account, but there is no "this one as opposed to that one" about any of them. Indeed, in the last usage, "any of them" sounded somehow wrong. There is no application of predicates to one as opposed to another dollar in my account. There are no determinate subsets of these dollars. There is no saying about one of these dollars, say the one I deposited on Tuesday, that *it* is the one I withdrew on Wednesday. Strict identity does not apply to these dollars. They are "things" without haecceities.

I have just used quotation marks around 'things' because a natural reaction to this example is to say that, strictly speaking, there are no *things* in my account. Many will insists that to speak accurately one needs to reexpress talk about dollars in my bank account as talk about properties of real things. The only real things that there are in this example are the bank, the people who work for it, the balance books, me, and the properties of all these things, saliently including the balance showing on the books and my creditworthiness, in the form of my right to withdraw up to $1,000 in bills.

We could, of course, decide to apply the word 'things' to dollars in a bank account in an extended or metaphorical way. But clearly it is best not to, to avoid any temptation to slip into applying strict identity where it does not apply.[5] Let us instead insist that, strictly speaking, 'thing', 'object', 'entity', and the devices of variables and quantification should be reserved for occasions on which strict identity can be presupposed and agree to withhold such linguistic devices when talking about—well, I can't say 'things'!—arrangements with no parts to which strict identity applies.

The analogy between dollars in a bank account and quanta in a box holds up surprisingly well, as we can see by examining an apparent disanalogy. Consider a quantum state with exactly one quantum in the left side of the box, l, and exactly one in the right, r. One might now suggest that these two properties, l and r, can serve to identify the quanta in a way in which properties cannot serve to identify individual dollars in a bank account. And if properties will serve to identify individual quanta, this might be taken to show that strict identity does, after all, apply—there is *the one on the left* and *the one on the right*, but no such identification by properties will work for dollars in a bank account.

Let's see how close an analogue we can get with a bank account. Suppose I set up a special account with exactly two dollars in it, and the stipulation that I may withdraw exactly one (of them?) from the West Branch and exactly one (of them? the other one??) from the East Branch. Of course there is a this one and a that one about the dollar bills once I have actually made my withdrawals, and sense to questions such as, Did I withdraw this bill from the West Branch, or Did it come from the East Branch? But all our alarm bells go off if, before making any withdrawals, someone suggests that there is one specific dollar waiting to be withdrawn at the West branch and a specific one waiting at the East Branch.

Now we see that to say anything different about the case of quanta in a box is to beg the question about the preobservation (prewithdrawal) applicability of strict identity. The argument was: one can use the property l to identify the quantum on the left. But of course, that is correct only if there *is* an identity to be identified, if there is a *thing* to which strict identity applies which can be pinned down with the property l. The facts about the quantum state function are consistent with a weaker set of assumptions, namely that if observations are made, exactly one will be observed on the left and exactly one on the right (exactly one dollar can be extracted from West Branch, exactly one from East Branch). There may be identities to the spots on the photographic plates (dollars received from the tellers). But it is a question-begging assumption that there are specific, identity bearing quanta (dollars) before the observations (withdrawals).[6]

To this point I have been using the expression '(minimalist) haecceity' to cover the idea that strict identity applies to a thing.[7] Why introduce this expression if 'applicability of strict identity' will do? I have two motivations.

First, 'haecceity' is traditional. In a widely cited contemporary article Adams speaks of a thing's 'thisness' which "is intended to be a synonym or translation of the traditional term 'haecceity'." Adams say that "A thisness is the property of being identical with a certain particular individual," but that "I want the word 'property' to carry as light a metaphysical load here as possible." In particular he comments that he does not mean to include the aspect of the conception of a haecceity which he attributes to Scotus, according to which a haecceity is "a special sort of metaphysical component of individuals."[8] Hence I take Adams to have in mind the sort of minimalist conception of haecceity that I have been attempting to develop here.

The second and more important reason for using the noun 'haecceity' is that in untangling the problem of quantum statistics there is at issue more than just the applicability of strict identity. There is also, in ways which I will explain, the use of labels and what I will call 'counterfactual switching'. I am confident that the three relevant characteristics need not be separately assumed. But I am not now in a position to argue this further claim and so need to list all three. I want then to use 'haecceity', a bit loosely, to cover what falls together under the three separately statable conditions.

To introduce labels, or more specifically nondescriptive reference, consider names, such as constants in first-order logic, that function by each being stipulated to refer to some one specific thing. For a fixed usage, the name 'a' refers to the very same thing when used repeatedly. If 'a' and 'b' are two such names there is one thing to which 'a' refers, one thing to which 'b' refers, and always a fact of the matter whether $a = b$ or $a \neq b$. The same kind of thing occurs with variables in first-order logic, as well as pronouns in natural languages. Although variables and pronouns function to refer to different things, in a fixed context the reference is fixed. When bound by a quantifier two occurrences of the variable 'x' refer to the same thing. Only in this way can first-order logic make the distinction between $(Ex)(Fx \ \& \ Gx)$ and $(Ex)(Fx) \ \& \ (Ex)(Gx)$.[9]

In preparation for the issue of counterfactual switching, let's quickly review some questions about counterfactuals, the notion of "possible worlds," and the issue of "transworld identity." Suppose I say: "Humphrey lost the 1968 presidential election. Had he won, many would have been surprised." Informally, we take the 'he' in the counter-factual to refer to the very same person as referred to by 'Humphrey'. Philosophers analyze such counterfactuals in terms of "possible worlds," thought of as alternative ways the world might have been. (Some philosophers take these possible worlds more literally than others.) To say that had he (Humphrey) won, that would have surprised many is to say that in all possible worlds in which Humphrey had won, and which are in other respect sufficiently similar to the actual world, many would have been surprised. But for this to capture what we intend with the counterfactual, there must be some sense in which the Humphrey in the possible worlds is the same individual as the real-world Humphrey. Many take this sense to be that of strict identity. Kripke says that we stipulate possible worlds, and in so stipulating them we refer to the very same individuals as occupy the real world.[10] Kaplan takes strict iden-tity across possible worlds to make antecedent sense and ties the notion explicitly to that of haecceities.[11]

With these ideas in place, I now offer what I will call three "tests" for whether a subject matter includes haecceities, or alternatively, three ways of indicating what is meant when we speak about entities with haecceities.

1. Strict identity: A subject matter comprises things with haecceities just in case the subject matter comprises things to which strict identity applies; that is, just in case there is a fact of the matter for two putatively distinct objects, either that they are distinct or, after all, that they are one and the same thing.

2. Labeling: A subject matter comprises things with haecceities just in case the subject matter comprises things that can be referred to with names directly attaching to the referents; that is just in case these things can be named, or labeled, or referred to with constants where the names, labels, or constants each pick out a unique referent, always the same on different occurrences of use, and the names, labels, or constants do not function by relying on properties of their referents.

3. Counterfactual switching: A subject matter comprises things with haecceities just in case the subject matter comprises things which can be counterfactually switched, that is just in case a being A and b being B is a distinct possible case from b being A and a being B, where A and B are complete rosters of, respectively, a's and b's properties in the actual world.[12]

Clearly there will be close connections between these tests. For example, it seems plausible that the applicability of the labeling test presupposes the applicability of strict identity. And one might argue that counterfactual switching presupposes labeling. In fact there is a lot of work to be done to sort out the relations between these tests, since one's conclusions may vary with one's views about naming and about counterfactuals. For example, if you believe, as does Lewis,[13] that possible worlds have no objects in common and that talk about transworld identity has to be reconstructed in terms of a counterpart relation, then one has better prospects for pulling apart the labeling and counterfactual switching tests. Before all this work is done it is not clear whether the three tests are different ways of getting at the same idea or whether the tests display different, if closely connected ideas. A full discussion of this subject will have to sort all of this out. I'm hoping that I've indicated enough about what haecceities are supposed to be to be able to support the critical examination that will follow.

I suggest that belief in haecceities, if only tacit and unacknowledged, plays a crucial role in the felt puzzles about quantum statistics. Haecceities in the counterfactual switching role are required in order to get the roster of possible cases which then leads one to expect classical statistics. For this reason it is hard to see how one would be particularly puzzled by the quantum statistics unless haecceities, at least in the counterfactual switching sense, are somehow assumed.

Haecceities also enter the discussion in their labeling role through the traditional use of particle labels in the traditional formulations of quantum mechanics.

3. Particle Labels in Quantum Mechanics

Let's start with quantum mechanical descriptions of a single particle.[14] In our stylized example from section 1, let's consider a single particle in our box, which can be in just one of two states, being in the left half of the box and being in the right half. In a widely used notation the first state is represented by $|l\rangle$ and the second by $|r\rangle$.

Now let's move to the case in which we have two qualitatively identical particles, two particles with the same mass and charge, but which can, of course, have different transient properties. Standard notation in use in physics is to combine particle labels, the numerals '1' and '2', with the state representations for single particles, $|l\rangle$ and $|r\rangle$. This gives the description of four cases:

$|l(1)\rangle|l(2)\rangle$ particle 1 on the left and particle 2 on the left;
$|l(1)\rangle|r(2)\rangle$ particle 1 on the left and particle 2 on the right;

$|r(1)\rangle|l(2)\rangle$ particle 1 on the right and particle 2 on the left;
$|r(1)\rangle|r(2)\rangle$ particle 1 on the right and particle 2 on the right.

Clearly the use of the labels in description of these four cases facilitates statement of the puzzle about quantum statistics. Maintain that the individual particles are independent and each equally likely to be on either side of the box, and we conclude that the above four cases are equally likely.

It is very tempting to see physicists' use of the particle labels as a behavioral manifestation of tacit belief that particles have haecceities in the minimalist sense that enables at least the labeling test. The labels are "pure." In the context in which we are considering a collection of qualitatively identical particles, we propose to call one of them 1, some second one 2, and so forth. Then '1', and '2', and so forth are then supposed each to refer to some one specific particle, and without the benefit of any property or feature that distinguishes the one from the other. But this is just to suppose that particles have haecceities according to the labeling test.

In more detail, I indulge in the following historical speculation. (I'm hoping that real historical work will be done to ascertain the degree of accuracy of this speculation!) The authors of quantum mechanics took themselves to be describing particles. In the process they found that they had to give up some of the aspects that classical physics ascribes to particles. Principally, they had to give up ascribing simultaneous position and momentum. But, perhaps unreflectively, they continued to think of particles as things that can be labeled in thought: we can talk about them as particle 1, particle 2, and so on, whether or not we have at our disposal some distinguishing features by which to assign the labels. But, of course, so doing is just to think of particles as having haecceities in the labeling sense, which then facilitates the application of counterfactual switching, which generates the four possible cases just described, and, finally, saddles us with the puzzle about quantum statistics.

4. Particle Labels, Superposition, and Symmetrization

Description of the possible states of two particles, as previously given in the four cases, can't be right—they lead to the wrong statistics! To get a correct description quantum mechanics has traditionally added another ingredient—symmetrization.[15] To explain symmetrization I need first to explain the more general idea of superposition.

Let $|m\rangle$ ascribe property m to a one-particle system. Similarly $|n\rangle$ ascribes n to a one-particle system, where, m and n are contrary properties, for example, two values of position or two values of momentum. The states $|m\rangle$ and $|n\rangle$ are represented by mathematical entities—vectors—which can be added. The principle of superposition says that the sum, $|m\rangle + |n\rangle$, represents a new state.[16] More generally, states are represented by weighted sums, $c_m|m\rangle + c_n|n\rangle$. $c_m|m\rangle + c_n|n\rangle$ represents a new property, which is related to the m and n properties as follows: if one examines the particle to see whether it has property m or n, one will find m and n with probabilities proportional to $|c_m|^2$ and $|c_n|^2$.

The principle of superposition applies very generally, so that we can add, or superimpose, the states $|l(1)\rangle|r(2)\rangle$ and $|r(1)\rangle|l(2)\rangle$. What does the superimposed state $|l(1)\rangle|r(2)\rangle + |r(1)\rangle|l(2)\rangle$ represent? We would like to say that this is a case in which there is an equal chance of observing the states $|l(1)\rangle|r(2)\rangle$ and $|r(1)\rangle|l(2)\rangle$. There are two problems with this.

First, for reasons which will emerge later, on many interpretations of quantum mechanics we seem forced to say that these two states, $|l(1)\rangle|r(2)\rangle$ and $|r(1)\rangle|l(2)\rangle$, never occur. Second, if l and r are maximal properties, completely describing individual particles, we cannot distinguish between these two states! If l and r are maximal properties, there are no properties left over to provide the wherewithall for identification, so that there is no way of determining *which* particle is l and which is r. All we can say about each of these states is that they describe a situation in which there is one l and one r. But this goes also for the superposition, $|l(1)\rangle|r(2)\rangle + |r(1)\rangle|l(2)\rangle$. If the superposition is understood as a case in which there will be one l and one r either because the system is in the $|l(1)\rangle|r(2)\rangle$ state or because the system is in the $|r(1)\rangle|l(2)\rangle$ state, then $|l(1)\rangle|r(2)\rangle + |r(1)\rangle|l(2)\rangle$ is a state in which there is sure to be one l and one r.

$|l(1)\rangle|r(2)\rangle + |r(1)\rangle|l(2)\rangle$ is said to be symmetric, or a symmetrized state, in the sense that if one exchanges the labels, '1' and '2', we get the original state back: exchanging the labels does not change the state. Quantum mechanics deals with the problem of quantum statistics by requiring states to be symmetric.[17] The symmetric states for the particles, 1, and 2, and two properties, l and r, are

$$|l(1)\rangle|l(2)\rangle,$$

$$|r(1)\rangle|r(2)\rangle,$$

$$|l(1)\rangle|r(2)\rangle + |r(1)\rangle|l(2)\rangle.$$

These cases exhaust the symmetric possibilities that quantum mechanics allows. Although we are not forced to assign these possibilities equal chances, equal chances is at least as plausible as any other distribution, and it gives the right statistics. The problem of quantum statistics has been resolved.

Let's state this a little more generally. In conventional quantum mechanics, an *observable* is said to be symmetric if it simply counts the number of particles that has each of a given list of properties. In other words, an observable is symmetric if it remains unchanged when particle labels are switched. A state is also said to be symmetric if it remains unchanged under any switching of particle labels. For example, the symmetric state for three particles, two of which are *l* and one of which is *r* is:

$$|l(1)\rangle|l(2)\rangle|r(3)\rangle + |l(3)\rangle|l(2)\rangle|r(1)\rangle + |l(1)\rangle|l(3)\rangle|r(2)\rangle.$$

Traditional quantum mechanics countenances only symmetric observables and restricts application of the formalism to the use of symmetric[18] states. With no nonsymmetric states describing cases that could arise, we have eliminated the cases such as $|l(1)\rangle|r(2)\rangle$ and $|l(2)\rangle|r(1)\rangle$ that pull us toward classical statistics. By using only symmetric observables we dodge the issue of how to understand the components, such as $|l(1)\rangle|r(2)\rangle$ and $|l(2)\rangle|r(1)\rangle$, in a symmetric state such as $|l(1)\rangle|r(2)\rangle + |r(1)\rangle|l(2)\rangle$. In many respects this is a natural solution: in treating a collection of particles that share all their fixed properties, the solution provides a way of paying attention only to the number of particles that exemplify each (maximal) variable property.

5. Problems with Interpreting the Labels

This is the end of the story with getting a correct description of the physics. But there is a lot to be said by way of evaluating just what has gone on in this story.

Now why was it that physics does not use nonsymmetric states in describing actual cases? The correct but superficial answer was: because only by never using nonsymmetric states do we get a correct description of the physics. Here is just a little more detail about how that works. Evolution of states in quantum mechanics—that is, the change of states over time—preserves the symmetry type of a state.[19] Consequently, if a system starts in a symmetric state, the new states into which it evolves must all be symmetric also. Consequently, no matter how a collection of symmetric states starts out, if it settles down to a thermodynamic equilibrium in which the available states are equally represented, this will be the collection of symmetric states exemplified in quantum statistics. Conversely, if a system were to start out in a nonsymmetric state we would expect thermodynamic equilibrium to result in classical statistics. But we never see classical statistics for quantum systems. We are driven to the conclusion that nonsymmetric states never occur.[20]

How we should understand this conclusion will depend on how we interpret quantum mechanical states. Let's start with the informal attitude often evinced in lectures and textbooks. Physicists describe themselves as describing some collection of particles. Let's, they say, take one of these particles to be particle 1, one of them to be particle 2, and so on. They then use these labels in writing down state descriptions, being careful never to use a nonsymmetric description.

On its face this widespread informal attitude expresses realism about particles and their capacity to bear labels. There really are these things, particles, out there. When we use a label, there is some one thing to which the label refers—each label has a referent, and use of the label refers to the distinguished referent and to nothing else. Permutation of the labels in a nonsymmetric state gives rise to a distinct nonsymmetric state, describing a distinct case—if this were not the case, symmetrization would not be necessary. Thus the informal realism about particles comes with the conditions of labeling and counterfactual switching.

In the context of thus thinking realistically about the quantum mechanical state descriptions, we are left with an unanswered question. We know what it means to say that a nonsymmetric state occurs: to say that the state $|l(1)\rangle|r(2)\rangle$ occurs is to say that there are two particles, one labeled 1, one labeled 2, and 1 is l and 2 is r. The unanswered question is, Why do these states never occur? Generally we expect that when a theory we accept describes states, we can actually find these states; or if not that at least the theory should tell us why not. Let me illustrate. Statistical mechanics describes a situation in which a cold cup of coffee spontaneously starts to boil, without the application of any heat source. We *never* see this happen. But statistical mechanics also tells us why this occurrence, clearly describable within the theory, is never found. Heat is random motion of molecules. Ordinarily the random motions are pretty evenly spread about. The theory allows for statistical flukes, in which the fast molecules all bunch up in a restricted area. This is the way an unheated cup of coffee could start to boil. But the theory also tells us that the probability for such a fluke is so fabulously small that we are morally certain never to see one.

Unlike statistical mechanics, quantum mechanics, realistically interpreted, gives us no account of why its never-occurring states don't occur.

I take the silence on this question to provide an at least small embarrassment for quantum mechanics, realistically construed. And there is a further embarrassment. On the realist reading we understand ascription of a nonsymmetric state. But how are we supposed to understand ascription of the superposition of nonsymmetric states that constitutes a symmetric state? If indeed the nonsymmetric states never occur, the superposition of such states can't be understood in terms of the probabilities for occurrence of the components of the superposition—in the case at hand, these components are just the never-occurring nonsymmetric states![21] It seems that the labels' real function is to aid in counting up how many times each property occurs, without regard to which particle has what property. But, on the realist label-bearing and counterfactual-switching way of thinking about the particles, it is at least not clear how this counting up functioning of the labels is supposed to work.[22]

These embarrassments might motivate an instrumentalist reinterpretation of physicists' casual way of talking. I am here not concerned with the general question of whether alpha particles, photons, and the like are "really out there," or even what that is supposed to mean. Instead I want to focus just on the way we understand the labels. On the instrumentalist reinterpretation, we refrain from taking the common informal presentation of the labels at face value—we refrain from thinking of '1', '2', and so forth as labels of anything. Instead the labels, functioning exclusively in writing down symmetric state descriptions, are just a formal device for getting the right quantum statistics by counting up the number of occurrences of each kind of property—so many l's, so many r's. In other worlds, we understand the labels as formal devices for presenting what are called *occupation numbers*.

This instrumentalist reinterpretation is consistent, coherent, and effective in getting the physics right. But there is still an embarrassment. In setting up the formal machinery we create excess formal descriptions. The formalism provides for the description of nonsymmetric states. These excess descriptions do not now saddle us with unanswered questions because these descriptions go completely uninterpreted. But they are still hanging around as excess baggage, idle cogs in the formal apparatus, which have no function in describing actually occurring states.

Let's take stock. On the realist reading we have a material question with no answer: why do nonsymmetric states never occur?[23] On what I've described as an instrumentalist alternative, we still have a formal dangler, an idle cog in the descriptive machinery. I submit that these are drawbacks in the larger theoretical structure.

Are these drawbacks serious? Most will say that they are no more than insignificant blemishes on what is otherwise a fabulously successful theory. The drawbacks should be evaluated as no more than a drop of methodological red ink in a sea of empirical and methodological black.

This sensible evaluation is compromised by the existence of an alternative formulation of the theory. Even small amounts of methodological red ink count when we have available an alternative with all the advantages and none of the disadvantages of the first formalism. And what is known as the Fock space representation provides just such an alternative. All we need say about this formalism is that it uses no particle labels. Instead it works directly from occupation numbers, the numbers describing how many times each maximal property is instantiated, with no regard to "which" particle has which of the properties. This account does not enable the description of nonsymmetric states, so that there is no question of why nonsymmetric states do not occur, nor is there any excess formal baggage.[24]

The Fock space formalism—or, as I will call it, the occupation number formalism—invites us to think about the subject matter of quantum mechanics as free of haecceities, free from things that support labels or admit of counterfactual switching. There are only amounts, or "heaps" of stuff, coming in discrete units, thought of in analogy to dollars in a bank account, with no this one or that one about ones with the same properties (ones in the same bank account), and no sense to a different case arising if they could somehow be switched. (Has anything changed if the bank redeposits one of your dollars in my account and one of my dollars in yours?)[25]

The issue of haecceities is certainly not forced by these considerations. One could insist that although there are no empirical repercussions, there must somehow be a "this one" and a "that one" about the individuals in a collection of quantum entities described with the occupation number formalism. But if one takes this insistence seriously, one ought to readmit a language in which such distinctions can be described, and get back the unanswered questions or empirically surplus formal baggage. These aren't compelling reasons for giving up haecceities. But I feel that they do provide good reasons.

6. Are Haecceities a Nonissue?

Bas van Fraassen appears to dispute my conclusions by claiming that the descriptions with particle labels and with occupation numbers are equivalent.[26] He and I agree that the empirical facts can be adequately described with either formalism,[27] but I feel that there is something at least misleading in his glossing this fact as an equivalence of sorts. So I want to go over his argument.

Van Fraassen's terminology differs from mine. He considers the questions, "Are there individual particles?" and "Is there a 'loss of individuality'?"[28] He speaks about "worlds [models] with and without individuals."[29] These expressions suggest that van Fraassen has in mind the minimalist conception of haecceities with which I have been concerned in this chapter. I assume this reading of van Fraassen in what follows, the point being to argue that his arguments do not show descriptions with and without haecceities to be equivalent if 'haecceities' is taken in my minimalist sense. This is, however, a reading of van Fraassen specifically driven by the needs of this chapter, and there is much to be said for a stronger reading. I see van Fraassen's arguments as entirely cogent if one reads him as taking minimalist haecceities for granted and then arguing that it makes no difference whether or not one assumes in addition what I have called haecceities in the metaphysically robust sense. I think readers will be able to see this difference for themselves, and since minimalist haecceities have been my concern, I will leave as an excercise the job of reevaluating the arguments on the stronger reading.

Van Fraassen and I agree that quantum mechanics does not force the issue. But in places van Fraassen maintains that there is no real contrast between the ways of thinking about our subject matter. He writes that

I shall argue that the 'loss of identity' dispute can be dissolved [by identifying a false presupposition without which the dispute dissolves]. The question rests on a mistake—or more precisely on a metaphysical position which has been moribund for centuries.[30]

He concludes that

The models required by semantic universalism [described later] are exactly those which can be described *equally* on either view [the views that take possible worlds to be, or not to be, populated by individuals]. Therefore every significant proposition can be restated entirely in terms of occupation numbers.[31]

Van Fraassen also talks about equivalence and isomorphism. He describes how, starting from the familiar label formalism, one constructs a Fock space representation. Then, citing the results of others, he claims that

All models of (elementary, non-relativistic) quantum field theory [which is label free] can be represented by (i.e. are isomorphic to) the sort of Fock space model constructions I have describe above. Since the latter are clearly carried out within a 'labelled particle' theory, we have a certain kind of demonstrated equivalence of the particle—and the particleless—picture.[32]

To make the form of his argument clear, van Fraassen provides an analogy:

The normal way, post-set-theory, of conceiving of geometry is as a theory of points and relations among points. Spheres, for example, are definable sets of points. But it is also possible to write an equivalent theory (in the above sense of equivalence)[33] in which the elements are spheres and relations among spheres. Points are then identifiable with a definable set of spheres (intuitively, the point *is* a set of spheres which contain it; points are introduced by limit construction). Obviously we are here offered two rival world-pictures for our consideration. But if someone likes to talk in terms of spheres, I can reconstrue his every assertion *salva veritate* (and saving also all valid inferential relations) as an assertion about points. And vice versa! This is not to deny that it is possible for a person to believe that points are the only real concrete individuals—what we cannot do is to say that geometry forces this view on us.[34]

I take van Fraassen to be claiming that a metaphysics with and without "individuals" is in every relevant way like this example. He first argues this point by examining the relation between the familiar label using quantum mechanics formalism and the label free formalism—Fock space—used in quantum field theory. He then argues the same point perfectly generally, without reference to any specific theory, by comparing label-burdened and label-free description of general facts conforming to the principle he calls "semantic universalism." I will outline both of these arguments and argue in my own turn that in each case the claim of equivalence is at least misleading. (The following material on Fock space presupposes a very small acquaintance with quantum mechanics. Readers without this background can still get the general drift of my thinking on this argument, or can skip immediately to the material on semantic universalism.)

6.1 *The Fock Space Argument*

To carry out the construction of Fock space "within a 'labelled particle' theory," van Fraassen follows a formulation frequently used in physics texts.[35] He starts with the Hilbert space for one particle, H. To treat n qualitatively identical particles, he considers the n-fold tensor product of H, H^n, from which he then "throws away" all the nonsymmetric states,[36] leaving only the symmetric states, which constitute a new Hilbert space, H_+^n. As so far described, H_+^n has been expressed with the aid of labels. But it can be equivalently described in terms of occupation numbers. For each symmetric state in the first formulation, just substitute a description expressed in terms of occupation numbers, exactly the occupation numbers that can easily be read off of the symmetric state.

In addition to the H_+^n for each positive integer, n, van Fraassen considers an "empty" Hilbert space, H^0, consisting of just one state, the vacuum, the state with no particles. Fock space is now the direct sum,

$H^0 + H^1_+ + H^2_+ + \cdots$, that is, the Hilbert space formed by taking super-positions of the states in any of the terms of the sum.

Finally, van Fraassen glosses this construction as follows: "Since [the Fock space model constructions] are clearly carried out within a 'labelled particle' theory, we have a certain kind of demonstrated equivalence of the particle—and the particleless picture."[37] I agree that the "pictures" are empirically equivalent, in the sense that all the facts that actually arise can, one way or another, be described in either framework.[38] But it seems to me misleading to parley the empirical equivalence into equiv-alence of the pictures across the board. In cutting down from H^n to H^n_+ we have lost expressive power. In H^n but not in H^n_+ we can describe cases that never occur. This fact, in turn, shows that although the cases that do occur can be described in either picture, there is an important sense in which the descriptions are not equivalent. One picture—the one using labels—describes the cases that do occur in terms of a conceptual framework that facilitates saying things that cannot be said with the re-sources of the other picture.

Alternatively, I can express my dissatisfaction in terms of a dilemma. Either the labels are to be thought of, instrumentalistically, as doing no more than providing a roundabout way of assigning occupation numbers—on this alternative the labels do not individually refer to particles and we have no more than a thinly disguised restatement of occupation numbers—or we take the labels seriously as labels. Then they facilitate description of cases which cannot be described with the Fock space formalism, the nonoccurring nonsymmetric cases. And ap-parently they do this in virtue of their haecceitistic function. Either we have one picture, described twice in essentially the same way, or we have two inequivalent pictures.

6.2 *Semantic Universalism*

Van Fraassen restates what I take to be essentially the same argu-ment, but put in much more general terms; and I have essentially the same complaint. To set the argument in its more general framework, van Fraassen considers what he calls *semantic universalism*, the position that "all factual description can be completely given in entirely general propositions."[39] To say that a proposition, understood as a set of pos-sible worlds, is general is to say that no change of truth value results from a permutation of individuals.[40] Whether or not semantic universal-ism holds across the board, it clearly holds for all facts (things expressed by true propositions) in the domain of quantum theory of identical par-ticles.

To set up his argument, van Fraassen tailors a specialized kind of semantic model. Each of these models includes a domain of individuals and a roster of "cells" that represent qualitative distinctions. Each cell corresponds to a maximally specific, consistent set of properties. Van Fraassen then takes a possible world to be an assignment of individuals to cells, and he assumes an access relation between possible worlds in terms of which modalities are described in the usual way. A model (also called a 'universe') is a collection of such possible worlds, all with a common domain and roster of cells and a fixed access relation. Finally, van Fraassen calls such a model *full* if it is closed under permutations on the individuals in the domain: if possible world w is in the model, so are all w' resulting by permuting the individuals, and the access relation is similarly preserved under permutations.

In terms of this formal machinery, van Fraassen offers the following argument:

> We return finally to the question of worlds with and without individuals. The models required by semantic universalism are exactly those which can be described *equally* on either view. So far we have described them in terms of individuals. But each world—a mapping of individuals into cells of a logical space—can be characterized simply as a set of occupation numbers for the cells. Closure under permutation of the access relation R entails that the R-modalities operate on fully general propositions without losing the generalities. Therefore every significant proposition can be restated entirely in terms of occupation numbers. This means that we can 'abstract' an individual-free model:
>
> A world is a mapping of cells into natural numbers,
>
> and this abstracted model corresponds to many models of the more ordinary sort, but (up to isomorphism) to a *unique* full model among them.[41]

Here is my gloss on this argument. 'Significant proposition' clearly means one that can be given by a "factual description," which in turn is one satisfying the principle of semantic universalism. Let us follow van Fraassen in taking propositions to be given by collections of possible worlds. Then semantic universalism plus fullness of a model—the fact that for each possible world, w, in the model, the model includes all w' resulting from permutation of the individuals in w—guarantees that if w is in a proposition, so is any w' resulting from w by permutation. Finally, in a full model a maximally strong, consistent nonmodal, general proposition can be characterized equally in terms of (1) an assignment of occupation numbers to cells, or (2) a collection of worlds that are related to one another by permutations of the individuals in the domain. Since permutation also preserves the access relation, a similar conclusion also holds for modal propositions.

I agree that the argument does show that what van Fraassen calls significant propositions (propositions that are general in the sense of being permutation invariant) are described equivalently in the two ways. And quantum mechanics deals only in propositions of this sort. But *if* one takes there to be things that have haecceities, these are not the only significant propositions! Haecceities in the sense of counterfactual switching simply says that we get a new possible case by a permutation. If cross world identity of individuals is assumed, as it is in van Fraassen's models, haecceities in the sense of labeling facilitate statement of counterfactual switching. And likewise, the applicability of strict identity involves distinctions between permuted worlds in any of van Fraassen's models, distinctions between things like this very thing being *A* and that one *B*, as opposed to the other way around. On all three tests for haeccceities, if haecceities are admitted there are more significant propositions than van Fraassen allows.

Van Fraassen's argument appears to pull in opposite directions at once. In order to set up the formalism with a domain of individuals that can be differentially assigned to cells, one must presuppose that these individuals have haecceities—for to suppose that there is a distinction as to *which* object gets mapped into which cell is just what we mean by having haecceities—at least in the counterfactual switching sense. But after presupposing individuals with haecceities, van Fraassen maintains that only general (in his sense) propositions are (in his sense) significant, which is to maintain that the distinction to which haecceities contribute are not significant. It feels like the left hand is taking away what the right hand granted.

Yet another way to express my reservations is to say that, in setting up the formalism in terms of functions that differentially map individuals into different cells, van Fraassen is presupposing what I will call a wider range of intelligible distinctions. The facts that can be expressed using only occupation numbers involve a narrower range of distinctions. While the facts of the narrow range can be expressed in terms of the wider range, this hardly shows that the two ranges of distinctions are equivalent. To put the same point in the formal mode: the occupation number language can express only the weaker range of distinctions. The stronger language, involving variables ranging over, functions on, and terms referring to members of a domain of individuals can be used to draw a wider range of distinctions. True, all distinctions made in the weaker language can be made in the stronger language. But that hardly justifies any conclusion about equivalence between what can be said with the two languages.

Van Fraassen had suggested his form of argument by reminding us that in geometry one can equivalently start from spheres or from points. In that subject "if someone likes to talk in terms of spheres, I can reconstrue his every assertion salva veritate (and saving also all valid inferential relations) as an assertion about points. And vice versa!"[42] In our present subject the "vice versa" fails.

Here is an alternative analogy. The theory of partial orders can be expressed in first-order logic. Ipso facto, it can also be expressed in second-order logic. Are the descriptions equivalent? In a sense, of course. Both languages give an exact description of the theory. But talk here of equivalence also threatens to be misleading—to say that the descriptions are equivalent suggests that there is no difference between first- and second-order logic.

Van Fraassen might well not be very moved by all of this. He is a constructive empiricist. That is, he believes only empirically significant propositions. And the wider set of distinctions that depends on haecceities does not expand the empirical content of what can be said with the narrower set. In quantum mechanics there is no observable that corresponds to one particle being l and the other r. More generally, inasmuch as distinctions that depend only on haecceities are in no way qualitative, they can make no observable difference. As a constructive empiricist, van Fraassen might appropriately insist that he takes no interest in any extra empirical inequivalence.

But constructive empiricism is not instrumentalism. Constructive empiricists allow that distinctions that go beyond empirical equivalence may be intelligible. Ascription of such distinction may have truth values. It's just that, for constructive empiricists, such ascriptions are not appropriate objects of belief.[43]

In other words, talk of equivalence of descriptions that do and do not presuppose haecceities seems right if we read 'equivalent' as 'empirically equivalent.' But full-blown equivalence has not been shown. Indeed, the argument appears to conflate the distinction, which constructive empiricism recognizes, between equivalence and empirical equivalence.

7. If Haecceities Are Taken Away, Are We Left with Fields?

I have argued that the subject matter of quantum theories should be taken not to have haecceities in even the minimalist sense and that, contrary to one reading of van Fraassen, this view provides a substantive position. Some will now argue that if we abandon particles with minimalist haecceities, we must be left with fields as the subject matter of quantum theories. The argument is that particles and fields provide exclusive and exhaustive alternatives for the subject matter of physical theories; that quantum field theory, which subsumes conventional quantum mechanics, bears a natural field theoretic interpretation; and that if we abandon a particle reading, the field reading is the only alternative remaining.

I don't agree with this further conclusion, but won't sort the issue out on this occasion because I have nothing to add to my earlier treatment.[44] But here is a synopsis. I reject the dichotomy of particles and fields. Instead I maintain that both particle and field concepts are imprecise conglomerates of more general component concepts. Once we analyze these components and reject the ones that are inconsistent with the facts of the quantum world, we discover that we can refashion a consistent new conceptual scheme, one that is not happily described in terms of either a particle or a field concept. Instead this scheme inherits parts of both antecedents. The quantal facts pressure us to give up the particle notions of exact trajectories and, if the argument of this chapter stands up, haecceities. But what remains is still susceptible to a high degree of localization; and, what is especially at odds with fields classically conceived, quanta are discrete. But to see how this all works in detail, one needs to lay everything out in the theoretical context of what is, I feel, somewhat misleadingly called "quantum field theory."

8. The Status of Haecceities

I have argued that there is a conceptually important difference in describing the facts of quantum mechanics with and without haecceities. But the Fock space formalism invites us to give up haecceities. That formalism gets all the work done with occupation numbers, eliminating the need to rely on anything that uses or presupposes haecceities. I agree with van Fraassen that quantum mechanics does not force the issue. A post-Lavoisier chemist could insist on embellishing the oxygen theory with talk of phlogiston: things burn as a result of combination with oxygen—but phlogiston also escapes! If all of mathematics could be done in first-order logic, so that second-order logic was entirely dispensible, one could nonetheless insist on using second-order logic and thinking in terms of its distinctions and the domains of its quantifiers. But why?

Some maintain that the parallel question for haecceities has a very pointed answer. These people maintain that haecceities are presupposed by the mechanisms of reference, without which language would not be possible. To review these arguments, to predicate property F of individual a requires that there be some distinct entity to which 'a' refers, an entity distinct from other entities, *b*, *c*, and so on. Again, the use of variables, both pronouns in natural language and variables in logic, requires that there be a domain of individuals with haecceities over which the pronouns or variables can range. This is required, it is maintained, for the repeated occurrence of the same pronoun or variable to pick out the same individual. For example, without this mechanism, there would be no distinction between $(Ex)(Fx \& Gx)$ and $(Ex)Fx \& (Ex)Gx$. In addition, without haecceities there is no set theory. Sets are individuated by their members. If there are no individuals with haecceities, the extensional notion of set collapse. And with the collapse of set theory, this apocalyptical story continues, all of mathematics goes. It seems we can't live without haecceities. The drop of methodological red ink mentioned earlier seems like a small price to pay for the very intelligibility of language and mathematics!

I have three responses to such apocalyptical fears.

1. The principle of identity of indiscernibles holds for much of our experience. In our normal experience things are completely individuated by their qualitative properties, by their size, shape, texture, and the like; by their space-time trajectories; and by their relations to other things. In short, the work of strict identity is accomplished by completely individuating properties. We can gloss these facts by saying that in our ordinary experience what will pass for things will have ersatz haecceities, so that we can think of them as things with haecceities, while this talk of haecceities can, in principle, be cashed in in favor of individuating properties and relations. Ersatz haecceities should suffice for the referential, predicational, and quantificational use of language. In the special case of the subject matter of quantum mechanics we need a different means of description. Where there are what people are in the habit of describing as collections of qualitatively identical particles, occupation numbers provide the most adequate description.

2. Individuation in mathematics may rest on a number of things other than extensionally characterized sets. For example, mathematical objects might themselves best be taken to be properties. To illustrate, we might take the number 3 to be the property of being three in number. Or, on a somewhat different alternative, structuralism characterizes mathematical objects entirely in terms of their relations. Set theory might indeed not be applicable to quantum objects—that is, there may well be no sense to talking about sets, extensionally understood, of quantum objects. But for ordinary objects, ersatz haecceities ought to suffice. And for set theory as a mathematical enterprise, all we need is for the empty set to be uniquely individuated by (the nonexistence of any of) its members.

3. A pluralist will find these issues not so pressing. I take science, as well as prescientific metaphysics, to be in the business of constructing

models. As a pluralist, I expect no one ultimate, complete, and perfectly accurate description of the world. Instead, we construct a variety of models, each of which shows us, with limited accuracy, what some aspects of the world are like. There are phenomena, including all of our common experience, accurately described in terms of haecceity-bearing individuals. And other phenomena, notably ones in the quantum domain, are best described without haecceities. In the enterprise of describing the great variety of phenomena that the world presents to us, we are well advised to have in store a spectrum of conceptual resources, to be applied variously in various situations.

One will sensibly ask how this pluralism squares with my first response, that where we appear to need haecceities, ersatz haecceities will do. The accuracy of the last claim rides on a closer reading of 'will do' than I have given. A satisfactory account must tell us, "Will do for what?" For this occasion I will only note that there are many situations in which arguably flawed descriptions are still the best descriptions to give. For example, in describing the fluid properties of water, water's capacities for flowing through pipes and supporting waves, one is best off describing water as an incompressible continuous medium. Though inaccurate in certain respects, such a description provides the best way to say many true things about water. More generally, a conceptual resource that we know to fail in important respects may nonetheless provide the best resource we have for getting at important things we want to say and to know.

So I conclude that quantum mechanics shows us that there are important facts about the world which can be adequately described without appeal to the conceptual resource of haecceities. And in so doing it demonstrates to us that thinking in terms of haecceities is not an unavoidable conceptual strategy, not a necessary part of any metaphysics of concreta. We have learned about limitations in thinking in terms of haecceities. But I refrain from concluding that therefore the concept has no usefulness whatsoever.

Notes

Comments on an earlier draft from Bas van Fraassen, Michael Jubian, and Nick Huggett have helped me immeasurably in clarifying the arguments in this chapter. This chapter further develops ideas originally in M. Redhead and P. Teller, "Particles, Particle Labels, and Quanta: The Toll of Unacknowledged Metaphysics," *Foundations of Physics* 21 (1991): 43–62, and M. Redhead and P. Teller, "Particle Labels and the Theory of Indistinguishable Particles in Quantum Mechanics," *British Journal for the Philosophy of Science* 43 (1992): 201–218. The argument of section 5 is essentially the same as the one worked out with Redhead, but there have been changes in formulation. Section 2 works at making the relevant notion of haecceity clearer than previously. Section 6 makes comparisons with relevant views of van Fraassen, and section 8 airs some new thoughts on repercussions of these arguments.

1. Davidson uses the expression 'identical in the necktie sense' for qualitatively identical, having leaned on the necktie example to bring out the intended usage.

2. Adams uses the expression 'primitive thisness'. (R. Adams, "Primitive This-ness and Primitive Identity," *Journal of Philosophy* 76 [1979]: 5–26.) Red-head and Teller ("Particles, Particle Labels, and Quanta," and "Particle Labels and the Theory of Indistinguishable Particles in Quantum Mechanics") adopt Heinz Post's expression, 'transcendental individuality'.

3. Adams characterizes a thing's 'thisness', which he intends as a synonym for 'haecceity', as the property of being identical with that thing. He attributes to Scotus the view that a thing's haecceity is a "special sort of metaphysical component" of that thing. See Adams, "Primitive Thisness," 6–7.

4. I develop this kind of example in P. Teller, "Quantum Physics, the Identity of Indiscernibles, and Some Unanswered Questions," *Philosophy of Science* 50 (1983): 309–319.

5. With my linguistic intuitions poisoned, perhaps, by too many decades of think-ing and talking about quanta, I did not find such an extended use particularly problematic. But, under protest from colleagues, I tried out the following lit-tle passage on some nonphilosophers: "Among the things I can credit to my name are the 1,000 dollars in my account. 238 of them were deposited on a Monday, the rest of them on a Tuesday. And you can't have any of them!" Nine protested strongly, largely for the same reasons cited by philosophers. One found the passage unproblematic, and one was unsure.

6. For a breathtaking moment one might think that in these facts we get some purchase on the issue of realism about quantum states. And this may be so if we are allowed use of an additional, very strong premise: for a determinate property to be instantiated it must be instantiated by some identity possessing *thing*. Now, if the larger argument of this chapter is correct, that the facts of quantum statistics mitigate against minimalist haecceities for quantum entities, we may be able to conclude that there can be no premeasurement determinate properties because there are no premeasurement entities-with-identity to have these properties. But I am skeptical about the additional premise. See this chapter's last section for speculation on the question of this additional premise.

7. Jubian introduces the primitive idea of 'stuff', which can have properties and from which things are constituted, and then he proposes to understand a thing's haecceity as the property of "being made of such and such stuff." (M. Jubian, *Ontonlogy, Modality, and the Fallacy of Reference* [Cambridge: Cam-bridge University Press, 1993], 45.) Although he does not say so explicitly, he clearly takes strict identity to apply to any bit of stuff, and, as far as I can see, it is the applicability of strict identity to any bit of stuff that gets the work of haecceities done for Jubian.

8. Adams, "Primitive Thisness," 6–7.

9. This aspect of variables in first-order logic is made precise with the idea of substitution instances and the satisfaction relation.

10. S. Kripke, *Naming and Necessity* (Oxford: Blackwell, 1980), 16–17.

11. D. Kaplan, "How to Russell a Frege-Church," *Journal of Philosophy* 72 (1976): 716–729. Not everyone shares this view of transworld identity. Lewis holds that, strictly speaking, each object occurs in only one world and analyzes identity across possible worlds in terms of counterpart relation. See D. Lewis, *On the Plurality of Worlds* (Oxford: Blackwell, 1986), esp. 1–5.

12. Properties *A* and *B* must be qualitative—that is, they must not include properties such as *being identical with a*, which would clearly undermine the possibility of the counterfactual switching. I know of no non-question-begging way of drawing a general distinction between qualitative and nonqualitative properties, so we will have to take this distinction as primitive and intuitive. Also, if *a* and *b* have essential properties, the counterfactual switching test will only work in the special case in which *a* and *b* have exactly the same essential (qualitative) properties. This condition will be met in all cases of interest in quantum mechanics.

13. Lewis, *On the Plurality of Worlds*.

14. I use the word 'quanta' for the particle-like concept for which haecceities have been rejected. If haecceities are assumed or if they are being debated, I will continue to use the conventional 'particle'.

15. I am going to streamline my exposition by ignoring antisymmetrization. Quantum mechanics describes two kinds of particles—bosons, described with symmetrized states, and fermions, decribed with antisymmetrized states. In ignoring fermions and antisymmetrization I neglect a bag of extremely important and interesting problems, of which I will give an outline in this note. I am headed for the recommendation that we drop haecceities. One would expect that quanta—what I am calling particles without haecceities—could occur in collections of 0, 1, 2, or any whole number all in exactly the same state. This is true of bosons only. There can be only 0 or 1 fermion in exactly the same state. Why? At present we can state only that nature presents us with these two different kinds of quanta. Something called the spin-statistics theorem shows that there are connections between facts of relativity, spin, and the contrast between bosons and fermions. Perhaps from this formal theorem one can extract a more robust physical understanding of why and how nature makes room for these two kinds of quanta.

16. I will ignore normalization throughout.

17. Symmetrized states are required for bosons. For fermions one requires states to be antisymmetrized, that is one requires that if two labels are switched one changes the sign for the whole state.

18. And antisymmetric states.

19. Because the Hamiltonian is symmetric.

20. A qualification on this conclusion occurs shortly, for interpretations that distinguish between "dynamic" and "value" states.

21. This claim is incorrect for certain interpretations of quantum mechanics, for which the claim that nonsymmetric states never occur also needs to be qualified. Let's distinguish between dynamic states, which evolve over time and are the ones ordinarily described in the quantum formalism, and value states, which give the values actually found on measurement and have no dynamics prescribed other than that they occur on measurement with the proba-

bilities prescribed by application of the usual algorithm to the dynamic states. These "hidden-variables" are not assumed everywhere—they occur only in sufficiently specialized cases so as not to run afoul of no hidden variable proofs. On such an account, nonsymmetric states never occur as dynamic states. But they do occur as value states—indeed, the symmetric dynamic states are understood as giving the probabilities for the occurrence on measurement of the nonsymmetric value states. In this brief summary I've followed the terminology of van Fraassen (B. van Fraassen, *Quantum Mechanics: An Empiricist View* [Oxford: Oxford University Press, 1991]), although van Fraassen does not describe himself as a realist. R. Healey, *The Philosophy of Quantum Mechanics: An Interactive Interpretation* (Cambridge: Cambridge University Press, 1989), develops similar ideas.

22. Again, this does not seems like such a problem for a realist view that distinguishes between dynamic and value states.

23. On an interpretation that distinguishes between dynamic and value states, nonsymmetric states still never occur as dynamic states.

24. The Fock space formalism actually has empirical content additional to that found in conventional quantum mechanics. It provides for description of indefinite particle numbers. I am not here counting this as an additional reason for preferring the Fock space form of description to the conventional one using labels because I suspect that indefinite particle number descriptions could be reconstructed within a label using formalism, though such a reconstruction might be quite complicated and awkward.

25. This description is intended for bosons. Since there can be only zero or one fermion in a state, there is a sense in which fermions are individuated by their properties. But we still need admit no sense to the thought that the "bare" fermions in different states have somehow been switched.

26. van Fraassen, *Quantum Mechanics*.

27. Again, the occupation number formalism naturally covers the case of indefinite particle number. I continue to assume that, if one really wanted to, one could reconstruct these facts in a formalism using labels.

28. van Fraassen, *Quantum Mechanics*, 435–436.

29. Ibid., 475.

30. Ibid., 434.

31. Ibid., 475.

32. Ibid., 448.

33. The referent of the "above sense of equivalence" is not clear. Two paragraph's back van Fraassen refers both to equivalence as "the representability of one sort of mathematical object as another sort" and to "a weaker but philosophical more interesting equivalence... [that] theories are necessarily empirically equivalent. Any possible phenomena that can be accommodated in the one sort of model can also be accommodated in a model of the other sort." (Ibid., 450.) Although he describes it as philosophically less interesting, I assume that van Fraassen must here have the stronger sort of equivalence in mind. In the next paragraph, the paragraph immediately preceding the quotation in the main body of my text, he refers to "strong equivalence," and this strong equivalence seems to be what is in question in the conclusion he draws

on p. 475, where he talks about alternative statements of "significant propositions," where by significant he means general, but not limited to empirical propositions.

34. van Fraassen, *Quantum Mechanics*, 451.
35. Ibid., 436–448.
36. And antisymmetric states.
37. van Fraassen, *Quantum Mechanics*, 448.
38. Once again waving any worries about inequivalence with respect to indefinite particle number.
39. van Fraassen, *Quantum Mechanics*, 465.
40. Ibid., 469.
41. Ibid., 475–476.
42. Ibid., 451.
43. See ibid., 4. For a more detailed exposition see B. van Fraassen, *The Scientific Image* (Oxford: Oxford University Press, 1980), 11–13.
44. P. Teller, *An Interpretive Introduction to Quantum Field Theory* (Princeton: Princeton University Press, 1995), chap. 5.

8

Quasiset Theories for Microobjects: A Comparison

M. L. Dalla Chiara, R. Giuntini, and D. Krause

1. Introduction

The basic aim of *quasiset theories* is to provide a mathematical framework for dealing with collections of microobjects, which seem to violate some characteristic properties of the classical identity relation.

Looking for axioms to describe the behavior of entities such as elementary particles is an important area of research into the foundations of mathematics, according to Manin. In his lecture at the 1974 Symposium of the American Mathematical Society, he observed:

> I would like to point out that it is rather an extrapolation of common-place physics, where we can distinguish things, count them, put them in some order, etc.... New quantum physics has shown us models of entities with quite different behaviour. Even [models] of photons in a looking-glass box, or of electrons in a nickel piece are much less Cantorian than the of grains of sand....
>
> The twentieth century return to Middle Age scholastics taught us a lot about formalism. Probably, it is time to look outside again. Meaning is what really matters.[1]

In this chapter we compare the basic ideas of two different approaches to the notion of *quasiset*. One has been developed (with different technical details) by da Costa, French, and Krause.[2] The other has been proposed and applied to semantic questions of quantum mechanics (QM) by Dalla Chiara and Toraldo di Francia.[3]

Our present analysis will mainly discuss the intuitive aspects of these different theories. For technical details, we refer to the original essays.

2. Schrödinger's Idea of Microobject and the Theory S^{**}

In some of his general writings, Schrödinger discussed the inconsistency between the classical concept of physical object as an individual entity and the behavior of particles in quantum mechanics. Quantum particles, he noticed, lack individuality, and the concept of identity cannot be applied to them in the same way as in classical physics. Schrödinger insisted that the language we normally use to deal with the objects of modern physics caused serious and apparently insuperable obstacles for an adequate discussion from the foundational point of view. Such a language "constantly drives our mind to ask for information which has obviously no significance. Its imaginative structure exhibits features which are alien to the real particles."[4]

One of the aims of our theory S^{**}[5] is to try and describe formally the following idea defended by Schrödinger: identity, which may hold for macroobjects, is not defined for microobjects. As a consequence, we cannot even assert that an "electron is identical with itself." In the realm of microobjects only an *indistinguishability relation* makes sense.

Let us now describe the basic features of S^{**}, which represents a generalization of a Zermelo-Fraenkel-like set theory with *ur-elements* (ZF). Differently from the classical case, here an ur-element may be either a *macroobject* or a *microobject*. Collections are represented by *quasisets* and classical sets turn out to be limit-cases of quasisets.

The theory is supposed formalized in a classical first-order logic *without* identity. The language contains the usual logical constants: the connectives ¬ (not), ∧ (and), ∨ (or), → (if...then) and the quantifiers ∀ (all), ∃ (at least one). The primitive concepts are the following:

1. three monadic predicates: *macroobject* (M), *microobject* (m), *set* (Z);
2. two binary relations: the *membership* relation (∈) and the *indistinguishability* relation ≡;
3. a 1-ary functional symbol: the *quasicardinal of* (*qcard*).

The notion of formula of S^{**} is defined in the standard way. We will use $x, y, z, \ldots$ as metavariables for individual variables, and $A, B, C, \ldots$ as metavariables for formulas. Restricted quantifiers will be used in the expected way. For instance, $\forall_Q xA$ will mean $\forall x[Q(x) \to A(x)]$. In this language, the following notions can be defined

Definition 2.1 *Ur-object*

An *ur-object* is either a macroobject or a microobject:

$$O(x) := M(x) \vee m(x).$$

Definition 2.2 *Quasiset*

A *quasiset* is something that is not an ur-object:

$$Q(x) := \neg O(x).$$

Definition 2.3 *Classical entity*

A *classical entity* is either a macroobject or a set:

$$C(x) := M(x) \vee Z(x).$$

Definition 2.4 *Subquasiset*

x is a *subquasiset* of y iff all the elements of x are elements of y:

$$\forall_Q xy[x \subseteq y := \forall z(z \in x \rightarrow z \in y)].$$

Definition 2.5 *Extensional equality*

(i) Two macroobjects are *extensionally equal* iff they are indistinguishable:

$$\forall_M xy[x =_e y \leftrightarrow x \equiv y].$$

(ii) Two quasisets are *extensionally equal* iff they have the same elements:

$$\forall_Q xy[x =_e y \leftrightarrow \forall z(z \in x \leftrightarrow z \in y)].$$

Following Schrödinger's suggestion, extensional equality has no definite meaning for microobjects. From a syntactical point of view, an expression like $m(x) \rightarrow x =_e x$ will be a well-formed formula, which is not a theorem. The basic axioms of S^{**} are the following:

(A1) The indistinguishability relation $\equiv$ is an equivalence relation (reflexive, symmetric, and transitive).

(A2) The extensional equality satisfies the substitutivity principle in the case of entities that are not microobjects. In other words, extensionally equal entities (that are not microobjects) can be mutually substituted in all contexts:

$$\forall xy[\neg m(x) \wedge \neg m(y) \wedge x =_e y \rightarrow (A(x, x) \leftrightarrow A(x, y))].$$

(with the usual syntactical restrictions).

As a consequence, one obtains that extensionally equal entities are indistinguishable, but not the other way around. At the same time, generally, extensional equality behaves like the standard first order equality only for entities that are not microobjects.

(**A3**) All that has elements is a quasiset:

$$\forall xy[y \in x \rightarrow Q(x)].$$

(**A4**) Sets are quasisets:

$$\forall x[Z(x) \rightarrow Q(x)].$$

(**A5**) Quasisets, containing microobjects, are not sets:

$$\forall_Q x[\exists_m y(y \in x \rightarrow \neg Z(x)].$$

(**A6**) Quasisets whose elements are either macroobjects or sets are sets:

$$\forall_Q x \, [\forall y \, (y \in x \rightarrow (M(y) \vee Z(y))) \rightarrow Z(x)].$$

(**A7**) All that is indistinguishable from a microobject is a microobject:

$$\forall x[m(x) \wedge x \equiv y \rightarrow m(y)].$$

(**A8**) Indistinguishable sets are extensionally equal:

$$\forall_Z xy[x \equiv y \rightarrow x =_e y].$$

Similarly to ZF, we assume some existence axioms:

(**A9**) *The null quasiset*
There exists a quasiset that has no elements:

$$\exists_Q y \forall x[\neg x \in y].$$

One can easily check that the null quasiset, for which we will use the symbol ∅, is unique and is a set.

(**A10**) *The pseudopair quasiset*
For any two entities there exists a quasiset which contains all the entities that are indistinguishable either from the first or from the second:

$$\forall xy \exists_Q z \forall u[u \in z \leftrightarrow u \equiv x \vee u \equiv y].$$

Let $[x, y]$ represent the pseudopair of x and y. The quasiset $[x, x]$ will represent the pseudosingleton of x. Notice that generally a pseudosingleton may contain more than one element. Similarly, pseudopairs may contain more than two elements.

(A11) *Separation*

Any property expressed in the language determines, for any quasiset, the subquasiset consisting of all elements that satisfy our property:

$$\forall_Q x \exists_Q y \forall z[z \in y \leftrightarrow z \in x \wedge A(z)]$$

(with the usual syntactical restrictions).

(A12) *Union*

For any quasiset x whose elements are quasisets, there exists a quasiset that contains all the elements of the elements of x:

$$\forall_Q x \left[\forall y(y \in x \rightarrow Q(y)) \rightarrow \exists_Q z(\forall t(t \in z \leftrightarrow \exists s(t \in s \wedge s \in x)))\right].$$

(A13) *Power set*

(only for sets!)

For any set x, there exists a set y that contains all the subsets of x:

$$\forall_Z x \exists_Z y \forall z[z \in y \leftrightarrow z \subseteq x].$$

(A14) *Infinity*

There exists a quasiset that contains the empty set and further contains the singletons of all its elements:

$$\exists_Q x[\emptyset \in x \wedge \forall y(y \in z \rightarrow [y] \in x].$$

(A15) *Replacement*

Let F represent a *pseudofunction* defined by a formula $A(x, y)$ of the language:

$$\forall x \exists y[A(x, y) \wedge \forall z(A(x, z) \rightarrow y \equiv z)].$$

If the domain of F is a quasiset, then also the range of F is a quasiset. Notice that F represents a pseudofunction: for, indistinguishability does not mean identity in the case of microobjects.

One can easily check that when applied to sets our axioms give rise to sets. Further, there exists at least an infinite quasiset that is a set.

(A16) *Foundation*

Sets are well founded. In other words, no infinite chain ... $x_n \in x_{n-1} \in$ ... x_1 is admitted for sets.

On this basis one can show that S^{**} contains a kind of "copy" of ZF theory. In other words, S^{**} is theoretically stronger than ZF. For the language of ZF can be naturally translated into the language of S^{**}. It is sufficient to interpret the membership predicate of ZF as the membership predicate of S^{**}, the equality predicate of ZF as the extensional equality of S^{**}, and the ur-element predicate of ZF as the macroobject predicate M of S^{**}. Finally, the quantifiers of any formula of ZF will be restricted to the predicate C ("being a classical entity"). On this basis, one can show that all theorems of ZF are translated into theorems of S^{**}. Since S^{**} is stronger than ZF, one can easily reconstruct, within S^{**}, the standard arithmetics of cardinal and ordinal numbers. Hence, in particular, the following notions will be definable: *ordinal (Ord)*, *cardinal (Card)*, the *cardinal of* x *(card(x))*, the *order* $\leq$ among ordinals and cardinals. Needless to stress that ordinals and cardinals are sets in S^{**}. As a consequence, they have a classical behavior. Let $\alpha, \beta, \ldots$ be metavariables for cardinal numbers. What about quasisets? Is it possible to associate a cardinal and an ordinal number even to proper quasisets that are not sets? As we have seen, quasisets of microobjects are not well behaved with respect to the standard identity relation. Indistinguishable microobjects lack an individual identity; as a consequence, they cannot be well ordered and labeled by a name, that should be uniquely attached to them. This is precisely what happens for collections of quantum indistinguishable particles. For instance, let us think of a collection of bosons. In spite of this, quantum physicists have tools for answering the question "how many particles belong to a certain collection?" In other words, differently from the classical case, the operation of *counting* seems to be, at least at a certain extent, independent of any individual identification and ordering. As a consequence, it seems reasonable to assume that even quasisets admit of a *quasicardinal*: the quasicardinal of a quasiset will be a cardinal number. Further, the notions of quasicardinality and of cardinality will coincide for sets. The axioms for the quasicardinality function (which is not derivable from ordinal notions and is primitive in our theory) are at least the following:

(A17) Any quasiset has a unique quasicardinal that is a cardinal number:

$$\forall_Q x \exists_Q y [qcard(x) =_e y \wedge Card(y) \wedge \forall_Z z (z =_e qcard(x) \rightarrow z =_e y)].$$

(A18) Quasicardinality and cardinality coincide for sets:

$$\forall_Z x [qcard(x) =_e card(x)].$$

(A19) Only the empty quasiset has quasicardinality zero.

$$\forall_Q x [qcard(x) = 0 \rightarrow x =_e \emptyset],$$

where 0 is the cardinal number of the empty set.

(A20) The quasicardinality of subquasisets is less or equal than the quasicardinality of the whole set:

$$\forall_Q xy [x \subseteq y \rightarrow qcard(x) \leq qcard(y)].$$

(A21) Suppose a quasiset x has a quasicardinal α. Then all the cardinals that are less or equal than α are the quasicardinal of a subquasiset of x:

$$\forall_Q x[qcard(x) = \alpha \wedge \beta \leq \alpha \rightarrow \exists_Q y(y \subseteq x \wedge qcard(y) =_e \beta)].$$

(A22) *Weak power set for quasisets*

$$\forall_Q y \forall \beta \{\beta =_e qcard(y) \rightarrow \exists_Q z[\forall u(u \in z \rightarrow u \subseteq x) \wedge \forall \alpha$$

$$\leq \beta \exists u(qcard(u) =_e \alpha \wedge u \in z)]\}$$

Finally, it is quite reasonable to require the following weak indistinguishability principle for quasisets:

(A23) *Weak indistinguishability*

$$\forall_Q xy [\forall u(u \in x \rightarrow \exists v(v \in y \wedge u \equiv v))$$

$$\wedge \forall u(u \in y \rightarrow \exists v(v \in x \wedge u \equiv v))$$

$$\wedge qcard(x) =_e qcard(y) \rightarrow x \equiv y].$$

Our theory S^{**} represents a kind of semiextensional theory. In a sense, collections of microobjects are not determined by their elements. For one cannot distinguish a given quasiset of microobjects from an other quasiset which has been obtained exchanging the elements of the previous collection with "other" indistinguishable elements. In spite of this, the extensionality principle is still valid: quasisets that have the same elements are by definition extensionally equal and hence indistinguishable. It is worthwhile noticing that a kind of Leibnizian equality turns out to be trivially definable for all entities (even for microobjects and for proper quasisets). It is sufficient to assume the following natural definition:

Definition 2.6 Two entities are *Leibniz-equal* iff they belong to the same quasisets:

$$\forall xy[x =_L y \leftrightarrow \forall z(x \in z \leftrightarrow y \in z)].$$

However, this Leibnizian equality does not correspond here to a strong identity relation. Differently from the classical case, x and y may represent two "different" microobjects that are indistinguishable. Both x and y may belong to $[x]$ and $[y]$: the strong singleton of x (which should distinguish x from y) does not necessarily exist.

3. Intensions in Microphysics and the Theory QST

Quantum compound systems of microobjects can be alternatively described as "intensional-like entities."[6] According to this approach, collections of microobjects have been shortly termed *quasets*. In spite of some similarities, the two notions of *quasiset* and *quaset* turn out to have some essentially different properties. The starting point of the theory of quasets (QST) is based on the following observation: physical kinds and compound systems in QM seem to share some features that are characteristic of intensional entities. Further, the relation between intensions and extensions turns out to behave quite differently from the classical semantic situations. Generally, one cannot say that a quantum intensional notion uniquely determines a corresponding extension. For instance, take the notion of *electron*, whose intension is well defined by the following physical property: mass $= 9.1 \times 10^{-28}$g, electron charge $= 4.8 \times 10^{-10}$e.s.u., spin $= 1/2$. Does this property determine a corresponding *set*, whose elements should be all and only the physical objects that satisfy our property at a certain time interval? The answer is negative. In fact, physicists have the possibility of recognizing, by theoretical or experimental means, whether a given physical system is an electron system or not. If yes, they can also enumerate all the quantum states available within it. But they can do so in a number of different ways. For example, take the spin. One can choose the x-axis and state how many electrons have spin-up and how many have spin-down. However, we could instead refer to the z-axis or any other direction, obtaining *different collections* of quantum states, all having the same cardinality. This seems to suggest that microobject systems present an irreducibly intensional behavior: generally they do not determine precise extensions and are not determined thereby. Accordingly, a basic feature of our QST will be a strong violation of the extensionality principle. Differently from S^{**}, QST is formalized in classical first-order logic *with* identity (which is dealt with as a logical constant). The language contains the following primitive concepts:

1. one monadic predicate: *ur-object* (O);
2. three binary predicates: the *positive membership* relation ($\in$), the *negative membership* relation ($\notin$), the *inclusion* relation ($\subseteq$). In the intuitive interpretation, $x \in y$ ($x \notin y$) will be read as "x certainly belongs to y" ("x certainly does not belong to y");
3. a 1-ary functional symbol: the *quasicardinal of* (*qcard*);
4. a binary functional symbol: the *quaset-theoretical intersection* ($\sqcap$).

We will use restricted quantifiers, like in S^{**}. Further, the quantifier "there exists exactly one" ($\exists!$) is supposedly defined in the usual way.

We will present here only a minimal axiomatic nucleous of our theory. The notion of *quaset* is defined (similarly to the definition of *quaset* in S^{**}).

Definition 3.1 *Quaset*

A *quaset* is something that is not an ur-object:

$$Q(x) := \neg O(x).$$

(**Ax1**) All that has elements is a quaset:

$$\forall xy[y \in x \to Q(x)].$$

(like in S^{**}).

(**Ax2**) The inclusion relation $\subseteq$ between quasets is a partial order (reflexive, symmetric, and transitive).

Recall that, differently from S^{**}, here the predicate $\subseteq$ is primitive. In the intended interpretation, $\subseteq$ has an intensional meaning: $x \subseteq y$ can be read as "the concept x *involves* (or *implies*) the concept y."

(**Ax3**) Suppose that something certainly does not belong to a given quaset. Then it is not the case that it certainly belongs to our quaset:

$$\forall xy[x \notin y \to \neg x \in y].$$

But generally it is not the case the other way around.

As a consequence, a strong *tertium non datur* principle $(x \in y \vee x \notin y)$ fails and indetermined membership relations are possible (in accordance with the quantum uncertainty relations).

(**Ax4**) Intensional inclusion implies extensional inclusion (but not the other way around):

$$\forall xy[x \subseteq y \to \forall z(z \in x \to z \in y) \wedge (z \notin y \to z \notin x)].$$

(**Ax5**) Any quaset has exactly one quasiextension, where the *quasiextension* of a quaset x is the unique quaset that certainly contains all the elements of x and certainly does not contain all the other entities:

$$\forall_Q x \exists!_Q y \forall z[(z \in y \leftrightarrow z \in x) \wedge (z \notin y \leftrightarrow \neg z \in x)].$$

Axiom 5 justifies the definition of a 1-ary functional symbol *ext* (the quasiextension of x).

Definition 3.2

$$\forall xy[y = ext(x) \leftrightarrow \forall z(z \in y \leftrightarrow z \in x) \wedge (z \notin y \leftrightarrow \neg z \in x)].$$

The predicate *set (Z)* is here defined (differently from *S***).

Definition 3.3 Sets are quasets that are identical with their quasiextension.

One can easily show that sets satisfy the extensionality principle. The extension of an empty quaset (which turns out to be trivially a set) is postulated.

(**Ax6**) *The empty quaset*
  There exists a quaset that necessarily does not contain any element:

$$\exists_Q y \forall x[x \notin y].$$

Similarly to *S***, we assume that *QST* contains a copy of *ZF*. For any formula *A* of *ZF*, let A^z be the corresponding formula of *QST* relativized to sets.

(**Ax7**) If *A* is any instance for an axiom of *ZF*, then A^z is an axiom of *QST*.

The notion of quasicardinality of a quaset is introduced exactly as in *S***:

(**Ax8**) Any quaset has a unique quasicardinal, which is a cardinal number:

$$\forall x \exists! y[card(y) \wedge qcard(x) = y].$$

(**Ax9**) Quasicardinals and cardinals coincide for sets:

$$\forall x[Z(x) \rightarrow qcard(x) = card(x)].$$

(**Ax10**) The quasicardinal of a subquaset is less or equal than the quasicardinal of the whole quaset:

$$\forall xy[x \subseteq y \rightarrow qcard(x) \leq qcard(y)].$$

(**Ax11**) The quasicardinal of a quaset is greater or equal than the quasicardinal of its quasiextension:

$$\forall x[qcard(x) \geq qcard(ext(x))].$$

(**Ax12**) $\sqcap$ represents a weak conjunction for quasets. This weak conjunction coincides with the usual set-theoretical intersection in the case of sets:

$$\forall xy[(x \sqcap y \subseteq x \wedge x \sqcap y \subseteq y) \wedge (Z(x) \wedge Z(y) \rightarrow x \sqcap y = x \cap y)].$$

As a consequence, a separation procedure may be applied. Notice that our axioms do not require that proper quasets (that are not sets) exist. From an intuitive point of view, the quasiextension of a proper quaset does not represent an adequate semantic counterpart for the usual notion of extension. Think for instance of the fact that the quasiextension a quaset, whose quasicardinal is greater than 0, might have an empty quasiextension.

What about the validity of Leibniz-indiscernible principle in this theory? Are nonidentical objects always distinguished by a property, represented by a quaset? As expected, the answer is negative. Namely, the implication

$$\neg x = y \rightarrow \exists_Q z[x \in z \land y \notin z]$$

generally fails. The classical set-theoretical argument founded on the theorem

$$\neg x = y \rightarrow x \in \{x\} \land y \notin \{x\}$$

cannot be obviously repeated here. Nothing guarantees that the singleton of x (which should be the characteristic property of x) exists and that x certainly belongs to it.

Notes

1. Y. I. Manin, "Problem of Present Day Mathematics: I (Foundations)," in F. E. Browder, ed., *Proceedings of the Symposia in Pure Mathematics*, vol. 28 (Providence, R.I.: American Mathematical Society, 1976).
2. See N.C.A. da Costa, S. French, and D. Krause, "The Schrödinger Problem," in M. Bibtol and O. Darrigol, eds., *Erwin Schrödinger: Philosophy and the Birth of Quantum Mechanics* (Paris: Editions Frontières, 1992), and D. Krause, "On a Quasi-Set Theory," *Notre Dame Journal of Formal Logic* 33 (1992): 402–411.
3. See M. L. Dalla Chiara and G. Toraldo di Francia, "Individuals, Kinds and Names in Physics," in G. Corsi, M. L. Dalla Chiara, and G. C. Ghirardi, eds., *Bridging the Gap: Philosophy, Mathematics, and Physics* (Dordrecht: Kluwer, 1993), and M. L. Dalla Chiara and G. Toraldo di Francia, "Identity Questions from Quantum Theory," in K. Gavroglu, J. Stachel, and M. W. Wartofski, eds., *Physics, Philosophy and the Scientific Community* (Dordrecht: Kluwer, 1995).
4. See E. Schrödinger, *Science and Humanism* (Cambridge: Cambridge University Press, 1952).
5. Such a theory is a slight modification of the theory presented in Krause, "On a Quasi-Set Theory."
6. See Dalla Chiara and Toraldo di Francia, "Individuals, Kinds and Names"; Dalla Chiara and Toraldo di Francia, "Identity Questions"; G. Toraldo di Francia, *Le cose e i loro nomi* (Rome: Laterza, 1986).

PART TWO

OBJECTS AND INVARIANCE

9

Physical Reality

Max Born

The notion of reality in the physical world has become, during the last century, somewhat problematic. The contrast between the simple and obvious reality of the innumerable instruments, machines, engines, and gadgets produced by our technological industry, which is applied physics, and of the vague and abstract reality of the fundamental concepts of physical science, as forces and fields, particles and quanta, is doubtless bewildering. There has already developed a gap between pure and applied science and between the groups of men devoted to the one or the other activity, a separation which may lead to a dangerous estrangement. Physics needs a unifying philosophy, expressible in ordinary language, to bridge this gulf between "reality" as thought of in practice and in theory. I am not a philosopher but a theoretical physicist. I cannot provide a well-balanced philosophy of science that would take due account of the ideas developed by differing schools, but I shall endeavour to formulate some ideas which have helped me in my own struggle with these problems.

There is a school of thought among theoretical physicists and scientific philosophers which advocates a standpoint radically abstract. This philosophy was expressed, for instance, in the notable lecture given by Professor H. Dingle to Section A of the British Association in Edinburgh (published in *Nature* 168 [1951]: 630), and I cannot explain my own standpoint better than by way of contrast. But in quoting extracts from Dingle's lecture I do not intend to conduct a personal controversy; these quotations serve only as examples suitable to develop my own differing views. Let us begin with the following sentence: "The quantities with which physics concerns itself are not evaluations of objective properties of parts of the external material world; they are simply the results we obtain when we perform certain operations." This looks like a denial of the existence of a preexisting material world; it suggests that the physicist does not care about the real world and makes an experiment solely in order to predict the results of yet another experiment. Why the physicist should take the trouble to make an experiment at all is not explained. This question is seemingly regarded as not worthy of a philosopher of science. Can we avoid asking what is the part played in this scheme of things by the instruments, made of steel, brass, glass, etc., carefully composed and adjusted for an experiment? Are they too no part of a preexisting external material world? Are they, like electrons, atoms and fields, merely abstract ideas used to predict the phenomena to be observed at the next experiment, which is again only an assembly of ghosts? We have before us a standpoint of extreme subjectivism, which may rightly be called "physical solipsism." It is well known that obstinately held solipsism cannot be refuted by logical argument. This much, however, can be said, that solipsism such as this does not solve but evades the problem. Logical coherence is a purely negative criterion; no system can be accepted without it, but no system is acceptable just because it is logically tenable. The only positive argument in support of this abstract type of ultrasubjectivism is a historical one. It is maintained that the belief in the existence of an external world is irrelevant and indeed detrimental to the progress of science, and that what the physicist is doing can be satisfactorily understood only in terms of "experiences," not of the external world.

The actual situation is very different. All great discoveries in experimental physics have been due to the intuition of men who made free use of models, which were for them not products of the imagination, but representatives of real things. How could an experimentalist work and communicate with his collaborators and his contemporaries without using models composed of particles, electrons, nucleons, photons, neutrinos, fields, and waves, the concepts of which are condemned as irrelevant and futile?

However, there is of course some reason for this extreme standpoint. We have learned that a certain caution is necessary in using these concepts. The naive approach to the problem of reality, which was so successful in the classical or Newtonian period, has been proved to be not satisfactory. Modern theories demand a reformulation. This new formulation is slowly evolving, but has probably not reached a final expression. I shall try to indicate the present tendencies.

The first point is to remember that the word "reality" is part of our ordinary language, and hence its meaning is ambiguous like that of most words. There are subjective philosophies which teach that only the mental world is real and the physical world merely an appearance, a shadow without substance. This standpoint, though of the greatest philosophical interest, is outside the scope of our discussion, which has to do only with physical reality. Still there remain enough other queries. The realities of a peasant or craftsman, a merchant or banker, a statesman or soldier have certainly little in common. For each of these the most real things are those which occupy the centre of his mind, the word "real" being used as almost synonymous with important. I wonder whether any philosophy can give a definition of the concept of reality that is untainted by some such subjective associations. The question concerning us is whether science can.

This leads to the second point, stressed by Dingle, whether the use of the concept and word "reality" can be discarded without detriment to science. My answer is that it could only be disregarded by men isolated in ivory towers, remote from all experience, from all actual doing and observing, the type of man who becomes extremely absorbed in pure mathematics, metaphysics, or logic. Niels Bohr, who has contributed more to the philosophy of modern science than anybody else, has repeatedly and emphatically said that it is impossible to describe any actual experiment without using ordinary language and the concepts of naive realism. Without this concession no communication about facts is conceivable, even between the most sublime minds. And it is an essential part of this procedure to distinguish between ideas, projects, theories, and formulae on the one side, and the real instruments and gadgets constructed according to those ideas. Here the naive use of the word real, the simple belief in the real existence of the material apparatus, is imperative. I presume that the abstract school represented by Dingle does not deny this, although he does not say so. He does, however, forbid the application of the concept of reality to atoms, electrons, fields, etc., terms used in the interpretation of observations. But where is the border between these two domains? Start with a piece of a crystal, which belongs to the domain of crude reality, and grind it into a powder, whose particles are too small to be seen by the unaided eye. You have to take a microscope: are the particles then less real? Still smaller particles, colloids, appear, properly illuminated, in the ultramicroscope, as bright points without structure. There is a continuous transition between these particles and single molecules or atoms. The ultramicroscope there deserts you. You then have the electron microscope with which you can see even large molecules. Where does that crude reality, in which the experimentalist lives, end, and where does the atomistic world, in which the idea of reality is illusion and anathema, begin?

There is, of course, no such border; if we are compelled to attribute reality to the ordinary things of everyday life including scientific instruments and materials used in experimenting, we cannot cease doing so for objects observable only with the help of instruments. To call these subjects real and part of the external world does not, however, commit us in any way to any definite description: a thing may be real though very different from other things we know.

Let me now discuss some examples which Dingle cites to show the failure in physics of the concept of an objective reality.

The first example is the kinetic theory of matter. Dingle discusses the statistical method, which is not concerned with the single orbits of the molecules and is content to calculate averages, in order to represent "observations (that is, appearances)" and he calls this attitude a "betrayal of the true mission of physics according to the accepted philosophy. They (the physicists) were dedicated to the investigation of reality, which had become the investigation of the nature and behaviour of molecules; and instead of pursuing that, they occupied themselves in showing how their ignorance of reality could be used in order to describe mere appearances." I have not been able to understand whether Dingle thinks the whole kinetic theory superfluous, or whether he suggests stripping the molecules of their reality by calling them "counters" or "dummies." For he makes no attempt to analyse the actual evidence provided by the kinetic theory for the existence of molecules. Let me sketch such an analysis in a few words.

The kinetic derivation of Boyle's law establishes only the possibility of an atomistic explanation, and can hardly be called evidence. However, the same derivation properly formulated leads to a definite value of the mean energy, hence of the specific heat ($\frac{3}{2}R$ for monatomic gases, R being the gas constant), which no phenomenological consideration could provide. The general formula for the mean energy contains numbers of degrees of freedom of the molecules—or "dummies," to use Dingle's expression. The kinetic interpretation of the deviations from Boyle's laws leads to an estimate of the size of the molecules, which is confirmed by a quite different set of phenomena, the irreversible processes of heat conduction, viscosity, diffusion. Many concepts first introduced in a theoretical way, like velocity distribution, free path, etc., have been confirmed and determined by direct measurements. The fluctuations predicted by the kinetic theory are observable in many ways, through the Brownian motion, the blue colour of the sky, etc. Of course, as Dingle says, these are all phenomena, "appearances," the molecules remaining in the background. But the essential point, not mentioned by Dingle, is that the kinetic theory leads to definite properties of the molecules, weight, size, shape (degrees of freedom), mutual interaction. A small number of molecular constants determines an unlimited number of phenomenological properties, in virtue of the molecular hypothesis. Therefore each new property predicted is a confirmation of the molecular hypothesis. Amongst these predictions are such amazing feats as von Laue's X-ray patterns produced by crystals, and the whole range of radioactive phenomena. Here the evidence of the reality of molecules is striking indeed, and to speak of a "dummy" producing a track in a Wilson chamber or a photographic emulsion seems to me—to say the least—inadequate. Compare this kind of reality with the following example. You see a gun fired and, a hundred yards away, a man breaking down. How do you know that the bullet sticking in the man's wound has actually flown from the gun to the body? Nobody has seen it, in fact nobody could have seen it, except a scientist after cumbersome preparations, e.g. through the installation of a complicated optical apparatus of the kind Ernst Mach invented for photographing flying projectiles. Yet I am sure you believe that the bullet has in the short interval between the firing of the gun and the wounding of the man performed a definite trajectory; you believe that it was really there during the interval; or are you content to say, "Oh, I don't know; it's enough to know the phenomena of the firing and wounding. All things between are theoretical imagination, the bullet in flight is merely a 'dummy' invented to account for the connection of the two phenomena by the laws of mechanics." I cannot refute this attitude by logical reasoning. I only wish to point out that if one denies the existential evidence of an atomic track which *can* be seen, one is committed to denying the existence of a bullet in flight which *cannot* be seen, and of numerous similar things.

The root of this strange denial of reality to things like molecules is the interpretation of the concept real as meaning "known in all detail." This does not agree with the usual application of the word. We think all the 500 millions of Chinese are real, although we know not a single one, or perhaps a few individuals, and have not the slightest knowledge of their whereabouts, activities, motions, reactions. We think the Romans of Caesar's time or the Chinese during the life of Confucious were real, although we have no possible means of verifying this in the way which Dingle demands in the case of molecules. Are these Romans or Chinese of the present or the past only dummies invented by the historians to connect phenomena? Which phenomena? Perhaps the words found in newspapers, in books, or on ancient tombstones?

All these considerations are rather on the surface and do not touch the actual difficulties which physics encounters, and which compel us to revise our fundamental notions. Dingle's next example, relativity, leads a little nearer to these problems. He asserts that "in accordance with the philosophy of the time, the real material world, whether regarded as consisting of molecules or of gross bodies, was conceived to possess its properties by intrinsic right. Thus its constituents had a size, a mass, a velocity, and so on." After elaborating this he continues: "Now the basic requirement of the theory of relativity was that all these properties were almost completely indefinite," and he exemplifies this by the notions of length and of mass, which according to relativity depend on the velocity of the observer. The same distance measured by different observers in relative motion may be anything between a maximum and nothing, the same mass anything between a minimum and infinity. He concludes that "by abandoning all attempts to assign any property at all to matter we can learn more and more about the relations of phenomena." Now this is a misrepresentation of the theory of relativity, which has never abandoned all attempts to assign properties to matter, but has refined the method of doing so in order to conform with certain new experiences, such as the famous Michelson-Morley experiment.

In fact this example is very well suited to get at the root of the matter. This root of the matter is a very simple logical distinction, which seems to be obvious to anybody not biased by a solipsistic metaphysics; namely this: that often a measurable quantity is not a property of a thing, but a property of its relation to other things. To give an example: cut out a figure, say a circle, of a piece of cardboard and observe its shadow thrown by a distant lamp on a plane wall. The shadow of the circle will appear in general as an ellipse, and by turning your cardboard figure you can give to the length of an axis of the elliptical shadow any value between almost zero and a maximum. That is the exact analogue of the behaviour of length in relativity, which in different states of motion may have any value between zero and a maximum. If you wish to have an analogue to the behaviour of mass which according to velocity may have any value between a minimum and infinity, take a long sausage and cut slices with different inclination which will be ellipses with one axis between a minimum and "practical" infinity. To return to the shadow of the circle, it is evident that the simultaneous observation of the shadows on several different planes suffices to ascertain the fact that the original cardboard figure is a circle and to determine uniquely its radius. This radius is what mathematicians call an invariant for the transformations produced by parallel projection. In the same way there is an invariant of all the cross sections of a sausage, that with the smallest area. Most measurements in physics are not directly concerned with the things which interest us, but with some kind of projection, this word taken in the widest possible sense. The expression coordinate or component can also be so used.

The projection (the shadow in our example) is defined in relation to a system of reference (the walls, on which the shadow may be thrown). There are in general many equivalent systems of reference. In every physical theory there is a rule which connects the projections of the same object on different systems of reference, called a law of transformation, and all these transformations have the property of forming a group, i.e. the sequence of two consecutive transformations is a transformation of the same kind. Invariants are quantities having the same value for any system of reference, hence they are independent of the transformations.

Now the main advances in the conceptual structure of physics consist in the discovery that some quantity which was regarded as the property of a thing is in fact only the property of a projection.

The development of the theory of gravity is an example. Using modern mathematical language, the primitive (pre-Newtonian) conception of gravity is connected with a group of transformations for which the vertical, the normal to the plane surface of the earth, is absolutely fixed. For these transformations the size and direction of the force of gravity is an invariant, which implies that the weight is an intrinsic property of the body which it carries along. The situation changed completely when Newton discovered gravity to be a special case of general gravitation. The group of transformations was extended in such a way that space became isotropic, with no fixed direction; gravity then became just a component of the gravitational force.

The theory of relativity has continued this development. The transformations of classical mechanics, often called Galilean transformations, kept space and time apart. The experiences condensed in the theory of relativity showed that this does not agree with facts. One has to use a wider group, called Lorentz transformations, in order to introduce an intimate connection between space coordinates and time. Naturally, quantities regarded by the older theory as invariants, like distances in rigid systems, time intervals shown by clocks in different positions, masses of bodies, are now found to be projections, components of invariant quantities not directly accessible. Still, as in the case of the shadow, by determining a number of these components, the invariants can be found. Thus it turns out that the maximum length and the minimum mass are relativistic invariants. It would perhaps have been preferable to call these invariants, which are properties of bodies, by the old names length, time, mass, and to invent new names for the projections. But science is strangely conservative in such matters, and it has been agreed to rename the invariants rest-length, proper-time, rest-mass, etc., and keep the old expressions for the components although these are now not properties of a body but of its relation to a system of reference.

I think the idea of invariant is the clue to a rational concept of reality, not only in physics but in every aspect of the world.

The theory of transformation groups and their invariants is a well-established part of mathematics. Already in 1872 the great mathematician Felix Klein discussed in his famous "Erlanger Programm" the classification of geometry according to this point of view; the theory of relativity can be regarded as an extension of this programme to the four-dimensional geometry of space-time. The question of reality in regard to gross matter has from this standpoint a clear and simple answer.

The situation is more difficult in atomic physics. It is well known that the laws of quantum mechanics lead to a kind of indeterminacy expressed by Heisenberg's uncertainty relations. Is not this vagueness, this impossibillty of answering definite questions about position and velocity of a particle, an argument against the reality of particles and altogether of the objective, real world? Here we have to reflect about what we mean by a particle, for instance a photon, an electron, a meson, a nucleon in regard to the experimental evidence; and again we find that these words signify definite invariants which can be unambiguously constructed by combining a number of observations.

The underlying transformation theory, however, is rather involved, and I can give here only a short, sketchy indication. The essence of the matter can be explained with the help of ordinary light.

The *wave* character of light was established by Young and Fresnel by showing that two beams of light, produced by splitting one beam, when reunited give interference fringes. Almost a hundred years later Einstein interpreted the photoelectric effect as the action of light quanta or photons which on hitting a metal surface knock out electrons. Thus light has in addition a corpuscular aspect, a fact confirmed by innumerable experiments. The strange thing is that between these apparently contradictory concepts there exists a simple quantitative relation, which Planck had derived already five years earlier from the behaviour of heat radiation, namely $E = h\nu$, where E is the energy of the photon, ν the frequency of the wave, and h a constant. The conceptual difficulty comes from the fact that the energy E is concentrated in a very small particle while the frequency ν, or better the wave length $\lambda = c/\nu$, needs for definition a (practically) infinite train of waves.

This paradox can only be solved by sacrificing some traditional concept. As we now know, what we have to give up is the idea that the particles, considered by themselves, follow deterministic laws similar to those of classical mechanics. The theory can predict only probabilities, and these are determined by the waves (they are the squares of the amplitudes). This is of course a decisive change in our attitude to nature. It calls for new ways of describing the physical world, but not the denial of its reality. The essence of the new method can be seen from a simple example.

Let a beam of light pass through a Nicol prism; it thus becomes linearly polarised. Let this primary beam, which may have the amplitude A, pass through a double-refractory crystal; there emerge two secondary beams, linearly polarised perpendicularly to one another. If θ is the angle between the direction of polarisation of the primary and of one of the secondary beams, the amplitudes of the latter are $A \cos\theta$ and $A \sin\theta$. Their intensities are therefore in the ratio $\cos^2\theta : \sin^2\theta$. If now the primary intensity is decreased until you see nothing with your eyes, you still can observe the arrival of photons with the help of a sensitive photocell and of proper amplification, and you can count the number of photons. Thus you will find that their average number in the two secondary beams is in the ratio of $\cos^2 \theta : \sin^2 \theta$. This is the simplest example of the statistical interpretation mentioned above, that probabilities are determined by the squares of the amplitudes of the waves. The point to which I wish to direct attention is that these secondary amplitudes are the projections of the primary amplitude in two directions determined by the instrument. The prediction made by the theory in regard to the intensities of the emerging beams, or the number of photons in these, has a meaning only in relation to the whole experimental arrangement, the Nicol prism and the crystal.

Now this example is typical for quantum phenomena. Take, for example, the corresponding experiment with electrons, known as the Stern-Gerlach effect, where the Nicol prism is replaced by a nonhomogeneous magnetic field and the polarisation by the direction of the spin. Again the observable part, the number of electrons of a given spin, depends on the special experimental arrangement in a way which can be described by saying that the instrument records projections of the actual state.

This description applies to any quantum effect. An observation or measurement does not refer to a natural phenomenon as such, but to its aspect from, or its projection on, a system of reference, which as a matter of fact is the whole apparatus used. Expressed in mathematical terms the word projection is perfectly justified since the main operation is a direct generalisation of the geometrical act of projecting, only in a space of many, often infinitely many, dimensions.

If these facts are analysed from the standpoint of particles alone, there appear those uncertainty relations, which I shall not discuss here, since they are now to be found in every textbook of quantum mechanics. Bohr has introduced the idea of complementarity to express the fact that the maximum knowledge of a physical entity cannot be obtained from a single observation or a single experimental arrangement, but that different experimental arrangements, mutually exclusive but complementary, are necessary. In the language proposed here this would mean that the maximum knowledge can only be obtained by a sufficient number of independent projections of the same physical entity, just as in the case of the circular piece of cardboard, where the shadows on several planes were necessary to determine its shape and invariant (radius). The observations of the different shadows on two perpendicular planes, used above to explain the concept of the invariant, also illustrate very well the essence of the idea of complementarity. The final result of complementary experiments is a set of invariants, characteristic of the entity.

The main invariants are called charge, mass (or rather rest-mass), spin, etc.; and in every instance, when we are able to determine these quantities, we decide we have to do with a definite particle. I maintain that we are justified in regarding these particles as real in a sense not essentially different from the usual meaning of the word.

Before defending this standpoint I wish to discuss in a few words the remark often repeated that quantum mechanics has destroyed the distinction between object and subject, since it cannot describe a situation in nature as such, but only that produced by a man-made experiment. This is perfectly true. The atomic physicist is very far removed from the idyllic attitude of the old-fashioned naturalist who, by watching butterflies in a meadow, hoped to penetrate into Nature's mysteries. The observation of atomic phenomena needs instruments of such sensitivity that their reaction in making measurements must be taken into account, and, as this reaction is subject to the same quantum laws as the particles observed, a degree of uncertainty is introduced, which prohibits deterministic prediction. It is therefore obviously futile to ponder about the situation which would have arisen without the interference of the observer, or independent of the observer. But in respect to a given interference of the observer, in a given experimental situation, quantum mechanics makes definite statements as to the maximum information obtainable. Although we cannot know everything, nor even approximate to a knowledge which is complete, by improving our instruments we can obtain certain restricted, but well-described, information which is independent of the observer and his apparatus, namely the invariant features of a number of properly devised experiments. The process of acquiring this information is certainly conditioned by the subject observing; but that does not mean that the results lack reality. For obviously the experimentalist with his apparatus is part of the real world, and even the mental processes used in designing his experiment are real. The boundary between the action of the subject and the reaction of the object is blurred indeed. But this does not prohibit us from using these concepts in a reasonable way. The boundary of a liquid and its vapour is also not sharp, as their atoms are permanently evaporating and condensing. Still we can speak of liquid and vapour.

Let us now return to the question of reality and recall the views of some modern philosophers on the subject.

In a recent book the American writer H. Margenau advocates the standpoint that reality consists of two layers; the immediate data of the senses and "constructs"; the latter include things of everyday life as well as scientific concepts, as far as they are verifiable by several independent experiments. The logical positivists who emphatically claim to possess the only rigorous scientific philosophy, as far as I understand, regard the constructs merely as conceptual tools for surveying and ordering the crude sense data which alone have the character of reality. These are minor variations of the same theme. These variations appear to me unimportant, as two essential points of reality are ignored. One such essential point is that it is psychologically and physiologically wrong to regard the crude sense impressions as the primary data; the other is that not every concept from the domain of scientific constructs has the character of a real thing, but only those which are invariant in regard to the transformations involved.

In respect of the first point, we have to remember that every human being has already acquired the ability to distinguish and recognize objects in his first childhood. As a result, the world of a normal human being is not a kaleidoscopic sequence of sensations but a comprehensible, continuously changing scene of events in which definite things preserve their identity, in spite of their ever changing aspects. This power of the mind to neglect the differences of sense impressions and to be aware only of their invariant features seems to me the most impressive fact of our mental structure. Imagine you are walking with your dog beside you. He sees a rabbit and follows it in a wild chase, and soon the dog will be a tiny spot in your field of vision. But all the time you see your dog, not a sequence of visual impressions of diminishing size. Modern psychology has recognized this fundamental situation; I mean the *Gestalt* psychology of Köhler, Hornbostel, Wertheimer, to name only a few German psychologists of this school whom I personally knew. I should like to translate the word *Gestalt* not as "shape" or "form" but as "invariant," and speak of "invariants of perception" as the elements of our mental world. The physiology and anatomy of the nervous system, of which I know a little from the writings of Professor B. D. Adrian and Professor J. Z. Young, are in full agreement with this result of psychological observation.

Each single nerve fibre, whether motor or sensor, and in the latter case whether carrying tactile, visual, auditory, or thermal messages, transfers a set of regular pulsations which have not the slightest similarity to the physical stimulus. The brain receives nothing but sequences of such pulsations, each propagated by a different fibre to a definite place in the cortex, and it has the amazing ability to disentangle these code messages almost instantaneously. What it does is the solution of an extremely difficult problem of algebra, determining the invariant feature in this welter of ever changing signals. These features thus determine not a blurred set of impressions but recognizable things.

If we attempted to build a philosophy of science on the assumption that our raw material is unordered sense impressions, we could not even describe our manipulations and simple instruments. Science must accept, as I said before, the concepts of ordinary life and the expressions of ordinary language. It transcends these by using magnifying devices, telescopes, microscopes, electromagnetic amplifiers, etc. Thus new situations are encountered where ordinary experience breaks down, and we are at a loss how to interpret the signals received. You will understand what I mean if you have ever looked through a microscope in which a medical friend is showing you some remarkable cells or microbes: you see nothing but a tangle of vague lines and colours and have to take his word for it that some oval yellow structure is the object of interest. Exactly the same happens in all branches of physics where amplification is used. We glimpse the unknown, and we are bewildered. For we are then not children any more; we have lost the power of unconsciously decoding the nerve messages we are receiving, and have to use our conscious technique of thinking, mathematics and all its tricks (we except a few men of rare genius like Faraday, who saw the inner connection of nature by intuition like a child). Thus we apply analysis to construct what is permanent in the flux of phenomena, the invariants. Invariants are the concepts of which science speaks in the same way as ordinary language speaks of "things," and which it provides with names as if they were ordinary things.

Of course, they are not. If we call an electron a particle we know very well that it is not exactly like a grain of sand or pollen. For instance, it has under certain circumstances not a distinct individuality: if you shoot an electron out of an atom by another electron, you can never tell which of the two electrons flying away is which. Still it has some properties in common with ordinary "particles," thus justifying its name. Such extensions of nomenclature are quite common in life as in science, and are systematically developed in mathematics. A number means originally an integer with which you can count a discrete set of objects. But the word is also used for fractions like $\frac{2}{3}$, radicals like $\sqrt{2}$, transcendentals like π, and imaginary numbers like $\sqrt{-1}$, although you cannot count with them. The justification is that they have some formal properties in common with integers, each type a little less, but enough to use a familiar word for them. The same principle is applied in analytical geometry, when we speak of the infinitely distant line in a plane, or of a four-dimensional sphere, and so on; and also in physics. We speak of infrared or ultraviolet light although we cannot see it, and of suprasonic sound although we cannot hear it. We are so accustomed to extrapolate into regions beyond our sense qualities that we have quite forgotten that we are extending concepts beyond their original domain of definition. The principle of doing this is always the same. Consider the concept of waves. We regard waves on a lake as real, though they are nothing material but only a certain shape of the surface of the water. The justification is that they can be characterized by certain invariant quantities, like frequency and wavelength, or a spectrum of these. Now the same holds for light waves; why then should we withhold the epithet "real," even if the waves represent in quantum theory only a distribution of probability? The feature which suggests reality is always some kind of invariance of a structure independent of the aspect, the projection. This feature, however, is the same in ordinary life and in science, and the continuity between the things of ordinary life and the things of science, however remote, compels us to use the same language. This is also the condition for preserving the unity of pure and applied science.

10

The Constitution of Objects in Kant's Philosophy and in Modern Physics

Peter Mittelstaedt

1. Introduction

From a historical point of view the main goal of Kant's transcendental deduction in the *Critique of Pure Reason* was to show that there are well-defined limits for the validity of the synthetic judgments a priori. Since objects of experience are constituted by means of certain categories, the general statements that follow from these categories are a priori valid for all objects of experience, but the a priori validity is also restricted to these objects. "Consequently, there can be no a priori knowledge, except of objects of possible experience."[1]

In contemporary physics and philosophy we find just the opposite problem. In contrast to Kant, who was deeply convinced that there is a "pure science," in the twentieth-century philosophy there are many doubts in the general validity of Kant's a priori judgments. Hence it is very important to investigate in detail the validity domain of the transcendental arguments. This means that one has to reconsider the necessary conditions of objective experience, the constitution of objects, and the question for which kinds of objects the a priori theorems really hold.

There is a general agreement that the most general laws of nature, which can be derived within the framework of transcendental philosophy as a priori judgments, are actually valid within the domain of classical physics. In this domain the formal as well as the material preconditions of experience are fulfilled and allow for a complete constitution of objects of experience in the sense of the Kantian philosophy. These objects are "completely determined" and subject to the a priori laws of nature.

In contrast to the situation in classical physics, in modern physics (relativity, quantum theory, cosmology) Kant's synthetic judgments a priori (conservation of substance, causality, and individuality) lose their general validity. The reason for this violation is closely connected with Einstein's observation that in modern physics the material preconditions of experience, that is, the physical laws of the measuring apparatus, are no longer independent constituents of the process of cognition, but follow from the physical laws of the domain of reality considered. These new and modified preconditions of experience provide some serious difficulties for the constitution of objects.

On the other hand, also in modern physics one requires objectivity of the cognition, which is one of the origins of Kant's system of categories. Hence we will investigate here the question whether objects of experience can also be constituted in modern physics at least in a somewhat modified way. This problem will be treated here within the framework of quantum physics. The investigations are based on the conviction that also in modern physics objective cognition can be achieved and that an adequate epistemology of contemporary physics must make use of a transcendental way of reasoning.

2. Objects in Classical Physics

In his *Treatise of Human Nature* (1739) David Hume emphasized that we never observe objects but only qualities and that it is nothing but imagination if we consider the observed qualities as properties of an object. Hence any scientific cognition begins with the observation of qualities, and it seems to be merely a question of interpretation whether in addition to the observed phenomena a fictitious object, "an unknown something," is used for the description of the experimental results. Obviously, there is no reason to expect that general laws like conservation of substance or causality hold in nature. The same problem was treated by Kant in his *Critique of Pure Reason* (1787). However, in contrast to Hume, Kant emphasized that "objects of experience" are not only arbitrary imaginations but entities that were constituted from the observational data by means of some well-defined conceptual prescriptions, the categories of substance and causality. Hence the interpretation of the empirical data as properties of an object can only be justified if the object as carrier of properties was constituted by the categories mentioned. Kant formulated necessary conditions that must be fulfilled by the observational data if the measuring results can be considered as properties of an "object of experience."

This program can be realized and specified within the framework of classical mechanics. If the cognition of the exterior reality is objective, then the observations in space and time must have been ordered and interpreted according to the categories of substance and causality. In this way objects of experience are constituted. It is not claimed by Kant that an interpretation of this kind is always possible. "If each representation were completely foreign to every other, ... no such thing as knowledge would ever arise."[2] If, however, the observations can be used for the constitution of objects, these objects fulfill the a priori laws of substance and causality.

In classical mechanics observations consist of mechanical data like position, momentum, and the like. More generally, observable qualities are given by subsets of the phase space. If the measuring results, for example, points in phase space at different time values can be related to an exterior entity as its properties, then the observed data must possess some invariance properties. Since each observer corresponds to a space-time coordinate system and since classical mechanics is invariant with respect to the group G_{10} of Galilei transformations, the objectivity of some mechanical experience, that is, its independence of various observers, corresponds to the covariance of the observed data with respect to Galilei transformations. It follows from these invariance properties of the observations that the object to which the qualities can be related as its properties must possess some invariant properties like mass, charge, spin, and so on. These time-independent "essential" properties E_k characterize a class of object systems—mass points and the like—but not an individual object.

In classical mechanics to the formal conditions of experience, that is, to the categories of substance and causality, *material* preconditions of experience must be added, which correspond to the material possibilities to measure mechanical observables. These material preconditions of possible experience extend and specify the possibilities for the constitution of objects. In the present case there are no obvious restrictions for measuring processes and hence all possible predicates $P_i (i = 1, 2, \ldots)$ can be measured jointly on a system. If the results (P_i or $\neg P_i$) are interpreted according to the categories of substance and causality, objects are constituted that are subject to the principle of complete determination: "Every *thing* as regards its possibility is likewise subject to the principle of *complete* determination according to which if all possible predicates are taken together with their contradictory opposites, then one of each pair of contradictory opposites must belong to it."[3] "Things" of this kind possess each possible "accidental" property P either positive (P) or negative ($\neg P$). Moreover, in this case the causality law leads to a strict and complete determinism. It follows from these results that the "accidental" properties P_i allow also to individualize systems if impenetrability is assumed as an additional condition.

The Kantian method of constituting objects is not only a philosophical conception but actually used in contemporary physics. This can easily be illustrated by an astrophysical example. In the planetary system there is

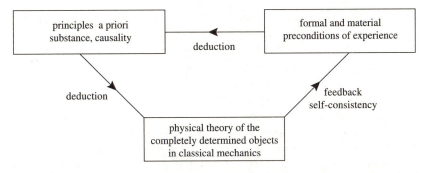

Fig. 10.1. The Kantian circle of self-consistency in classical mechanics. The physical laws of the measuring apparatus are material preconditions of experience *and* consequences of the physical theory of the domain of reality considered.

a large number of circa 2,000 small planets that can be observed only occasionally. Hence the observations that can be obtained during a time interval of several months, say, consist of a large number of isolated light points without any obvious connection. However, if impenetrability is assumed, one can try to interpret these light points by means of the laws of classical mechanics as points on some mechanical trajectories. These trajectories will then constitute individual small planets, which are, of course, subject to the laws of substance and causality, which are specified here by the Galilei-invariant laws of classical mechanics.

Finally it should be emphasized that Einstein's observation mentioned is almost self-evident in classical mechanics. In fact the material preconditions of experience, that is, the physical laws of the measuring instruments, are independent constituents of the process of cognition, but also in complete accordance with the laws of classical mechanics, that is, with the laws of that field that should be tested by means of the measuring apparatus considered. This self-referential consistency or *"self-consistency"* is obviously not a new requirement but already fulfilled in classical mechanics and for mechanical measuring instruments (figure 10.1). It is, however, no longer trivial if one considers theories of modern physics, for example, relativity and quantum mechanics.

3. Quantum Physics I: Incompletely Determined Objects

According to Kant the conservation of substance and the causality are laws that are a priori valid for any object of experience. This statement can be confirmed within the framework of classical mechanics. Moreover, the most general preconditions of experience are then in accordance with the physical laws of the considered domain of reality. These preconditions allow for the constitution of objects that are "completely determined" and subject to the laws of conservation of substance, causality, and individuality. These results reproduce the self-consistency schema mentioned earlier.

However, in quantum mechanics one finds that for quantum systems (atoms, electrons, neutrons, etc.) the mentioned laws of causality, conservation of substance, and individuality are no longer generally valid. In particular, it turns out that it is in general not possible to relate all the experimental data obtained by measurement to an object as their carrier. Furthermore, observational data subsequently obtained on a system by measurements cannot be connected in general by causality-like laws. Hence it is not possible to determine individuals with temporal identity.

In principle there are two possible ways to react to this discrepancy between transcendental arguments and quantum mechanics. First, one could restrict the physical investigations to the observed data and measuring results, thus avoiding the inconsistencies mentioned. This approach corresponds to the empiricism of David Hume or to the positivism of Ernst Mach and was first applied to quantum mechanics by Niels Bohr.[4] This way of reasoning, which was later elaborated in all detail by G. Ludwig,[5] leads to a consistent description of the measuring results without any reference to objects and their properties.

Second, one could try to resolve the inconsistencies in some way and to interpret also quantum physics in terms of objects and their properties. This attempt goes back to Heisenberg[6] and was further discussed by G. Hermann,[7] C. F. von Weizsäcker,[8] E. Cassirer,[9] and others.[10] For an adequate treatment of this approach one has to analyze carefully the statements of transcendental philosophy and quantum mechanics and to investigate the corresponding justifications. According to Kant the synthetic judgments a priori are not considered to be true in an apodictic sense, but their validity is restricted to objects of experience the constitution of which has been performed by means of the categories of substance and causality.

On the other hand, it is by no means clear whether the entities, which appear as quantum objects (atoms, electrons, neutrons, etc.), are objects of experience in the transcendental sense, that is, constituted by means of the Kantian categories. At first glance one finds that quantum systems are not "subject to the principle of complete determination," which means that every property P or its contrary $\neg P$ pertains to the system. Whether this deficiency of quantum objects is actually the reason for the violation of the a priori judgments in quantum physics must be further investigated.

In quantum mechanics the material preconditions of experience, that is, the physical laws of the measuring apparatus, do not allow to determine all possible properties $P_i \in \mathbf{P}$ (with values P_i and $\neg P_i$) of a given system. In any contingent situation ψ only a subset $\mathbf{P}_\psi \subset \mathbf{P}$ of properties can be measured jointly on the system S. The properties $P_i \in \mathbf{P}_\psi$ are mutually "commensurable," which means that they can be measured in arbitrary sequence without thereby changing the results of the measurements. The totality $\mathbf{P}_\psi$ of commensurable properties corresponds to the state of the system S, which is denoted here by W_ψ. For any state W_ψ of S there are also properties $P_k \in \mathbf{P}/\mathbf{P}_\psi$ which are *incommensurable* with the properties $\mathbf{P}_\psi$, that is, the measurement of a property of this kind would provide a material change of the state W_ψ of S.

If in a state W_ψ of S the properties $P_i \in \mathbf{P}_\psi$ are measured, the measuring results (P_i or $\neg P_i$) can be related to the object system S as its properties just as in classical mechanics. Hence we refer to these properties $P_i \in \mathbf{P}_\psi$ as the "objective" properties of S in the state W_ψ. Clearly, the object system obtained in this way is not "completely determined" since only the subset of objective properties pertains to the object with state W_ψ. However, the quantum object system S may be considered as constituted by means of the objective properties P_ψ in a consistent way.

It should be noted that the properties that are not objective in a given state do not pertain to the system in any consistent way. Since the experimental determination of these properties provides a change of the state of the system, it is obviously not possible to measure simultaneously the totality of all properties. Moreover, it is even impossible to assume that in a given state W_ψ a property P that is not objective in W_ψ pertains objectively to the system and that it is merely subjectively unknown to the observer whether S possesses the property P or its contrary $\neg P$. It can be shown that this hypothetical attribution of P-values to the system leads to a contradiction with the statistical predictions of quantum mechanics and can thus be disproved even experimentally.

The temporal development of the state W_ψ is completely determined by the Schrödinger equation in a causal way, that is, the state $W_\psi(t)$ at time t determines the state $W_\psi(t')$ at a later time $t' \geq t$. However, since a state W_ψ corresponds to a set $\mathbf{P}_\psi$ of objective properties, at different time values $t', t'', \ldots$ we have different sets $\mathbf{P}_{\psi'}, \mathbf{P}_{\psi''}, \ldots$ of objective properties. Hence it will in general not be possible to establish a causal connection between a property $P_a(t)$ at time t and the same property $P_a(t')$ at a later time t'. There is only a very limited causality law between the objective properties $\mathbf{P}_\psi$ and $\mathbf{P}_{\psi'}$ at different time values t and t'. Consequently, individualization of quantum objects is not possible since it would require a complete causal determination of all possible properties. Thus for quantum objects the laws of substance and causality are restricted to the respective objective properties, and individuality cannot be achieved.

Hence it seems that for quantum physics a new and somewhat relaxed self-consistency scheme can be established that connects the quantum physical concepts in the sense of the transcendental way of reasoning. The restricted laws of substance and causality allow to constitute quantum objects by means of objective properties. These quantum objects are then subject to the laws of quantum substance and quantum causality, which may be considered as synthetic judgments a priori in a modified sense. Moreover, the material preconditions of experience, that is, the laws of the measuring apparatus, are nothing else but the laws of quantum physics and thus in complete accordance with the a priori laws of quantum substance and quantum causality. Hence the full transcendental way of reasoning is again self-consistent.

4. The Range of Kant's Transcendental Arguments

It is an important question whether this relaxed and modified post-Kantian self-consistency scheme is still in the spirit of Kant's transcendental philosophy. This problem has been investigated in full detail by I. Strohmeyer.[11] The starting point of these investigations is the original Kantian basic requirement of objectivity. The cognition of the exterior reality should be objective, that is, perceptions and observations must not be merely subjective imaginations but qualities that can be considered as properties of an exterior object. According to Kant the objectivity of our cognition implies the existence of categories by means of which the objects of experience are constituted from observational data. This means in particular, that the categories of substance and causality are necessary conditions of objectivity. Together with space and time, we refer to the categories as the *formal* preconditions of objective experience. Since also quantum physics is usually assumed to provide objective experience, it follows that the categories of substance and causality are also necessary preconditions of quantum physical experience.

For the constitution of objects the formal preconditions mentioned are necessary but in general not sufficient. They must be completed by *material* preconditions, which are given by the specific material possibilities to obtain experimental data. Within the framework of quantum physics, these material preconditions are given by the physical properties of the measuring instruments. It is well known that the physical laws of quantum measurements lead to some restrictions on the acquisition of experimental information. In particular, it will not be possible to obtain a sufficiently large number of data which can be used for the constitution of objects that are "subject to the principle of complete determination." Hence this principle must be abandoned in the case of quantum physics.

It has been shown by Strohmeyer in a careful analysis of Kant's writings that this principle does not belong to the necessary preconditions of experience. There is no proof of the a priori validity of this principle in the *Critique of Pure Reason*. According to the table of judgments, which contains not only affirmative and negative judgments but also indefinite (or infinite) ones, we have to distinguish three cases for qualities or properties. An object of experience may posses a property P (affirmative case), it may possess the counterproperty $\neg P$ (negative case), or the property P does not pertain to the system in any sense (indefinite case). This third possibility shows that (in principle) objective experience can also be achieved for objects that are not "completely determined" such that only a limited number of properties pertain to the object.

Since quantum objects possess only the objective properties, the laws of substance and causality can only be applied to these properties. However, according to the investigations mentioned, this restriction does not invalidate the possibility of objective experience. Hence one gets the impression that not only the classical (completely determined) objects but also incomplete quantum objects can be incorporated into the transcendental way of reasoning. The range of transcendental philosophy seems to be large enough to cover classical physics as well as quantum physics. Objective cognition of objects of experience implies the validity of a priori judgments, the specific structure of which depends on additional material preconditions of experience.

5. Quantum Physics II: Unsharp Objects

The result of the preceeding discussion can be summarized in the post-Kantian circle of self-consistency (figure 10.2). Quantum objects that possess only the restricted class of "objective" properties are subject to the restricted laws of substance and causality, which can be established as synthetic judgments a priori. The objectivity of quantum physical experience and the general applicability of transcendental philosophy can be demonstrated in this way. However, the specific solution of the object problem used here looks rather artificial and unphysical for the following three reasons.

First, the case of an indetermined quality in Kant's argumentation is somewhat different from the situation in quantum physics. In the indetermined case a property P does not belong at all to the object, that is, even a measurement of P will not lead to a definite result (P or $\neg P$). In quantum mechanics, however, a nonobjective property Q is "nonobjective" only with respect to a well-defined contingent situation which is

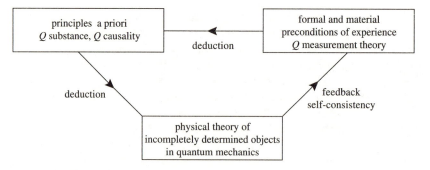

Fig. 10.2. The post-Kantian circle of self-consistency in quantum physics for incompletely determined quantum objects.

described by the state W_ψ of the system. If the property Q is measured on the system, one will always get a definite result (Q or $\neg Q$), which means that the state of the system has been changed into a new state W_ψ by this measurement.

Second, if a quantum system is in the state W_ψ say, then the knowledge of the totality of all objective properties $\mathbf{P}_\psi$ is not the complete information about the system. In addition to the values (P, $\neg P$ or 1, 0) of the objective properties, for the nonobjective properties $Q \in \mathbf{P}/\mathbf{P}_\psi$ one knows the probabilities $p_\psi(Q)$ to obtain the value 1 if Q is measured on the system in the state W_ψ. This probabilistic information, which is also contained in the state, is, however, not contained in the interpretation with incomplete objects.

Third, the causality law that can be justified for incomplete quantum objects is rather artificial and intractable from an experimental point of view. If at a certain time t_a a complete set of objective properties $\mathbf{P}^a$ was measured such that the corresponding results can be related to a quantum object and if corresponding measurements were performed at a time t_b with $t_b \geq t_a$, then there are two possibilities. (I) One measures just those properties $\mathbf{P}^b$, which are objective in the state $W_\psi(b)$ which has evolved from $W_\psi(a)$ according to the Schrödinger equation. These properties can be tested without disturbing the system, and they can be used for the constitution of an incompletely determined quantum object with temporal identity. (II) However, from an experimental point of view it is unknown which properties $\mathbf{P}^b$ are objective at the time t_b. If, on the other hand, one measures the properties $\mathbf{P}^a$ already tested at t_a also at t_b, then the state of the system will in general be changed and the measuring outcomes at t_a and t_b cannot be connected by a causality law. In this case it is not possible to constitute an object system with temporal identity.

These three objections against the solution of the object problem in quantum physics by means of incomplete objects show that a convincing theory should not only be in accordance with Kant's philosophy but also tractable in the sense of quantum physics. As mentioned earlier, the interpretation of quantum mechanics that was used as a basis of the incomplete object solution is not sufficient. The full set $\mathbf{P}_\psi$ of objective and mutual commensurable properties is incomplete since it must be completed by the probabilities of the nonobjective properties $\mathbf{P}/\mathbf{P}_\psi$. The informational incompleteness of the set $\mathbf{P}_\psi$ means that it is not possible to infer from the simultaneous measuring results of a set $\mathbf{P}_\psi$ to the state of the system before the measurements.

There is, however, a description of quantum systems that avoids these problems. Instead of using the restricted class of objective properties (with truth-values 0 and 1), one could start with the totality of all measurable properties, if these properties are understood in the sense of unsharp propositions. In recent years "unsharp" or "fuzzy" properties have been studied in quantum mechanics in full detail. Two incommensurable properties P and Q—for example, momentum and position values—can be related to a single quantum system if these properties are understood as unsharp observables, the unsharpness of which is described by some parameter ϵ with $0 \leq \epsilon \leq 1$. If this unsharpness parameter is sufficiently large, two complementary but unsharp properties P_ϵ and Q_ϵ can not only be related to a system in a consistent way but they can also be measured jointly on the system.[12]

The use of unsharp properties offers new possibilities for the constitution of quantum objects. If the full set $\mathbf{P}$ of possible properties is measured ϵ-unsharply, then it can happen that the results of these measurements can consistently be related to an object as the carrier of these properties. At the time t_1 of these joint measurements, an object is constituted by the unsharp properties as an ϵ-unsharp object S_ϵ. (For objects the unsharpness ϵ means that there is a probability $p \approx \epsilon$ to confuse two objects S_ϵ and S'_ϵ which are constituted by the same unsharp properties.) Also the concept of causality can be applied to unsharp properties. If the full set of ϵ-unsharp properties is measured again at some time $t_2 \geq t_1$, the system will be disturbed only in such a weak sense that the unsharp properties $P^\epsilon(t_1)$ and $P^\epsilon(t_2)$ can at least be connected by a probability law. Hence unsharp measuring results at successive time values t_1 and t_2 can be used for the constitution of an object with approximate temporal identity. In this way the application of the categories of substance and causality to ϵ-unsharp measuring outcomes provides the constitution of an ϵ-unsharp object S_ϵ.

The approximate constitution of objects by means of unsharp properties and some general laws, which is described here in terms of fuzzy

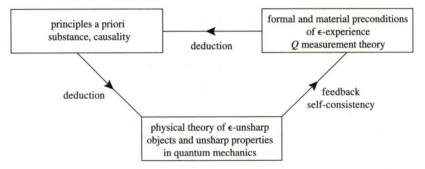

Fig. 10.3. The post-Kantian circle of self-consistency in quantum mechanics for ϵ-unsharp quantum objects.

observables, is not new and very familiar from an experimental point of view. In cloud chamber experiments one performs a large number of unsharp position measurements with outcomes $\tilde{x}(t_1), \tilde{x}(t_2), \ldots$ at successive time values $t_1, t_2, \ldots$. Similarly as in celestial mechanics these experimental data (and the corresponding unsharp momenta) are then interpreted by means of the laws of classical mechanics, which should be considered here as a specification of the general principles of substance and causality. In this way the unsharp position values can be interpreted as unsharp elements of a mechanical trajectory of an unsharp quantum object that is constituted in this way.

The applicability of the categories of substance and causality to unsharp observations has the consequence that individual objects with temporal identity can at least be constituted in an approximate sense. On the other hand the categories of substance and causality are necessary conditions for the objectivity of our cognition. Hence we find that even in quantum physics one can achieve objective cognition of objects that are, however, constituted in an unsharp way. In other words quantum physics provides objective knowledge of unsharply constituted objects. Consequently the quantum mechanical unsharpness of measurable properties must be considered as an objective indeterminateness and not as a mere subjective ignorance of the observer. (See figure 10.3.)

6. Conclusion

Kant's reaction to the skepticism of David Hume shows how in principle knowledge about objects can be obtained. Objects of experience must be constituted from the observations by means of categories, in particular by substance and causality. In physics we are confronted with the same problem. For classical as well as for modern physics, Kant has shown in which way one has to proceed. The necessary conditions of objectivity, that is, the categories, must be used for the constitution of objects.

In classical mechanics Kant's procedure can be made more explicit. Here objects are "completely determined," they are subject to a strict determinism and they can be individualized. The advantage of this method can be demonstrated by astrophysical examples. Moreover, one recognizes that the material preconditions of experience are in complete accordance with the laws of classical physics, a condition that was self-evident for Kant but which is an important requirement in modern physics.

It turns out that *objective* cognition in the Kantian sense is possible also in quantum physics. Here, however, the material preconditions of experience, which follow from the quantum theory of measurement, allow to constitute only two kinds of objects: either objects that are *incomplete* and thus *intractable* or objects that are *unsharp* and constituted only approximately. The latter case, which is in accordance with the contemporary interpretation of quantum mechanics, shows that it is in general not possible to begin with the assumption that we actually have some cognition of an object of experience. In many cases these objects are only unsharp objects that are constituted in an approximate way. However, this restriction does not invalidate the objectivity of physical knowledge.

Notes

1. I. Kant, *The Critique of Pure Reason* (1787), B 166.
2. Ibid., A 97.
3. Ibid., B 600.
4. N. Bohr, "The Quantum Postulate and the Recent Development of Atomic Theory," *Nature* 121 (1928): 580–590.
5. G. Ludwig, *Foundations of Quantum Mechanics I and II* (Berlin: Springer, 1983, 1985).
6. W. Heisenberg, *Die physikalische Prinzipien der Quantentheorie* (Leipzig: Hirzel, 1930).
7. G. Hermann, "Die naturphilosophischen Grundlagen der Quantenmechanik," *Abhandlungen der Fries'schen Schule*, new ser. 6, vol. 2 (Berlin: Verlag Öffentliches Leben, 1935).
8. C. F. von Weizsäcker, "Das Verhaltnis der Quantenmechanik zur Philosophie Kants" (1941), in *Zum Weltbild der Physik* (Stuttgart: Hirzel, 1943).
9. E. Cassirer, *Determinismus und Indeterminismus in der modernen Physik* (Göteborg: Elanders Boktryckeri Aktiebolag, 1937).
10. P. Mittelstaedt, *Philosophical Problems of Modern Physics* (Dordrecht: Reidel, 1976).
11. I. Strohmeyer, "Tragweite und Grenze der Transzendentalphilosophie zur Grundlegung der Quantenphysik," *Zeitschrift für allgemeine Wissenschaftstheorie* 18 (1987): 239–275.

12. See: P. Busch and P. Lahti, "A Note on Quantum Theory, Complementarity, and Uncertainty," *Philosophy of Science* 52 (1985): 64–77; P. Busch, "Unsharp Reality and Joint Measurements for Spin Observables," *Physical Review D* 33 (1986): 2253–2261; P. Busch, "Unsharp Reality and the Question of Quantum Systems," in P. Lahti and P. Mittelstaedt, eds., *Symposium on the Foundations of Modern Physics, 1987* (Singapore: World Scientific, 1987); P. Mittelstaedt, A. Prieur, and R. Schieder, "Unsharp Particle-Wave Duality in a Photon Split Beam Experiment," *Foundations of Physics* 19 (1987): 891–903; P. Mittelstaedt, "Unsharp Particle-Wave Duality in Double-Split Experiments," in F. Selleri, ed., *Wave-Particle Duality* (New York: Plenum, 1992). It should be emphasized that the "unsharpness" meant here corresponds to an objective indeterminateness which must not be confused with the observer's subjective ignorance or the experimental inaccuracy, respectively.

11

Galilean Particles: An Example of Constitution of Objects

Elena Castellani

1. Objects and Physics

Speaking of physical objects, we can momentarily bracket the general question, What is there? and focus on the more specific problem of what could reasonably be taken as an *object* in the context under examination.

As regards physics, it is uncontroversial that subatomic and subnuclear particles—the best candidates for qualifying as "objects" in the realm of contemporary physics—are quite different from the ordinary physical objects we can see or touch in our everyday experience. Microphysical entities are surely not immediate data of perception: they can be "observed" only with the help of instruments, and sometimes they are even in principle unobservable as free particles, as in the well-known case of quarks. Moreover, properties traditionally attributed to an object, such as its persistence through time or the possibility of distinguishing it from another similar object, are not easily available in quantum contexts.

These are some of the reasons why it is often claimed that the classical conception of physical objects, that is, the conception grounded on everyday experience, is no longer appropriate when we turn to microphysics. Quantum objects are remote from the objects of our common understanding; if microparticles are to be taken as *objects*, this has to be justified: they must be constructed or "constituted" as objects, depending on the theoretical framework being used as well as on the experimental data at disposition.

Supporters of this view usually assume that, with regard to the characterization of objects of physics, difficulties typically arise in the case of nonclassical entities; in classical physics, objects are not really different from ordinary "material beings."

But one can arrive at a more radical position. Physics, classical or not, does not speak immediately of the objects that populate our external world. Classical mechanics, for example, is formulated in terms of *mass points*, which are obviously of quite another nature than everyday things. Mass points can be seen as "ideal objects," taken to represent some main features of ordinary physical objects.[1] How the properties symbolized through such ideal entities can actually be related to some "real" macroscopic object has to be clarified. And those same classical material beings, whose physical behavior is described using the mass points representation, are not always directly given to us as definite objects: take the case of some astronomic bodies of which we have only a series of isolated light points as empirical evidence.[2]

Arguments of this kind may lead to the following position: *all* the objects of physics, classical as well as nonclassical, must be "constituted."[3] Such a position can be taken as the starting point of what we shall call the *group-theoretic approach* to the problem of physical objects. This chapter provides a brief overview of this approach, entering into some more details for the case of nonrelativistic objects, the so-called *Galilean particles*.

2. Invariance, Symmetry Groups, and Constitution of Objects

In the framework of the "constitution view," objects are generally conceived as "carriers of properties." What the carriers themselves are is not a relevant point on this view. What matters is specifying which the constituting properties of an object are and how these properties are related to their carrier. The question is then, What is it that confers to the carrier of a set of properties the dignity of an *object*? or, more specifically, What kind of properties and prescriptions do we need in order to construct an object?

The group-theoretic approach to the problem of physical objects is grounded on the idea of *invariance*. More precisely, the basic consideration is that the fundamental role that the notions of *invariance* and *symmetry* (i.e., invariance with respect to a group of transformations) have acquired in contemporary physics can also provide a key for addressing the problem of the constitution of physical objects.

Arguments attributing special significance to the notion of invariance with regard to the object question are surely not new. Permanence or invariance in time is a classical requirement for defining the identity of an object in the philosophical tradition. More generally, invariance with respect to change in space and time, when interpreted as invariance under change of reference frames or "observers," is a typical *objectivity condition* required in determining physical objects: we would have difficulties in speaking of "objects" in the case of entities that were not recognizable as the same ones under a simple change of the spatiotemporal perspective.[4]

What is specific to the group-theoretic approach for constituting objects is the exploitation of the invariance idea by using the results of the application of group theory in physics. The theory of *groups of transformations* and their *representations* constitutes the appropriate mathematical tool for investigating the consequences of the symmetry characteristics of physical systems, that is, the characteristics that are usually formulated in terms of *invariance principles*. In physics, the significance of symmetry groups started to be realized around the middle 1920s (with the consequent introduction of group-theoretic techniques in theoretical elaboration) mainly thanks to the fundamental contributions of Hermann Weyl and Eugene Wigner on the application of group theory in quantum mechanics. Today, we can say that symmetry groups are among the basic ingredients of theoretical physics: the so-called elementary particles and their interactions are described essentially with the help of symmetry arguments and group-theoretic methods.

What can the theory of symmetry groups actually tell us in relation to the constitution of objects? Let us begin by recalling the well-known work of Wigner (1939) on the representations of the inhomogeneous Lorentz group (also called "Poincaré group"), a locus classicus for the above approach to the object question.[5] Wigner's aim was to determine all the unitary representations of the Poincaré group (the space-time symmetry group of special relativity), investigating the connection of the representations with quantum mechanical wave equations. Among the far-reaching results he obtained, we find the possibility of arriving at a complete classification of free relativistic elementary systems in association with the study of the *irreducible representations* of the Poincaré group. This possibility of establishing a correspondence between "irreducible representations" of the symmetry group and "elementary physical systems" furnishes the basic motivation for the group-theoretic approach to the constitution of objects. Since the fundamental work of Wigner, indeed, it has become quite usual to classify elementary particles on the basis of their correspondence with the irreducible representations of symmetry groups. In fact, with respect to the characterization of objects it is possible to use symmetry considerations in more than one way. Let us focus on the two main arguments we can find in the literature.

A first possibility is to say, following Wigner's 1939 results, that each "elementary particle" is associated with an irreducible representation of the symmetry group.[6] This implies that the particle has a given number of invariant properties, and that these are exactly the properties which characterize the kind of particle in question.[7] Note, however, that what we obtain in this way is no more than a *class of objects*. The invariant properties that we ascribe to a "particle" on the ground of group-theoretic considerations are not sufficient for constituting the particle as an individual object, distinct from other similar particles. These properties are "necessary" or "essential," in the sense that the given "particle-object" couldn't be determined as such without them (an electron couldn't be an *electron* without given values of mass and spin). But an object cannot be determined as an *individual object* just on the basis of such "essential" properties (also called, in the literature, "intrinsic" properties). One has therefore to deal with the well-known problem of how to obtain "individuating" properties, or properties that can confer individuality on the particle in question, distinguishing it from the other similar particles, that is, from the other particles having the same invariant "essential" properties.

Here comes to the rescue another kind of symmetry argument, which goes back to the notion of a *system of imprimitivity* associated with a symmetry group. The method of imprimitivity systems, which proved to be of great importance for the theory of group representations and the application of this theory to the domain of quantum physics, was systematically developed in a series of papers of G. W. Mackey in the 1950s.[8] In the literature, the explicit use of the notion of an imprimitivity system with regard to the definition of a "particle" is due especially to Piron.[9] The basic idea is to obtain a definition of a particle by employing physical quantities or *observables*, such as for example the position observable, through which the particle could be determined also as an individual object. As we shall see, the method of Mackey's imprimitivity systems for the space-time symmetry group provides a way of approaching the object problem from this point of view.

As an example of the foregoing ideas, let us now have a closer look at how "particles" can in fact be constituted in the case of nonrelativistic (classical and quantum) mechanics.

3. Galilean Particles (1)

The space-time symmetry group of nonrelativistic physics is the so-called *Galilei group*, the group of transformations relating classical frames of reference. According to the group-theoretic approach, for arriving at obtaining "Galilean particles" one has therefore to investigate the irreducible representations of this symmetry group. This has to be done in the formalism of either classical or quantum mechanics, depending on what kind of objects—classical or quantum objects—we want to obtain. But let us start with some general features for discussing physical systems, space-time symmetry, and the consequences of this symmetry for the physical description.

A *physical system*, to begin with, is generally described by first specifying its observable properties or *observables* and its possible modes of preparation or *states*. Observables are physical quantities that are measurable attributes of the system, that is, their values can be measured on the system. The result of such measurements depends on the conditions in which the system is, that is, its modes of preparation or states. Given observables and states, the description of the system at a certain moment is then completed by assigning a rule that tells us the expected value or *expectation value* of every observable for every state of the system. A "dynamical description" will be obtained by specifying, in addition to the foregoing "instantaneous description," how the system evolves with time, that is, by furnishing its *evolution law* (or *dynamical law*).

Space-time symmetry is the symmetry postulated through the *principle of relativity* (the laws of physics are invariant under changes of reference frame). In group-theoretic terms, this means the invariance of physical laws with respect to the group of transformations of reference frames, that is, the group of space-time symmetry consisting of the translations in space and time, the spatial rotations, and the transitions to a uniformly moving coordinate system. In the case of so-called nonrelativistic physics,[10] this is precisely the group of Galilean space-time transformations or Galilei group.

Given a physical system and its description, what follows from the assumption of space-time symmetry? In general, if G is a symmetry group of a theory describing some physical system (i.e., the fundamental equations of the theory are invariant under the transformations of group G) this implies, among other things, that the states of the system transform into each other according to some *representation* of the group: the group operations are "represented" in the states' space by operations relating the states, one to each other.[11]

Accordingly, if the Galilei group is the symmetry group of the theory, the set of the states of the system should provide a "representation space" for this group. We shall therefore proceed toward the determination of "Galilean particles" by first defining as a *Galilean system* a physical system whose states form a representation space for the Galilei group.[12] In order to arrive at *Galilean particles*, the strategy will then be to select, among Galilean systems, those which are *elementary*. For a physical system representing a single particle, elementarity is a quite natural requirement already from an intuitive point of view. In group-theoretic terms, however, this assumes the following precise meaning: an *elementary system* is a system whose set of states constitutes a representation space for an *irreducible representation* of the space-time symmetry group.[13] This means, in other words, that there is a correspondence between the "elementarity" of the system and the "irreducibility" of the representation associated with the system: for such a system, it will not be possible that the set of its states decomposes into (linear) subsets that are each invariant under the transformations of the space-time symmetry group.[14]

At this point, we can propose the following definition of a *Galilean particle*:

> **Definition (1)**: a Galilean particle is an elementary Galilean system, that is, a physical system whose states form an irreducible representation space for the Galilei group.

Actually, in order to take into account the fact that we have a whole class of equivalent elementary Galilean systems, we should more accurately say:

> **Definition (1′)**: a Galilean particle is a physical equivalence class of elementary Galilean systems.

What have we obtained so far? Let us see what the preceding definition implies. For this purpose, we shall have to return to a more abstract level and introduce some aspects of the *Lie group formalism* appropriate for investigating the irreducible representations of space-time symmetry groups.

Space-time transformation groups such as the invariance groups of Galilean and Lorentz relativity are *continuous groups* and, in particular, *Lie groups*. This means, first of all, that the group elements are functions of a certain number r of continuous parameters a_l ($l = 1, 2, \ldots r$), which are characteristic of the group. Without entering into further mathematical technicalities, let us just recall that such group elements can be written in terms of a corresponding number r of infinitesimal operators X_l, the *generators* of the group, which satisfy the "multiplication law" represented by the "Lie brackets"

$$[X_s, X_t] = c_{st}^q X_q,$$

so forming what is called the *Lie algebra* of the group. The coefficients c_{st}^q are constants that characterize the particular structure of the group, whence their being named the *structure constants* of the Lie group. For every Lie group, we can construct operators that are scalar quadratic in the infinitesimal operators X_l, the so-called *Casimir operators*

$$C = \sum_{s,t} (g^{-1})_{st} X_s X_t$$

with the matrix elements g_{st} given by $g_{st} = \sum_{pq} c_{sq}^p c_{tp}^q$. Casimir operators have the special property of commuting with all the infinitesimal operators X_l (the *generators* of the Lie group), that is, they are the fundamental *invariants* of the group. In the context of the theory of group representations, this implies, in particular, that in an irreducible representation the Casimir operators are simple multiples of the unit operator, whence the possibility of labeling the representations directly in terms of the eigenvalues of these operators. The eigenvalue spectra of the invariants of the group therefore provide the labels for classifying the irreducible representations of the group: on this fact is grounded the possibility of associating the values of the invariant properties characterizing physical systems with the labels of the irreducible representations of symmetry groups.

This is a very general scheme, which can be abstracted from the usual way of proceeding in quantum relativistic physics (quantum field theory) for classifying elementary particles. In order to see how and to what extent such a scheme can be applied to the case of classical and quantum "Galilean particles," one has to consider how classical and quantum physical systems can respectively provide a representation space for irreducible representations of the Galilei (Lie) group, and what kind of invariant properties can be consequently attributed to either class of systems. In this chapter we shall give only a very brief account of this program, by summarizing it in the following points.[15]

3.1 *Galilei Group*

The (proper) Galilei group $\mathcal{G}$ is a 10-parameter Lie group, containing the translations in space $\mathbf{x}' = \mathbf{x} + \mathbf{a}$ (3 translation parameters $\mathbf{a}$), the translations in time $t' = t + b$ (1 time parameter b), the space rotations $\mathbf{x}' = R\mathbf{x}$ (3 rotation parameters in rotation matrix R), and the transitions to a uniformly moving coordinate system $\mathbf{x}' = \mathbf{x} + \mathbf{v}t$, also called "pure Galilei transformations" or "Galilei boosts" (3 velocity parameters $\mathbf{v}$).

A general inhomogeneous Galilei transformation is then a map

$$g : (\mathbf{x}, t) \mapsto (\mathbf{x}', t')$$

of the form

$$\mathbf{x}' = R\mathbf{x} + \mathbf{v}t + \mathbf{a},$$
$$t' = t + b.$$

In terms of the group parameters, a generic element g of $\mathcal{G}$ is denoted by

$$g = (b, \mathbf{a}, \mathbf{v}, R)$$

and the multiplication law of $\mathcal{G}$ is

$$gg' = (b, \mathbf{a}, \mathbf{v}, R)(b', \mathbf{a}', \mathbf{v}', R') = (b + b', \mathbf{a} + R\mathbf{a}' + b'\mathbf{v}, \mathbf{v} + R\mathbf{v}', RR').$$

Given the multiplication law or "group law" and hence the group structure,[16] one can compute the Lie brackets for the infinitesimal generators and consequently obtain the Lie algebra of the Galilei group $\mathcal{G}$.

3.2 Galilei Group and Classical Mechanics

The usual way of describing a classical mechanical system with n degrees of freedom is to associate with it a space Ω, called the *phase space* of the system, whose points are defined by the values $x_1, \ldots, x_n, p_1, \ldots, p_n$ of the configuration $(x_1, \ldots, x_n)$ and the momentum vector $(p_1, \ldots, p_n)$ of the system at a given instant of time. These points represent the *states* of the system and the physical quantities or *observables* are described by real valued functions defined on the phase space. The system we are considering here is a massive point particle.

In the phase space of the states of the system, the symmetry transformations are "represented" by so-called *canonical transformations*, which are the transformations of the phase space leaving invariant the equations of motion. Canonical transformations can be regarded as infinite sequences of *infinitesimal canonical transformations* $x'_i = x_i + \delta x_i$, $p'_i = p_i + \delta p_i$. The quantities W_s, such that $\delta x_i = (\partial W_s/\partial p_i)\delta\alpha_s$ and $\delta p_i = -(\partial W_s/\partial x_i)\delta\alpha_s$, are called the *generators* of the infinitesimal transformations.

The Lie algebra of the Galilei group has a physically significant realization in terms of the ten quantities H, $\mathbf{P}$, $\mathbf{J}$, and $\mathbf{K}$, which are the generators of the canonical transformations representing the time translations, space translations, space rotations, and Galilean boosts, respectively. H can be identified with the Hamiltonian or energy of the system, the three generators P_i with the components of the momentum $\mathbf{P}$, the three generators J_i with the components of the angular momentum $\mathbf{J}$, and the three generators K_i with the components of the quantity $\mathbf{K} = t\mathbf{P} - m\mathbf{X}$. The Lie brackets characterizing the Lie algebra of the Galilei group are represented by the following *Poisson brackets*:

$$\{P_i, P_j\} = 0, \ \{P_i, H\} = 0, \ \{J_i, H\} = 0,$$

$$\{J_i, J_j\} = \epsilon_{ijk}J_k, \ \{J_i, P_j\} = \epsilon_{ijk}P_k,$$

$$\{J_i, K_j\} = \epsilon_{ijk}K_k, \ \{K_i, H\} = P_i, \ \{K_i, K_j\} = 0,$$

and

$$\{K_i, P_j\} = 0 + m\delta_{ij},$$

where the constant m can be identified with the mass of the system.[17]

3.3 Galilei Group and Quantum Mechanics

In the usual formulation of quantum mechanics, the behavior of a system with n degrees of freedom is described in terms of a *wave function* or *state vector* $\psi(\mathbf{x}, t)$, where $\mathbf{x} = (x_1, \ldots, x_n)$ is the configuration of the system at the instant of time t. To each physical system is associated a *Hilbert space* $\mathcal{H}$, which is the space of the wave functions or vectors corresponding to the (pure) states of the system, and to each observable quantity corresponds a self-adjoint operator acting in this Hilbert space.

Any group G of coordinate transformations $(\mathbf{x}, t) \mapsto (\mathbf{x}', t')$ defines a corresponding set of induced trasformations $\psi \mapsto \psi'$ in the Hilbert space of the wave functions. If G is a symmetry group of the system description, the transformations of the group are represented in the space $\mathcal{H}$ by *unitary* operators acting on the wave functions. In the case of a Lie group, the unitary operators can be regarded as generated by infinitesimal (self-adjoint) operators.

A physically significant representation of the Galilei Lie group is given in terms of the infinitesimal generators which are the quantum-mechanical analogues of the classical quantities generating the Galilean transformations: that is, the operator Hamiltonian H for time translations, the three operators components of the momentum $\mathbf{P}$ for space translations, the three operators components of the angular momentum $\mathbf{J}$ for space rotations, and the three operators components of the quantity $\mathbf{K}$ for Galilei boosts. The Lie brackets of the Galilei Lie algebra are here represented by the *commutation relations*

$$[P_i, P_j] = 0, \quad [P_i, H] = 0, \quad [J_i, H] = 0,$$
$$[J_i, J_j] = \epsilon_{ijk} J_k, \quad [J_i, P_j] = \epsilon_{ijk} P_k,$$
$$[J_i, K_j] = \epsilon_{ijk} K_k, \quad [K_i, H] = P_i, \quad [K_i, K_j] = 0,$$

and

$$[K_i, P_j] = 0 + m\delta_{ij} I,$$

where m is the mass of the system.[18]

3.4 *Invariant Properties*

As we have seen, the values of the fundamental invariants of the symmetry group label the irreducible representations and hence the elementary physical systems.

Both in the classical and quantum case, the characteristic invariant quantities one obtains by constructing the irreducible (projective) representations of the Galilei group are the mass m (as can be seen also from the relations representing the Lie brackets for the Galilei group generators), the "internal energy" (i.e., the energy in the rest system) and a quantity which can be interpreted as an "internal angular momentum" or "spin."[19]

From the group-theoretic point of view, "Galilean particles" are therefore characterized in terms of the invariant properties of mass, internal energy, and spin.[20]

4. Galilean Particles (2)

As already remarked, another way of addressing the object question from the group-theoretical point of view is that grounded on the notion of a *system of imprimitivity* for the space-time symmetry group. The aim is to arrive at a definition of a particle by determining "individuating" observable quantities (such as the position and momentum) with the help of the method of imprimitivity systems.[21]

The framework is that of the *logical formulation* of a physical system: according to the viewpoint introduced in the 1930s by John von Neumann, to each physical system is associated a (orthocomplemented) lattice $\mathcal{L}$ of "propositions," which is called the *logic* of the system. These propositions are the most general experimentally verifiable statements which can be made about the system, that is the statements of the form "the value of a certain observable quantity lies in a given set of real numbers." In such a framework, if $A\{E\}$ is the proposition that, for a given system S, the value a of the observable quantity A lies in the set $E \in \mathcal{B}(\mathcal{R})$ (where $\mathcal{B}(\mathcal{R})$ is the family of the Borel subsets of the real line $\mathcal{R}$), the observable A may be identified with the mapping

$$A : E \mapsto A\{E\}$$

of $\mathcal{B}(\mathcal{R})$ into the lattice $\mathcal{L}$.

Given the triplet $(A, \mathcal{B}, \mathcal{L})$ for a physical system, the method is then to consider what the action of the space-time symmetry group G implies for the elements of the triplet. To state it very shortly, we shall have, in

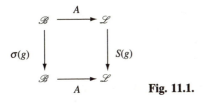

Fig. 11.1.

particular, a *condition of covariance* for the observable A, which can be expressed in terms of the following *imprimitivity condition:*

$$A\{\sigma(g)[\mathscr{B}]\} = S(g)[A\{E\}]. \qquad (*)$$

Graphically, the condition corresponds to the commutativity shown in figure 11.1, where $\sigma(g)$ and $S(g)$ are representations of the group G in terms of the automorphisms of $\mathscr{B}$ and of $\mathscr{L}$, respectively. A triplet $(A, \mathscr{B}, \mathscr{L})$ that satisfies this condition is called a *system of imprimitivity* for the group G.[22]

Now, if one takes as "identifying" properties the observables position, momentum, and time, a "Galilean particle" can be defined in the following way:[23]

> **Definition (2):** A Galilean particle is an elementary system for which the observables position, momentum, and time are defined.

that is, in the described framework,

> **Definition (2′):** A Galilean particle is a system of propositions $\mathscr{L}$, for which an irreducible representation of the Galilei group is defined such that the observable quantities $A_i \colon \mathscr{B}(\mathscr{R}) \mapsto \mathscr{L}$, where $i = 1, 2, 3$ and A_1, A_2, A_3 are, respectively, the position, momentum, and time, satisfy the imprimitivity condition $(*)$.

Galilean particles can then be explicitly constructed by studying the imprimitivity systems for the observables position, momentum, and time both in the classical and in the quantum case.[24]

Notes

Support for this work was provided in 1993–1994 by the N.A.T.O.–C.N.R. Advanced Fellowships Program.

1. See, for example, W.V.O. Quine, *Word and Object* (Cambridge, Mass.: MIT Press, 1960), 248–251.
2. This astrophysical example is treated, in particular, in P. Mittelstaedt, "The Constitution of Objects in Kant's Philosophy and in Modern Physics," in P. Parrini, ed., *Kant and Contemporary Epistemology* (Dordrecht: Kluwer Academic Publishers, 1994), which is reprinted here as chapter 10.

3. The term *constitution* is here used to indicate that objects are determined as such by using some conceptual prescription—in the case of physics objects, some physical laws. For arguments supporting such a use of this "Kantian" terminology with regard to the problem of physical objects, see Mittelstaedt, "The Constitution of Objects," and also, by same author, *Sprache und Realität in der modernen Physik* (Mannheim: Bibliographisches Institut-Wissenschaftsverlag, 1986). The view according to which all the objects of physics need to be constituted is more extensively presented in E. Castellani, "Sulla nozione di oggetto nella fisica classica e quantistica," in C. Cellucci, M. C. Di Maio, and G. Roncaglia, eds., *Logica e filosofia della scienza: problemi e prospettive* (Proceedings Soc. Italiana di Logica e Filos. delle Scienze, Lucca 1993) (Pisa: Edizioni ETS, 1994).

4. For a more extensive treatment of this point, see, for example, E. Castellani, "Quantum Mechanics, Objects and Objectivity," in C. Garola and A. Rossi, eds., *The Foundations of Quantum Mechanics-Historical Analysis and Open Questions* (Dordrecht: Kluwer, 1995).

5. E. Wigner, "On Unitary Representations of the Inhomogeneous Lorentz Group," *Annals of Mathematics* 40 (1939): 149–204.

6. More precisely, an elementary particle is described as a physical system whose states transform under the operations of the symmetry group according to a definite irreducible representation.

7. In the context of relativistic quantum physics, for example, an elementary particle is associated with an irreducible representation of the Poincaré group: within such a determination, the invariant properties characterizing a (free) particle are its rest-mass and spin.

8. For a clear and detailed account of Mackey's contributes on the notion of an imprimitivity system and a discussion of the impressive variety of applications of the notion, see in particular the chapter on systems of imprimitivity and "Mackey's machine" in V. S. Varadarajan, *Geometry of Quantum Theory*, 2d ed. (New York: Springer, 1985).

9. See C. Piron, *Foundations of Quantum Physics* (Reading, Mass.: W. A. Benjamin, 1976), 93. A definition of an "elementary particle" making use of the notion of a system of imprimitivity can be found already in J. M. Jauch, *Foundations of Quantum Mechanics* (Reading, Mass.: Addison-Wesley, 1968), 205.

10. Which it would be more appropriate to call "Galilean relativistic physics," as it has been particularly stressed by J.-M. Lévy-Leblond. See for instance his "Galilei Group and Galilean Invariance," in E. M. Loebl, ed., *Group Theory and Its Applications,* vol. 2 (New York: Academic Press, 1971).

11. Where the basic condition for a set of operations $T(g_i)$, $T(g_j)$, ... to form a "representation" of the group G (with elements $g_i, g_j, ...$) is that $T(g_i)T(g_j) = T(g_ig_j)$.

12. In an analogous way, a *Lorentz system* will be defined as a system whose states form a representation space for the Lorentz group.

13. The concept of an "elementary system" is actually broader than the intuitive concept of an "elementary particle" (an elementary particle requiring the further condition of being structureless). On this point see in particular T. D.

Newton and E. P. Wigner, "Localized States for Elementary Systems," *Review of Modern Physics* 21 (1949): 400–406.

14. That is, subsets whose component states are transformed into states of the same subset by the transformations of the symmetry group.

15. For simplicity sake, in considering the states of a physical system we shall take into account only the case of "pure state." (A more general treatment of representation spaces for the space-time symmetry group, which includes also the case of "mixed states," is of course possible; see, for example, D. G. Currie, T. F. Jordan, and E.C.G. Sudarshan, "Relativistic Invariance and Hamiltonian Theories of Interacting Particles," *Review of Modern Physics* 35 [1963]: 350–375.)

16. The multiplication law is easily obtained by applying successively two Galilei transformations g and g'.

17. The generators H, **P**, **J**, **K** so constitute a "Poisson algebra" of the form

$$\{W_i, W_j\} = c_{ij}^k W_k + \beta_{ij}$$

(where β_{ij} are constants), that is a so-called *projective* realization of the Galilei group Lie algebra in the phase space.

18. The infinitesimal operators H, **P**, **J**, **K** so form an algebra

$$[W_i, W_j] = c_{ij}^k W_k + \beta_{ij} I,$$

which is a projective representation of the Galilei group Lie algebra in the Hilbert space.

19. For more details on how to obtain these invariant quantities and their meaning in the classical case and in the quantum nonrelativistic case—and, in general, for what regards the Galilei group and its irreducible representations—see in particular Lévy-Leblond, "Galilei Group and Galilean Invariance," and also, by the same author, "Galilei Group and Nonrelativistic Quantum Mechanics," *Journal of Mathematical Physics* 4 (1963): 776–788.

20. Recalling, however, that for a nonrelativistic isolated particle the internal energy has indeed no physical significance. See Lévy-Leblond, "Galilei Group and Nonrelativistic Quantum Mechanics," 782.

21. The view at issue here is that according to which individuality is conferred upon an object by some of its properties and, in particular, by space-time properties.

22. See for example Piron, *Foundations of Quantum Physics*, 94–95.

23. We follow here the approach of Piron.

24. Some examples of such a way of constructing classical and quantum Galilean particles can be found in Piron, *Foundations of Quantum Physics*. In the quantum case, the problem of the incommensurability of properties such as position and momentum can be approached by using so-called "unsharp observables." The problem of the (approximate) constitution of objects by means of unsharp properties, that is, by means of the study of imprimitivity systems in the case of unsharp observables, is investigated in particular by P. Mittelstaedt, P. Busch,

and P. Lahti. On the subject see, for instance, Mittelstaedt, "The Constitution of Objects"; P. Mittelstaedt, "The Constitution of Objects in Classical Mechanics and in Quantum Mechanics," *International Journal of Theoretical Physics* 34 (1995): 1615–1626; P. Busch, "Unsharp Reality and the Question of Quantum Systems," in P. Lahti and P. Mittelstaedt, eds., *Symposium on the Foundations of Modern Physics, 1987* (Singapore: World Scientific, 1987), and P. Busch, "Macroscopic Quantum Systems and the Objectification Problem," in P. Lahti and P. Mittelstaedt, eds., *Symposium on the Foundations of Modern Physics, 1990* (Singapore: World Scientific, 1990).

PART THREE

OBJECTS AND MEASUREMENT

12

What Is an Elementary Particle?

Erwin Schrödinger

Increasing knowledge has in some ways made us not more certain but less certain of the nature of matter. Whereas Dalton and his school had a clear picture of the existence of atomic particles as real and indestructible solid particles, modern wave mechanics implies very clearly that, in fact, they are not identifiable individuals at all. Although Daltonian conceptions provide convenient terms in which to describe the properties of matter, their original significance has undergone very important changes during the last thirty years.

1. A Particle Is Not an Individual

Atomism in its latest form is called quantum mechanics. It has extended its range to comprise, besides ordinary matter, all kinds of radiation, including light—in brief, all forms of energy, ordinary matter being one of them. In the present form of the theory the "atoms" are electrons, protons, photons, mesons, etc. The generic name is elementary particle, or merely particle. The term atom has very wisely been retained for chemical atoms, though it has become a misnomer.

This essay deals with the elementary particle, more particularly with a certain feature that this concept has acquired—or rather lost—in quantum mechanics. I mean this: that the elementary particle is not an individual; it cannot be identified, it lacks "sameness." The fact is known to every physicist, but is rarely given any prominence in surveys readable by nonspecialists. In technical language it is covered by saying that the particles "obey" a newfangled statistics, either Einstein-Bose or Fermi-Dirac statistics. The implication, far from obvious, is that the unsuspected epithet "this" is not quite properly applicable to, say, an electron, except with caution, in a restricted sense, and sometimes not at all. My objective here is to explain this point and to give it the thought it deserves. In order to create a foil for the discussion, let me summarize in sections 2–5 what we are usually told about particles and waves in the new physics.

2. Current Views: The Amalgamation of Particles and Waves

Our image of the material world had been made up of two kinds of "fittings": waves and particles. The former were instanced mainly, if not exclusively, by Maxwell's waves of electromagnetic energy, comprising such as are used in radio, light, X-rays, and gamma-rays. Material bodies were said to consist of particles. One was also familiar with jets of particles, called corpuscular rays, such as cathode rays, beta rays, alpha rays, and anode rays. Particles would emit and absorb waves. For instance, cathode rays (electrons), when slowed down by colliding with atoms, emit X-rays. The distinction between particles and waves was, however, considered as clear-cut as that between a violin and its sound. An examinee who alleged cathode rays to be waves, or X-rays to be jets of particles, would have got very bad marks.

In the new setting of ideas the distinction has vanished, because it was discovered that all particles have also wave properties, and vice versa. Neither of the two concepts must be discarded; they must be amalgamated. Which aspect obtrudes itself depends not on the physical object, but on the experimental device set up to examine it. A jet of cathode rays, for example, produces in a Wilson cloud chamber discrete tracks of water droplets—curved tracks if there is a magnetic field to deflect the electrons, otherwise straight alignments of droplets. We cannot but interpret them as traces of the paths of single electrons. Yet the same jet, after crossing a narrow tube placed at right angles to it and containing crystal powder, will produce on a photographic plate at some distance behind the tube a pattern of concentric circles. This pattern can be understood in all its details when looked upon as the interference pattern of waves, and in no other way. Indeed, it bears a close resemblance to similarly produced X-ray patterns.

The suspicion arises: are the conical jets that impinge on the photographic plate and form the pattern of circles really cathode rays; are they not perhaps secondary X-rays? The suspicion has to be dismissed, for the whole system of circles can be displaced by a magnet, while X-rays can not; moreover, by putting a lead screen with a small hole in it in the place of the photographic plate, a jetlet can be isolated from one of the conical jets and made to display any of the typical particle characters of cathode rays: it will produce discrete tracks in a cloud chamber; bring about discrete discharges in a Geiger-Müller counter; and charge up a Faraday cage in which it is intercepted.

A vast amount of experimental evidence clinches the conviction that wave characteristics and particle characteristics are never encountered singly, but always in a union; they form different aspects of the same phenomenon, and indeed of all physical phenomena. The union is not a loose or superficial one. It would be quite unsatisfactory to consider cathode rays to consist both of particles and of waves. In the early days of the new theory it was suggested that the particles might be singular spots within the waves, actually singularities in the meaning of the mathematician. The white crests on a moderately rough sea would be a fairly adequate simile. The idea was very soon abandoned. It seems that both concepts, that of waves and that of particles, have to be modified considerably, so as to attain a true amalgamation.

3. Current Views: The Nature of Waves

The waves, so we are told, must not be regarded as quite real waves. It is true that they produce interference patterns—which is the crucial test that in the case of light had removed all doubts as to the reality of the waves. However, we are now told that all waves, including light, ought rather to be looked upon as "probability waves." They are only a mathematical device for computing the probability of finding a particle in certain conditions, for instance (in the above example), the probability of an electron hitting the photographic plate within a small specified area. There it is registered by acting on a grain of silver bromide. The interference pattern is to be regarded as a statistical registration of the impinging electrons. The waves are in this context sometimes referred to as guiding waves—guiding or directing the particles on their paths. The guidance is not to be regarded as a rigid one; it merely constitutes a probability. The clear-cut pattern is a statistical result, its definiteness being due to the enormous number of particles.

Here I cannot refrain from mentioning an objection which is too obvious not to occur to the reader. Something that influences the physical behaviour of something else must not in any respect be called less real than the something it influences—whatever meaning we may give to the dangerous epithet "real." It is certainly useful to recall at times that all quantitative models or images conceived by the physicist are, epistemologically, only mathematical devices for computing observable events, but I cannot see that this applies more to, say, light waves than to, say, oxygen molecules.

4. Current Views: The Nature of Particles (Uncertainty Relation)

As regards the modification required in the concept of a particle, the stress is on Heisenberg's uncertainty relation. The so-called classical mechanics hinged on Galileo's and Newton's discovery that the thing which in a moving body is determined at any instant by the other bodies in its environment is only and precisely its acceleration, or, in mathematical terms, the second derivatives with respect to time of the coordinates. The first derivatives, commonly called the velocity, are therefore to be included in the description of the momentary state of the body, together with the coordinates themselves, which label its momentary place in space or "whereness" (or ubiety, to use an antiquated but convenient word). Thus, to describe the momentary state of a particle, two independent data were required: its coordinates and their first time derivatives, or ubiety and velocity. According to the new theory less is required, and less is obtainable. Either of the two data can be given with arbitrary accuracy, provided that no store is set on the other, but both cannot be known together with absolute precision. One may not even conceive of both as having absolutely sharp values at the same instant. They mutually blur each other, as it were. Broadly speaking, the product of the latitudes of their respective inaccuracies cannot be reduced below a fixed constant. For an electron, this constant happens to be about 1 if the units centimetre and second are used. Thus, if the velocity of an electron is considered sharp with a latitude of only 1 cm/sec, its location has to be considered as blurred within the latitude of 1 cm. The strangeness does not lie in the mere existence of inaccuracies, for the particle might be a thing of vague and changeable extension, within which slightly different velocities prevailed at different spots. Then, however, a sharp location or ubiety would probably entail a sharply defined velocity and vice versa. Actually it is just the other way round.

5. Current Views: The Meaning of the Uncertainty Relation

Two links connect this strange and certainly very fundamental statement to other parts of the theory. It can be arrived at by declaring that a particle is equivalent to its guiding wave, and has no characteristics save those indicated by the guiding wave according to a certain code. The code is simple enough. The ubiety is indicated by the extension of the wave, the latitude in the velocity by the range of wave numbers. "Wave number" is short for reciprocal of the wavelength. Each wave number corresponds to a certain velocity proportional to it. That is the code. It is a mathematical truism that the smaller a wave group, the wider is the (minimum) spread of its wave numbers.

Alternatively, we may scrutinize the experimental procedure for determining either the ubiety or the velocity. Any such measuring device implies a transfer of energy between the particle and some measuring instrument—eventually the observer himself, who has to take a reading.

This means an actual physical interference with the particle. The disturbance cannot be arbitrarily reduced, because energy is not exchanged continuously but in portions. We are given to understand that, when measuring one of the two items, ubiety or velocity, we interfere with the other the more violently the higher the precision we aim at. We blur its value within a latitude inversely proportional to the latitude of error allowed in the first.

In both explanations the wording seems to suggest that the uncertainty or lack of precision refers to the attainable knowledge about a particle rather than to its nature. Indeed, by saying that we disturb or change a measurable physical quantity we logically imply that it has certain values before and after our interference, whether we know them or not. And in the first explanation, involving the wave, if we call it a guiding wave, how should it guide the particle on its path, if the particle has not got a path? If we say the wave indicates the probability of finding the particle at A, or at B, or at C . . . this seems to imply that the particle is at one, and one only, of these places; and similarly for the velocity. (Actually the wave does indicate both probabilities simultaneously, one by its extension, the other by its wave numbers.) However, the current view does not accept either ubiety or velocity as permanent objective realities. It stresses the word "finding." Finding the particle at point A does not imply that it has been there before. We are more or less given to understand that our measuring device has brought it there or "concentrated" it at that point, while at the same time we have disturbed its velocity. And this does not imply that the velocity "had" a value. We have only disturbed or changed the probability of finding this or that value of the velocity if we measure it. The implications as to "being" or "having" are misconceptions, to be blamed on language. Positivist philosophy is invoked to tell us that we must not distinguish between the knowledge we can obtain of a physical object and its actual state. The two are one.

6. Criticism of the Uncertainty Relation

I will not discuss here that tenet of positivist philosophy. I fully agree that the uncertainty relation has nothing to do with incomplete knowledge. It does reduce the amount of information attainable about a particle as compared with views held previously. The conclusion is that these views were wrong and we must give them up. We must not believe that the completer description they demanded about what is really going on in the physical world is conceivable, but in practice unobtainable. This would mean clinging to the old view. Still, it does not necessarily follow that we must give up speaking and thinking in terms of what is really going on in the physical world. It has become a convenient habit to picture it as a reality. In everyday life we all follow this habit, even those philosophers who opposed it theoretically, such as Bishop Berkeley. Such theoretical controversy is on a different plane. Physics has nothing to do with it. Physics takes its start from everyday experience, which it continues by more subtle means. It remains akin to it, does not transcend it generically, it cannot enter into another realm. Discoveries in physics cannot in themselves—so I believe—have the authority of forcing us to put an end to the habit of picturing the physical world as a reality.

I believe the situation is this. We have taken over from previous theory the idea of a particle and all the technical language concerning it. This idea is inadequate. It constantly drives our mind to ask for information which has obviously no significance. Its imaginative structure exhibits features which are alien to the real particle. An adequate picture must not trouble us with this disquieting urge; it must be incapable of picturing more than there is; it must refuse any further addition. Most people seem to think that no such picture can be found. One may, of course, point to the circumstantial evidence (which I am sorry to say is not changed by this essay) that in fact none has been found. I can, however, think of some reasons for this, apart from the genuine intricacy of the case. The palliative, taken from positivist philosophy and purporting to be a reasonable way out, was administered fairly early and authoritatively. It seemed to relieve us from the search for what I should call real understanding; it even rendered the endeavour suspect, as betraying an unphilosophical mind—the mind of a child who regretted the loss of its favourite toy (the picture or model) and would not realize that it was gone forever. As a second point, I submit that the difficulty may be intimately connected with the principal subject of this paper, to which I shall now turn without further delay. The uncertainty relation refers to the particle. The particle, as we shall see, is not an identifiable individual. It may indeed well be that no individual entity can be conceived which would answer the requirements of the adequate picture stated above.

It is not at all easy to realize this lack of individuality and to find words for it. A symptom is that the probability interpretation, unless it is expressed in the most highly technical language of mathematics, seems to be vague as to whether the wave gives information about one particle or about an *ensemble* of particles. It is not always quite clear whether it indicates the probability of finding "the" particle or of finding "a" particle, or indicates the likely or average number of particles in, say, a given small volume. Moreover the most popular view on probability tends to obliterate these differences. It is true that exact mathematical tools are available to distinguish between them. A point of general interest is involved, which I will explain. A method of dealing with the problem of many particles was indicated in 1926 by the present writer. The method uses waves in many dimensional space, in a manifold of $3N$ dimensions, N being the number of particles. Deeper insight led to its improvement. The step leading to this improvement is of momentous significance. The many-dimensional treatment has been superseded by so-called second quantization, which is mathematically equivalent to uniting into one three-dimensional formulation the cases $N = 0, 1, 2, 3 \ldots$ (to infinity) of the many-dimensional treatment. This highly ingenious device includes the so-called new statistics, with which we shall have to deal below in much simpler terms. It is the only precise formulation of the views now held, and the one that is always used. What is so very significant in our present context is that one cannot avoid leaving indeterminate the number of the particles dealt with. It is thus obvious that they are not individuals.

7. The Notion of a Piece of Matter

I wish to set forth a view on matter and the material universe, to which Ernst Mach,[1] Bertrand Russell,[2] and others were led by a careful analysis of concepts. It differs from the popular view. We are, however, not concerned with the psychological origin of the concept of matter but with its epistemological analysis. The attitude is so simple that it can hardly claim complete novelty; some pre-Socratics, including the materialist Democritus,[3] were nearer to it than were the great men who resuscitated science and moulded it during the seventeenth to nineteenth century.

According to this view, a piece of matter is the name we give to a continuous string of events that succeed each other in time, immediately successive ones being as a rule closely similar. The single event is an inextricable complex of sensates, of associated memory images, and of expectations associated with the former two. The sensates prevail in the case of an unkhown object, say a distant white patch on the road, which might be a stone, snow, salt, a cat or a dog, a white shirt or blouse, a handkerchief. Even so, within the ensuing string of events we usually know from general experience how to discount the changes caused by motions of our own body, in particular of our direction of sight. As soon as the nature of the object is recognized, images and expectations begin to prevail. The latter concern sensations as hard, soft, heavy, flexible, rough, smooth, cold, salty, etc., associated with the image of touching and handling; they also concern spontaneous movements or noises such as barking, mewing, shouting. It should be noted that I am not speaking of our thoughts or considerations about the object, but of what forms part and parcel of our perception of it—of what it is to us. However, the limit is not sharp. As our familiarity with a piece of matter grows, and in particular as we approach its scientific aspect, the range of expectations in regard to it widens, eventually to include all the information science has ascertained, e.g. melting point, solubility, electric conductivity, density, chemical and crystalline structure, and so on. At the same time, the momentary sensational core recedes in relevance the more the object becomes familiar to us, whether by scientific knowledge or by everyday use.

8. Individuality or "Sameness"

After a certain wealth of association has come to outshine the core of sensates, the latter is no longer needed to keep the complex together. It persists even when the contact of our senses with the object temporarily ceases. And more than that: the complex is latently conserved even when the whole string is interrupted by our turning away from the object to others and forgetting all about it. Indeed, this is not exceptional, but a rule which—since we sometimes sleep—has no exception. But we have adopted the useful device of filling these gaps. We supplement the missing parts of the strings relating to pieces of matter in our nearer and farther surroundings, to cover the periods when we neither watch them nor think of them. When a familiar object reenters our ken, it is usually recognized as a continuation of previous appearances, as being the same thing. The relative permanence of individual pieces of matter is the most momentous feature of both everyday life and scientific experience. If a familiar article, say an earthenware jug, disappears from your room, you are quite sure somebody must have taken it away. If after a time it reappears, you may doubt whether it really is the same one—breakable objects in such circumstances are often not. You may not be able to decide the issue, but you will have no doubt that the doubtful sameness has an indisputable meaning—that there is an unambiguous answer to your query. So firm is our belief in the continuity of the unobserved parts of the strings!

No doubt the notion of individuality of pieces of matter dates from time immemorial. I suppose animals must have it in some way, and a dog, when seeking for his ball that has been hidden, displays it very plainly. Science has taken it over as a matter of course. It has refined it so as safely to embrace all cases of apparent disappearance of matter. The idea that a log which burns away first turns into fire, then into ashes and smoke, is not alien to the primitive mind. Science has substantiated it; though the appearance in bulk may change, the ultimate constituents of the matter do not. This was (in spite of his occasional scepticism mentioned above) the teaching of Democritus. Neither he nor Dalton doubted that an atom which was originally present in the block of wood is afterward either in the ashes or in the smoke.

9. The Bearing of Atomism

In the new turn of atomism that began with the papers of Heisenberg and of de Broglie in 1925, such an attitude has to be abandoned. This is the most startling revelation emerging from the ensuing development, and the feature which in the long run is bound to have the most important consequences. If we wish to retain atomism we are forced by observed facts to deny the ultimate constituents of matter the character of identifiable individuals. Up to recently, atomists of all ages, for all I know, had transferred that characteristic from visible and palpable pieces of matter to the atoms, which they could not see or touch or observe singly. Now we do observe single particles; we see their tracks in the cloud chamber and in photographic emulsions; we register the practically simultaneous discharges caused by a single swift particle in two or three Geiger counters placed at several yards' distance from each other. Yet we must deny the particle the dignity of being an absolutely identifiable individual. Formerly, if a physicist were asked what stuff the atoms themselves were made of, he might smile and shirk the answer. If the inquirer insisted on the question whether he might imagine them as small unchangeable bits of ordinary matter, he would get the smiling reply that there was no point in doing so but that it would do no harm. The formerly meaningless question has now gained significance. The answer is definitely in the negative. An atom lacks the most primitive property we associate with a piece of matter in ordinary life. Some philosophers of the past, if the case could be put to them, would say that the modern atom consists of no stuff at all but is pure shape.

10. The Meaning of the New Statistic

We must at last proceed to give the reasons for this change of attitude in a more comprehensible form than at the end of section 6. It rests on the so-called new statistics. There are two of them. One is the Bose-Einstein statistics, whose novelty and relevance were first stressed by Einstein. The other is the Fermi-Dirac statistics, of which the most pregnant expression is Pauli's exclusion principle. I shall try to explain the new statistics, and its relation to the old classical or Boltzmann statistics, to those who have never heard about such things and perhaps may be puzzled by what "statistics" means in this context. I shall use an instance from everyday life. It may seem childishly simple, particularly because we have to choose small numbers—actually 2 and 3—in order to make the arithmetic surveyable. Apart from this, the illustration is completely adequate and covers the actual situation.

Three schoolboys, Tom, Dick, and Harry, deserve a reward. The teacher has two rewards to distribute among them. Before doing so, he wishes to realize for himself how many different distributions are at all possible. This is the only question we investigate (we are not interested in his eventual decision). It is a statistical question: to count the number of different distributions. The point is that the answer depends on the nature of the rewards. Three different kinds of reward will illustrate the three kinds of statistics.

(*a*) The two rewards are two memorial coins with portraits of Newton and Shakespeare respectively. The teacher may give Newton either to Tom or to Dick or to Harry, and Shakespeare either to Tom or to Dick or to Harry. Thus there are three times three, that is nine, different distributions (classical statistics).

(*b*) The two rewards are two shilling-pieces (which, for our purpose, we must regard as indivisible quantities). They can be given to two different boys, the third going without. In addition to these three possibilities there are three more: either Tom or Dick or Harry receives two shillings. Thus there are six different distributions (Bose-Einstein statistics).

(*c*) The two rewards are two vacancies in the football team that is to play for the school. In this case two boys can join the team, and one of the three is left out. Thus there are three different distributions (Fermi-Dirac statistics).

Let me mention right away: the *rewards* represent the particles, two of the same kind in every case; the boys represent states the particle can assume. Thus, "Newton is given to Dick" means: the particle Newton takes on the state Dick.

Notice that the counting is natural, logical, and indisputable in every case. It is uniquely determined by the nature of the objects: memorial coins, shillings, memberships. They are of different categories. Memorial coins are individuals distinguished from one another. Shillings, for all intents and purposes, are not, but they are still capable of being owned in the plural. It makes a difference whether you have one shilling, or two, or three. There is no point in two boys exchanging their shillings. It does change the situation, however, if one boy gives up his shilling to another. With memberships, neither has a meaning. You can either belong to a team or not. You cannot belong to it twice over.

Experimental evidence proves that statistical counts referring to elementary particles must never follow the pattern (*a*), but must follow either (*b*) or (*c*). Some hold that for all genuinely elementary particles (*c*) is competent. Such particles, electrons for instance, correspond to membership in a club; I mean to the abstract notion of membership, not to the members. Any person eligible to membership in that club represents a well-defined state an electron can take on. If the person is a member, that means there is an electron in that particular state. According to Pauli's exclusion principle, there can never be more than one electron in a particular state. Our simile renders this by declaring double membership meaningless—as in most clubs it would be. In the course of time the list of members changes, and membership is now attached to other persons: the electrons have gone over into other states. Whether you can, in a loose way, speak of a certain membership going over from Dick to Tom, thence from Tom to Harry, etc., depends on the circumstances. They may suggest this view, or they may not, but never in an absolute fashion. In this our simile is perfect, for it is the same with an electron. Moreover, it is quite appropriate to consider the number of members as fluctuating. Indeed, electrons too are created and annihilated.

The example may seem odd and inverted. One might think, "Why cannot the people be the electrons and various clubs their states? That would be so much more natural." The physicist regrets, but he cannot oblige. And this is just the salient point: the actual statistical behaviour of electrons cannot be illustrated by any simile that represents them by identifiable things. That is why it follows from their actual statistical behaviour that they are not identifiable things.

The case (*b*), illustrating Einstein-Bose statistics, is competent for light quanta (photons), inter alia. It hardly needs discussion. It does not strike us as so strange for the very reason that it includes light, i.e. electromagnetic energy; and energy, in prequantum times, had always been thought of in very much the way our simile represents it, viz. as having quantity but no individuality.

11. Restricted Notion of Identity

The most delicate question is that of the states of, say, an electron. They are, of course, to be defined not classically, but in the light of the uncertainty relation. The rigorous treatment referred to at the end of section 6 is not really based on the notion of "state of one electron" but on that of "state of the assembly of electrons." The whole list of members of the club, as it were, has to be envisaged together—or rather several membership lists, corresponding to the several kinds of particles that go to compose the physical system under consideration. I mention this, not to go into details about it, but because, taken rigorously, the club simile has two flaws. First, the possible states of an electron (which we had assimilated to the persons eligible for membership) are not absolutely defined; they depend on the arrangement of the—actual or imagined—experiment. Given this arrangement, the states are well-defined individuals, which the electrons are not. They also form—and this is the second flaw of the simile—a well-ordered manifold. That is, there is a meaning in speaking of neighbouring states as against such as are farther remote from each other. Moreover, I believe it is true to say that this order can be conceived in such a fashion that, as a rule, whenever one occupied state ceases to be occupied, a neighbouring state becomes occupied.

This explains that, in favourable circumstances, long strings of successively occupied states may be produced, similar to those contemplated in sections 7 and 8. Such a string gives the impression of an identifiable individual, just as in the case of any object in our daily surrounding. It is in this way that we must look upon the tracks in the cloud chamber or in a photographic emulsion, and on the (practically) simultaneous discharges of Geiger counters set in a line, which discharges we say are caused by the same particle passing one counter after another. In such cases it would be extremely inconvenient to discard this terminology. There is, indeed, no reason to ban it, provided we are aware that, on sober experimental grounds, the sameness of a particle is not an absolute concept. It has only a restricted significance and breaks down completely in some cases.

In what circumstances this restricted sameness will manifest itself is fairly obvious: namely, when only few states are occupied in the region of the state manifold with which we are concerned, or, in other words, when the occupied states are not too crowded in that region, or when occupation is a rare event—the terms "few," "crowded," and "rare" all referring to the state manifold. Otherwise, the strings intermingle inextricably and reveal the true situation. In the last section we shall formulate the quantitative condition for the prevailing of restricted individuality. Now we ask what happens when it is obliterated.

12. Crowdedness and Wave Aspect

One gains the impression that according as the individuality of the particles is wiped out by crowding, the particle aspect becomes altogether less and less expedient and has to be replaced by the wave aspect. For instance, in the electronic shell of an atom or molecule the crowding is extreme, almost all the states within a certain region being occupied by electrons. The same holds for the so-called free electrons inside a metal. Indeed, in both cases the particle aspect becomes entirely incompetent. On the other hand, in an ordinary gas the molecules are extremely rare in the wide region of states over which they spread. No more than 1 state in 10,000 or so is occupied. And, indeed, the theory of gases, based on the particle aspect, was able to attain great perfection long before the wave nature of ordinary matter was discovered. (In the last remark I have been speaking of the molecules as if they were ultimate particles; this is legitimate as far as their translatory motion is concerned.)

It is tempting to assign to the two rivals, the particle aspect and the wave aspect, full competences in the limiting cases of extreme "rarefaction" and extreme "crowding" respectively. This would separate them, as it were, with only some sort of transition required for the intermediate region. This idea is not entirely wrong, but it is also far from correct. One may remember the interference patterns referred to in section 2 in evidence of the wave nature of the electron. They can be obtained with an arbitrarily faint bundle of cathode rays, provided the exposure is prolonged. Thus a typical wave phenomenon is produced here, irrespective of crowding. Another instance is this. A competent theoretical investigation of the collision of two particles, whether of the same or of different kind, has to take account of their wave nature. The results are duly applied to the collisions of cosmic ray particles with atomic nuclei in the atmosphere, both being extremely rarefied in every sense of the word. But perhaps this is trivial; it only means that even an isolated particle, which gives us the illusion of transitory individuality, must yet not be likened to a classical particle. It remains subject to the uncertainty relation, of which the only tolerable image is the guiding wave group.

13. The Conditions for the Particle Aspect

The following is the quantitative condition for strings to develop which counterfeit individuals and suggest the particle aspect: the product of the momentum p and the average distance l between neighbouring particles must be fairly large compared with Planck's constant h; thus

$$pl \gg h.$$

(The momentum p—and not the velocity—is the thing we should really have referred to when, in sections 4 and 5, we dealt with the uncertainty relation; p is simply the product of the mass and the velocity, unless the latter is comparable with that of light.)

A large l means a low density in ordinary space. What matters, however, is the density in the manifold of states—or phase space, to use the technical term. That is why the momentum p comes in. It is gratifying to remember that those very obvious strings—visible tracks in the cloud chamber or in the photographic emulsion, and simultaneous discharges of aligned counters—are all produced by particles with comparatively very large momentum.

The above relation is familiar from the theory of gases, where it expresses the condition which must be fulfilled in very good approximation in order that the old classical particle theory of gases should apply in very good approximation. This theory has to be modified according to quantum theory when the temperature is very low and at the same time the density very high, so that the product pl is no longer very large compared with h. This modification is called the theory of degenerate gases, of which the most famous application is that by A. Sommerfeld to the electrons inside a metal; we have mentioned them before as an instance of extreme crowding.

There is the following connection between our relation and the uncertainty relation. The latter allows one at any moment to distinguish a particle from its neighbours by locating it with an error considerably smaller than the average distance l. But this entails an uncertainty in p. On account of it, as the particle moves on, the uncertainty in the location grows. If one demands that it still remains well below l after the particle has covered the distance l, one arrives precisely at the above relation.

But again I must warn of a misconception which the preceding sentences might suggest, viz. that crowding only prevents us from registering the identity of a particle, and that we mistake one particle for the other. The point is that they are not individuals which could be confused or mistaken one for another. Such statements are meaningless.

Notes

1. E. Mach, *Erkenntnis und Irrtum* (Leipzig: J. A. Barth, 1905).
2. B. Russell, *Human Knowledge: Its Scope and Limits* (London: Allen and Unwin, 1948).
3. H. Diels, *Die Fragmente der Vorsokratiker* (Berlin: Weidmannsche Buchhandlung, 1903). (The reference is mainly to the fragment 125 of Democritus.)

13

The Nature of Elementary Particles

Werner Heisenberg

The question, What is an elementary particle? must find its answer primarily in experiment, although it must also be confronted with philosophical considerations. I will therefore begin by giving a short survey of the important experimental results of the past fifty years. This survey will show that a critical unbiased study of these results already gives an answer to the question; theory, as we shall see, cannot add much to this answer.

Next I will deal with the philosophical problems that arise in connection with the concept of an elementary particle. It may be objected that in this question we should concentrate on physics rather than on philosophy. But this separation is not so simple. In fact, I believe that certain erroneous developments in particle theory—and I am afraid that such developments do exist—are caused by a misconception by some physicists that it is possible to avoid philosophical arguments altogether. Starting with poor philosophy, they pose the wrong questions. It is only a slight exaggeration to say that good physics has at times been spoiled by poor philosophy.

Finally I will discuss these problematic developments. Having witnessed similar mistakes in the development of quantum mechanics fifty years ago, I am in a position to make some suggestions to avoid such errors in the future. This will lead us to conclude on an optimistic note.

1. Particle Number Not Conserved

Let us start with the experimental facts. Nearly fifty years have passed since P.A.M. Dirac predicted, on the basis of his theory of the electron, the existence of its antiparticle, the positron: a few years later the existence of these was demonstrated experimentally by Carl Anderson and P.M.S. Blackett. They produced them in pair creation, forming what we now call "antimatter" artificially.

This was a discovery of prime importance. Before this time it was assumed that there were two fundamental kinds of particles, electrons and protons, which, unlike most other particles, were immutable. Therefore their number was fixed and they were referred to as "elementary" particles. Matter was seen as being ultimately constructed of electrons and protons. The experiments of Anderson and Blackett provided definite proof that this hypothesis was wrong. Electrons can be created and annihilated; their number is not constant; they are not "elementary" in the original meaning of the word.

The next important step was the discovery of artificial radioactivity by Frederic Joliot and Irène Curie. From many experiments it became clear that an atomic nucleus can be transmuted into another nucleus by emitting particles, provided the laws of conservation of energy, angular momentum, electric charge, and so on permit this transmutation. The transformation of energy into matter, predicted as a possibility very early in the theory of special relativity, has become recognized as a rather common phenomenon. Contrary to the earlier views, there was no conservation of particle number. However, there are physical properties that can be characterized by quantum numbers, for instance, angular momentum and electric charge; these quantum numbers may assume positive or negative values and are subject to laws of conservation.

The 1930s brought a few other important experimental discoveries. In cosmic radiation very energetic particles are observed. A cosmic-ray particle, when colliding with another particle such as an atomic nucleus in a photographic emulsion, can produce a shower of many secondary particles. For some time many physicists believed that such showers can be produced only by cascades in the interior of a heavy nucleus. Later it became clear, however, that even the collision of two protons can lead to the production of many secondaries in one step. In the late 1940s Cecil Powell discovered the pions, which play the main role in this process of multiple production of particles. His results emphasized again that the transmutation of energy into matter is the decisive process, and that it would be meaningless to speak about the "division" of the original particles. Experimentally, the concept of "dividing" had lost its meaning.

This new situation was confirmed again and again in the experiments of the 1950s and 1960s; many new particles of various lifetimes were discovered and there was no answer to the question, What do these particles consist of? A proton could be obtained from a neutron and a pion, or from a Λ hyperon and a kaon, or from two nucleons and one antinucleon, and so on. Could we therefore simply say a proton consists of continuous matter? Such a statement would be neither right nor wrong: there is no difference in principle between elementary particles and compound systems. This is probably the most important experimental result of the past fifty years.

This development convincingly suggests the following analogy: let us compare the so-called elementary particles with the stationary states of an atom or a molecule. We may think of these as various states of one single molecule or as the many different molecules of chemistry. One may therefore speak simply of the "spectrum of matter." Experiments in the 1960s and 1970s with large accelerators have demonstrated that this picture fits elementary particles as well. Like the stationary states of atoms, the elementary particles can be characterized by quantum numbers—that is, by their behavior under certain transformations. The corresponding laws of conservation determine what transmutations are possible. For an excited hydrogen atom it is its behavior under rotation that determines whether it can fall into a lower state with the emission of a light quantum. In the same way for a ϕ boson, it is its symmetry properties that determine whether it can disintegrate with the emission of a pion into a ρ boson.

The stationary states of atoms have very different lifetimes, and the same is true of particles. The ground state of an atom has an infinite lifetime, and there are many particles with this property, including the electron, the proton, and the deuteron. But these stable particles are no more elementary than the unstable ones—the ground state of hydrogen is a solution of the same Schrödinger equation as any of the excited states. Similarly, the electron and the proton are no more elementary than the Λ hyperon.

During recent years experimental particle physics has applied itself to tasks similar to those of spectroscopy in the early 1920s. Just as at that time all stationary states of atoms were collected in large tables, the "Paschen-Götze" volumes, so nowadays the "reviews of particle properties" every year collect new or improved data on the masses and quantum numbers of particles. This kind of work corresponds to the astronomical surveys, and obviously every observer hopes occasionally to find an especially interesting object in his sector.

2. Two Types of Broken Symmetry

There are nevertheless certain characteristic differences between the physics of atomic shells and particle physics. In the shells the relevant energies are small, so the typical features of the theory of relativity play no role, and their behavior can be described at least approximately by nonrelativistic quantum mechanics. Hence in shell physics and in particle physics the underlying symmetry groups are different. The Galilean group in shell physics is replaced by the Lorentz group in particle physics.

Other groups have to be added, such as the isospin group, which is isomorphic to SU_2; then SU_3, the scale group, and others. It is an important experimental proposition to determine all groups relevant to particle physics, and this problem has been solved to a large extent during the past twenty years.

From the physics of atomic shells we can learn that among those groups that describe only approximate symmetries in nature, two essentially different types can be distinguished. In the optical spectra of atoms, for example, the groups O_3 and $O_3 \times O_3$ play very different roles. The fundamental equations of quantum mechanics of atoms are strictly invariant under O_3. Therefore the stationary states with higher angular momenta are strictly degenerate, so that there are always several states with exactly the same energy. Only when external electromagnetic fields are applied do these states split and the well-known fine structures of the Zeeman effect or the Stark effect appear.

A similar effect can be produced in systems in which the ground state is not invariant under rotations, such as a crystal or a ferromagnet: the two directions of electron spin in a ferromagnet do not belong to precisely the same energy. In this case, according to a well-known theorem of Goldstone, bosons must also exist, with energy that tends to zero with increasing wavelength; in the case of ferromagnetism the spin waves of Bloch, "magnons," take the place of the Goldstone waves.

An entirely different situation is met with in the group $O_3 \times O_3$, which produces the well-known multiplets in the optical spectra. The group $O_3 \times O_3$ is only an approximate symmetry, which comes about if the spin-orbit interactions become small in a certain part of the spectrum, so that the orbits and the spins of the particles can be rotated almost independently. The symmetry $O_3 \times O_3$ results from the dynamics of the system and is useful only in certain parts of the spectrum. Empirically the two types of broken symmetries can be distinguished by the existence or nonexistence of the Goldstone modes. If they are found, it is plausible to assume that the degeneracy of the ground state plays an important role.

If we apply the experience of shell physics to particle physics, the experiments suggest that the Lorentz group and the group SU_2 should be interpreted as fundamental symmetries of the underlying natural law. Electromagnetism and gravitation then appear as those long-range effects that, according to Goldstone, are connected with the broken symmetry of the ground state. The more complicated groups such as SU_3, SU_4, SU_6, $SU_2 \times SU_2$, or $SU_3 \times SU_3$ should be taken as dynamical symmetries, like $O_3 \times O_3$ in shell physics. We may doubt whether the dilatation group or scale group should be counted among the fundamental symmetries. They are disturbed by the existence of particles of finite mass and by the gravitational influence of the big masses in the universe. Because of their close mathematical relation to the Lorentz group they probably belong to the fundamental group. This particular coordination of the broken empirical symmetries with the two proposed types of broken symmetries is suggested by the existing experimental evidence, but it may not yet be finally settled. It is important to emphasize that, for any symmetry group arising out of the phenomenological analysis of the spectrum, the question must be asked—and if possible answered—to which of these two types it belongs.

Another special feature of shell physics should be mentioned. Among the optical spectra there are noncombining or weakly combining term systems such as the parahelium and the orthohelium spectrum. In particle physics one could perhaps compare the division of the fermion spectrum into baryons and leptons with this feature.

It is evident that the analogy between the stationary states of an atom or molecule on one hand and the particles of high-energy physics on the other is nearly complete, qualitatively answering the question about the nature of the "elementary particle." But only qualitatively. For the theoretician the further question arises whether this interpretation can be based on quantitative calculations. Here we must answer a preliminary question: what does it mean to "understand a spectrum quantitatively"?

3. Dynamics and Contingent Conditions

A number of examples, both from classical physics and from quantum mechanics, teach the general procedure by which we acquire physical understanding. Let us think of the spectrum of elastic vibrations of a steel plate. If a qualitative theoretical interpretation is not enough, we have to start with the elastic properties of the steel plate, which should be represented mathematically. When this has been done, we must add boundary conditions that tell whether the plate is a circle or a square, and whether it is stretched in a frame or free. This knowledge then should be sufficient to calculate, at least in principle, the spectrum of the acoustic vibrations. It is true that, because of the high degree of complication, we are frequently unable to calculate all the vibrational frequencies precisely, but only the lowest ones—those with the smallest number of nodes.

Hence two elements are prerequisite to a quantitative understanding of the spectrum: a precise knowledge of the elastic behavior of the plate and the boundary conditions. The latter may be called "contingent" because they depend on the particular circumstances—the plate could have been cut differently. A similar case would be the electrodynamical vibrations of a cavity. Maxwell's equations determine the dynamical

216 - WERNER HEISENBERG

behavior, and the shape of the cavity defines the boundary conditions. Another comparable situation is met with in the optical spectrum of the iron atom. The Schrödinger equation for a system of one nucleus and twenty-six electrons determines the dynamical behavior, while the boundary condition determines that the wave function vanishes at infinity. If the atom were enclosed in a small box the spectrum would be different.

To relate these results to particle physics, our first problem must be to determine experimentally the dynamical properties of the system "matter" and to formulate them mathematically. Then the contingent element, the boundary conditions, has to be added. These contain in this case statements about the so-called empty space: the cosmos and its symmetry properties. In other words: the first step must be the attempt to formulate mathematically a natural law that defines the dynamics of matter. The second step is to determine the boundary conditions. Without such conditions the spectrum can not be defined. We might, for example, conjecture that inside a black hole the spectrum of elementary particles would be quite different from the spectrum in normal space—but unfortunately we can not check this by experiments!

Let me add a word about that decisive first step, the formulation of the governing dynamical law. There are pessimists among the particle physicists who believe that no such natural law defining the dynamics of matter exists. This view appears quite absurd to me. There must be some clearly defined dynamics of matter or there could be no spectrum; therefore a mathematical description should be possible. The pessimistic view implies that particle physics aims at no other goal than presenting an immense volume of particle data, a "super review of particle properties." In this supervolume nothing could be understood since there is no dynamics of matter and so the volume would scarcely be read.

But I would like to emphasize strongly that I can not see any reason for such pessimism. A spectrum with sharp lines is observed, and this should imply a well-defined dynamics of matter. The experimental results mentioned in the beginning give definite hints as to the fundamental invariances of the underlying natural law, and from the dispersion relations a lot is known about the degree of causality this law contains. Essential parts of the natural law therefore already belong to our definite knowledge. Because so many other spectra in physics have finally come to be understood quantitatively, this should be also possible here—in spite of the high degree of complication. It is on account of this intricacy that I will not discuss here the special proposal that Wolfgang Pauli and I made many years ago for the mathematical formulation of the underlying law, one I still regard as having the best chance of being the correct one. It is more important to emphasize that the formulation of such a law is the unavoidable precondition for a quantitative understanding of the spectrum. Anything else can not be called understanding; it would be scarcely more than looking up the table of data, and theoreticians at least should not be content with that.

4. Philosophical Problems

Taking this point of view, I am now going to discuss the philosophy that, whether consciously or unintentionally, has determined the direction of particle physics. For 2,500 years philosophers and scientists have pondered the questions: What happens if one tries to divide matter again and again? What are the smallest particles of matter? Different philosophers have given very different answers, all of which have influenced the history of natural science. The best-known answer is that of the philosopher Democritos: in the attempt to divide again and again one finally ends up with indivisible, unchangeable units, called atoms, of which all matter is composed. The positions and motions of the atoms determine the qualities of matter.

For Aristotle and his medieval successors, on the other hand, the concept of the smallest particle is not so well defined. It is true that for every kind of matter smallest particles are assumed—further division would change the characteristic qualities of the substance—but these smallest particles can be changed continuously like the substances themselves. Mathematically the substances can be divided ad infinitum. Matter is taken as continuous.

The clearest position against Democritos was taken by Plato. In his opinion the attempt to divide again and again results in mathematical forms: the regular bodies of stereometry, defined by their symmetry, and the triangles of which they are composed. These forms are not matter themselves, but they make up matter. For the element earth, for example, the characteristic body is the cube; for fire, the tetrahedron. What all these philosophies have in common is the attempt to deal with the antinomy of the infinitely small, which was discussed extensively by Immanuel Kant.

There have been even more naive attempts to rationalize this paradox. Some biologists have proposed the idea that within the seed of an apple an invisibly small apple tree is concealed, which again bears flowers and fruits; that the fruits again contain a still much smaller apple tree and so on ad infinitum. It was an amusing game, in the early days of the Bohr-Rutherford theory of the atom as small planetary system, to develop a similar idea: the electrons, the planets of the system, are inhabited by very small living bodies who build houses, cultivate their soil, and study atomic physics—and find that their atoms are again small planetary systems, and so on in an unending progression.

In the background of every such fiction is Kant's antinomy: it is difficult to imagine that matter can be divided again and again, but it is equally difficult to imagine that this division must necessarily come to an end. As we now know, the paradox is caused by the erroneous assumption that our intuition can be applied to the smallest dimensions.

The strongest influence on the physics and chemistry of the past century undoubtedly came from the atomism of Democritos. This view allows an intuitive description of chemical processes on a small scale. Atoms can be compared with the mass points of Newtonian mechanics, and from this a satisfactory statistical theory of heat was developed. It is true that the atoms of the chemists turned out to be not mass points but small planetary systems, and the atomic nucleus likewise was a compound system formed of protons and neutrons. Nevertheless, the electron, the proton, and possibly the neutron could, it seemed, be considered as the genuine atoms, the indivisible building blocks, of matter. In this way the atomism of Democritos became an essential part of the materialistic interpretation of the world during the past century: easily understood and intuitively plausible, it determined the way of thinking of even those physicists who insisted on not dealing with philosophy. At this point let me substantiate my earlier statement that, in the physics of elementary particles of our time, good physics has sometimes been unconsciously spoiled by poor philosophy.

We can not avoid using a language bound up with the traditional philosophy. We ask, What does a proton consist of? Can an electron be divided or is it indivisible? Is a photon simple or compound? But all these questions are wrongly put, because words such as "divide" or "consist of" have to a large extent lost their meaning. It must be our task to adapt our thinking and speaking—indeed, our scientific philosophy—to the new situation created by the experimental evidence. Unfortunately this is very difficult. Wrong questions and wrong pictures creep automatically into particle physics and lead to developments that do not fit the real situation in nature. We will discuss these fallacies below.

But first a word should be added concerning the postulate that understanding requires a visual picture of the phenomena. Some philosophers have claimed that such pictures are the preconditions for understanding. For example, the philosopher Hugo Dingler of Munich held the view, in regard to the theory of relativity, that Euclidean geometry was the only possible correct geometry because we assume its correctness in building our measuring apparatus; on this latter point Dingler was right, of course. Therefore he argued that the experimental facts that are at the foundation of general relativity can not be described by a Riemannian, non-Euclidean geometry, because this would lead to contradictions. Here the postulate seems to be overdrawn: it is enough to know that, within the dimensions of our apparatus, Euclidean geometry applies with sufficient accuracy.

We will have to accept the fact that experimental data on a very large or a very small scale do not necessarily produce pictures, and we must learn to do without them. We then come to recognize that the antinomy of the smallest dimensions is solved in particle physics in a very subtle manner, of which neither Kant nor the ancient philosophers could have thought: the word "dividing" loses its meaning.

If we wish to compare the results of present-day particle physics with any of the old philosophies, the philosophy of Plato appears to be the most adequate. The particles of modern physics are representations of symmetry groups and to that extent they resemble the symmetrical bodies of Plato's philosophy.

5. Wrong Questions

My intention, however, is not to deal with philosophy but with physics. Therefore I will now discuss that development of theoretical particle physics that, I believe, begins with the wrong questions. First of all, there is the thesis that the observed particles such as the proton, the pion, the hyperon consist of smaller particles: quarks, partons, gluons, charmed particles, or whatever else, none of which have been observed. Apparently here the question was asked, What does a proton consist of? But the questioners appear to have forgotten that the phrase "consist of" has a tolerably clear meaning only if the particle can be divided into pieces with a small amount of energy, much smaller than the rest mass of the particle itself.

To demonstrate how a word that seems to be well defined can lose its meaning in special situations, I can not resist repeating a story that Niels Bohr loved to quote. A small boy enters a shop with two pennies in his hand and asks the grocer for two pennies' worth of mixed sweets. The grocer hands him two sweets and adds: "You can do the mixing yourself." The concept "consist of" for a proton has just as much meaning as the concept of "mixing" in the story of the boy.

At this point many readers may object that the quark hypothesis was derived from experimental material, namely from the empirical relevance of the SU_3 group, and that it has been successfully applied in the interpretation of many experiments even beyond the use of the SU_3 group. This can not be denied. But I would like to mention a counterexample from the history of quantum mechanics, of which l have been a witness— a counterexample that shows clearly the weakness of such arguments.

Before Bohr's theory of the atom many physicists believed that an atom must consist of harmonic oscillators. The optical spectrum contains sharp lines, and sharp lines can only be emitted by harmonic oscillators. The charges in these oscillators, however, would correspond to e/m values different from those of the electron, and it would be necessary to assume very many different oscillators, since there are very many lines in the spectrum.

Notwithstanding these difficulties, Woldemar Voigt in Göttingen in 1912 developed a theory of the anomalous Zeeman effect of the D lines in the optical spectrum of sodium, on the following basis: he assumed two coupled oscillators which, without an external magnetic field, emitted the frequencies of the two D lines. The interaction between the two oscillators, and their coupling with an outer magnetic field could be arranged such that for weak fields the anomalous Zeeman effect, and for strong fields the Paschen-Back effect, could be reproduced correctly. For the intermediate range of medium fields the frequencies and intensities were described by long and complicated square roots, which, however, seemed to fit the experiments well.

Fifteen years later Pascual Jordan and I took the trouble to treat the same problem on the basis of the perturbational calculus of quantum mechanics. It came as a great surprise to us that we got exactly the old formulas of Voigt both for the frequencies and for the intensities, even for intermediate fields. The reason, as we understood later, was of purely formal mathematical nature. The perturbational calculus leads to a system of coupled linear equations and the frequencies are determined by the eigenvalues of the system. A system of coupled harmonic oscillators in classical theory also leads to a system of coupled linear equations. Since in Voigt's theory the important parameters had been adjusted to the empirical situation, it was not strange that the results ended up the same in the two cases. But Voigt's theory has not contributed to our understanding of atomic structure.

Why was this attempt by Voigt so successful, on the one hand, and so useless, on the other hand? Because Voigt intended to discuss only the D lines, without taking notice of the complete spectrum. Voigt phenomenologically used a special aspect of the oscillator hypothesis but ignored the other problems and difficulties of the model; at least he consciously left them undefined. He did not really take the oscillator hypothesis seriously. In the same way I am afraid that the quark hypothesis is not really taken seriously today by its proponents. Questions dealing with the statistics of quarks, the forces that keep them together, the reason why the quarks are never seen as free particles, the creation of pairs of quarks inside an elementary particle, are all left more or less undefined. If the quark hypothesis is really to be taken seriously it is necessary to formulate precise mathematical assumptions for the quarks and for the forces that keep them together and to show, at least qualitatively, that all these assumptions reproduce the known features of particle physics.

There should be no problem in particle physics to which these assumptions can not be applied. I do not know of any such attempt, and I am afraid that every attempt written down in precise mathematical language would easily lead to contradictions. My objections to the quark hypothesis can therefore be put in the form of questions:

- Does the quark hypothesis contribute more to the understanding of the particle spectrum than Voigt's hypothesis contributed in an earlier period to the understanding of the structure of the atomic shells?
- Do we not find behind the quark hypothesis the old idea—refuted long ago by experiments—that simple and compound particles can be distinguished?

6. No More Surprises?

Let us now discuss a few special questions. If the SU_3 group plays an important role in the structure of the spectrum—and this appears from the experimental evidence to be the case—then it is important to decide whether SU_3 is a fundamental symmetry of the underlying natural law or a dynamical symmetry, which by its very nature can be only approximately valid. If this decision is left open, the other assumptions concerning the dynamics of the system are left open as well, and no understanding can be gained. The higher symmetries such as SU_4, SU_6, SU_{12}, $SU_2 \times SU_2$ probably belong to the dynamical symmetries that may be useful for a phenomenological description; but their heuristic value could probably be compared with the heuristic value of cycles and epicycles in Ptolemy's astronomy. They give only very indirect information about the underlying natural law.

In the most important experimental results of recent years, bosons with relatively high mass (3–4 GeV) and long life have been discovered. Such states were to be expected, as has been especially emphasized by Hans-Peter Dürr. Whether their long life allows us to interpret them as being "composed of" other particles of long life is a difficult dynamical question, in which all the complications of many-body physics play a role. I would, however, consider it as useless speculation once again to introduce ad hoc new particles, of which these objects are assumed to "consist." This would again be the wrong question, which can not contribute to the understanding of the spectrum.

Recently, in the storage rings in Geneva and in the Batavia machine, total cross sections of collisions between protons at extremely high energies were measured. The result was that at very high energies the cross sections increase roughly as the square of the logarithm of the energy. This behavior had long ago been conjectured in the theory for the asymptotic region, independent of the nature of the particles. In the meantime the collisions between other particles have led to similar results, and this outcome strongly suggests that in the big accelerators the asymptotic region has already been reached—that even at highest energies no more surprises are to be expected.

New experiments generally can not be expected to yield a deus ex machina that suddenly leads to an understanding of the spectrum. The experiments of the past fifty years have already given qualitatively a quite satisfactory, consistent, and complete answer to the question of the nature of elementary particles. The quantitative details can—as in quantum chemistry—be analyzed only in the course of years by much detailed work in physics and mathematics, far from being solved in a single step.

Therefore this article can be concluded with a more optimistic view of those developments in particle physics that promise success. New experimental results are always valuable, even if they only enlarge the data table; but they are especially interesting if they answer critical questions of the theory. In the theory one should try to make precise assumptions concerning the dynamics of matter, without any philosophical prejudices. The dynamics must be taken seriously, and we should not be content with vaguely defined hypotheses that leave essential points open. Everything outside of the dynamics is just a verbal description of the table of data, and even then the data table probably yields more information than the verbal description can. The particle spectrum can be understood only if the underlying dynamics of matter is known; dynamics is the central problem.

14

The Entity and Modern Physics:
The Creation-Discovery View
of Reality

Diederik Aerts

The classical concept of "physical entity," be it particle, wave, field, or system, has become a problematic concept since the advent of relativity theory and quantum mechanics. The recent developments in modern quantum mechanics, with the performance of delicate and precise experiments involving single quantum entities, manifesting explicit nonlocal behavior for these entities, brings essential new information about the nature of the concept of entity. Such fundamental categories as space and time are put into question, and only a recourse to more axiomatic descriptions seems possible. In this contribution we want to put forward a "picture" of what an "entity" might be, taking into account these recent experimental and theoretical results, and using fundamental results of the axiomatic physical theories (describing classical as well as quantum entities) such as they have been developed during the past decade. We call our approach the "creation-discovery view" because it considers measurements as physical interactions that in general entail two aspects: (1) a discovery of an already existing reality and (2) a creation of new aspects of reality during the act of measurement. We analyze the paradoxes of orthodox quantum mechanics in this creation-discovery view and point out the prescientific preconceptions that are contained in the well-known orthodox interpretations of quantum mechanics. Finally we identify orthodox quantum mechanics as a first-order nonclassical theory, and explain in this way why it is so successful in its numerical predictions.

1. Introduction

If one takes into consideration the general approach to relativity theory[1] it seems to be all inclusive. It starts by stating that *"reality is the collection of all events,"* and this is indeed a starting point that should incorporate all possible physical descriptions. It does not seem to be possible to imagine a more general setting, and yet it is well known that quantum mechanics cannot be incorporated in relativity theory. All attempts to describe a quantum entity by means of a collection of events in the relativity framework have failed till now and, in our opinion, are bound to fail. We should, of course, add that after this very general opening of most approaches to general relativity theory (shortly stated here by the phrase: *"reality is the collection of all events"*), the theory proceeds by taking a second step, that reduces it to a much more specific and limited framework. This second step consists in stating that *"each event is identified with a point of space-time."* This second step takes for granted that *an event can always be characterized by the place and the time at which it happens.* It puts forward the space-time structure as the setting in which all of reality exists. One of the consequences of the analysis that we put forward in this chapter is that this second step is at the origin of the incompatibility between the relativity approach and quantum mechanics. Indeed, we shall be led to conclude that a quantum entity cannot be squeezed into the space-time frame introduced by relativity theory, because of its profound quantum nature itself. This impossibility for a quantum entity to be described within a space-time framework has during the past decades been exhibited in a dramatic way by delicate experiments on single quantum entities. The most spectacular effect connected with this impossibility is the property of *nonlocality* entailed by quantum entities. This nonlocality effect is well known, in the sense that quantum theory incorporates states for a quantum entity that entail nonlocal effects; but the effect has now also been demonstrated in various experimental settings, which shows that it really exists and is not just due to a theoretical interpretation of the quantum formalism.

We shall also indicate in which way the concept of *entity* (which we must define, as we shall do later) is brought into difficulties by the nature of these typical quantum effects and their experimental identification. Then we shall argue that in relation with this problem we are on the verge of an important paradigmatical choice: do we save the primitive notion of *quantum entity,* existing within a reality independent of the observer, or, do we opt for a much more subjective worldview, where the concept of entity, existing independently of the observer, looses its meaning and sense? We propose a solution that is very specific, and that we shall call the *creation-discovery view*, where the concept of quantum entity, existing independently of the observer, remains a primary concept (and hence we make the paradigmatical choice to save the primary concept of *entity*). We analyze in detail the prescientific preconceptions that we have to drop in order to fit the recent experimental results into this creation-discovery view.

Before we proceed, we want to stress that the creation-discovery view proposed here has not grown out of philosophical speculations. On the contrary, it is the result of an attempt to consolidate the most recent experimental and theoretical findings into a consistent view on reality.

2. Ptolomaeus and Copernicus, Waves, Particles, and Quantum Entities

The Ptolomean system for our universe was not abandoned by reason of experimental errors, for it fitted very well with all existing experimental results. To incorporate the descriptions of the known phenomena, it only had to introduce additional constructions, called epicycles, which gave rise to many complications but gave a good fit to the experimental observations. But since the primary hypotheses (a) *the earth is the center of the universe,* and (b) *all celestial bodies move in circles around the earth,* were felt to be absolutely essential, the complications could be interpreted as being due to specific properties of the planets. Copernicus (and Greek scientists long before him) dropped hypothesis (a), substituting it with a new one, (c) *the sun is the center of the universe.* Clearly this new hypothesis gave rise to a model that is much simpler than the original Ptolomaeus model. Till the theoretical findings of Kepler, using the refined experimental results of Brahe, hypothesis (b), the circle as the basic motion for the celestial objects, remained unaltered, and Kepler was very unhappy when it became clear to him that it was a wrong hypothesis. Now that we know the motion of the planets around the sun as a general solution of Newton's mechanics, the fact that these motions proceed along ellipses does not bother us anymore. On the contrary, the elliptic orbits become a part of a much greater whole, Newtonian mechanics, which incorporates more beauty and symmetry than the original two axioms that were of primary importance to Ptolomaeus.

The change from Ptolomaeus to Copernicus is typical in the evolution of scientific theories. Usually one is not conscious of the concepts that prevent scientific theories from evolving in a fruitful direction. We claim that we have now a similar situation for quantum mechanics, and that the concept of quantum entity, and its meaning, is at the heart of it. We believe that the prescientific preconception that has to be abandoned can be compared with that of the earth being the center of the universe. It is a preconception that is due to the specific nature of our human interaction with the rest of reality, and of the subjective perspective following from this human interaction. We can only observe the universe from the earth, and this gave us the perspective that the earth plays a central role.

In an analogous way we can only observe the microworld from our position in the macroworld; this forces us to extend the concepts of the worldview constructed for this macroworld into the worldview that we try to construct for the microworld. That space-time is the global setting for reality is such a concept.

The two successful pictures that have been put forward to describe quantum entities make use of two basic prototypes: particles and waves. The particle is identified by the fact that upon detection it leaves a spot on the detection screen, while waves are to be recognized by their characteristic interference patterns. Certain experiments with quantum entities give results that are characteristic for particles, other experiments reveal the presence of waves. This is the reason why we use in physics the concepts of particles and waves to represent quantum entities.

2.1 *De Broglie and Bohm: An Attempt to Fit Quantum Entities into the Space-Time Setting*

There exists a representation using waves and particles together, introduced by Louis de Broglie in the early years of quantum mechanics,[2] and which, after a long period of neglect, was rediscovered by David Bohm and Jean Pierre Vigier[3] and which is still now the object of active study in different research centers. In this representation, it is assumed that a quantum entity is at the same time always both a particle and a wave. The particle has the properties of a small projectile, but is accompanied by a wave which is responsible for the interference patterns. This representation of de Broglie and Bohm incorporates the observed quantum phenomena and attempts to change as little as possible at the level of the underlying reality where these quantum entities exist and interact. This reality is the ordinary three-dimensional Euclidean space; the quantum entity is considered to be a wave and a particle, existing, moving, and changing in this three-dimensional Euclidean space. The specific quantum effects are accounted for by a quantum potential that is effective in this three-dimensional Euclidean space, and brings about the quantum nonlocal effects. The quantum potential is the entity that carries most of the strange quantum behavior. The quantum probabilities appear in the de Broglie–Bohm picture as ordinary classical probabilities, resulting from a lack of knowledge about where the point particle associated with the quantum entity is, exactly as the probabilities of a classical statistical theory, due to a lack of knowledge about the microstates of the atoms and molecules of the substance considered. The de Broglie–Bohm picture is thus a hidden variable theory. The variables describing the state of the point particle are the hidden variables, and the lack of knowledge about these hidden variables is the origin of the probability.

There is, however, a serious problem with the de Broglie–Bohm theory when one attempts to describe more than one quantum entity. Indeed, for the example of two quantum entities, the wave corresponding to the entity consisting of the two quantum entities is a wave in the six-dimensional configuration space, and not a wave in the three-dimensional Euclidean space, and the quantum potential acts in this six-dimensional configuration space and not in the three-dimensional Euclidean space. Moreover, when the entity consisting of the two quantum entities is in a so-called *nonproduct state*, this wave in the six-dimensional configuration space cannot be written as the product of two waves in the three-dimensional space (hence the reason for naming these states nonproduct states). These nonproduct states give rise to the typical quantum mechanical Einstein-Podolsky-Rosen-like correlations between the two subentities. The existence of these correlations has meanwhile been experimentally verified by different experiments, so that the reality of the nonproduct states, and consequently the impossibility to define the de Broglie–Bohm theory in three-dimensional space, is firmly established. This important conceptual failure of the de Broglie–Bohm theory is certainly also one of the main reasons that Bohm himself considered the theory as being a preliminary version of yet another theory to come.[4]

2.2 *Bohr: Vanishing Reality, and a Definite Movement toward Subjectivism*

There exists also the representation using either a wave or a particle, associated with the Copenhagen school, but which was also present in quantum mechanics from the very start. In this picture it is considered that the quantum entity can behave in two ways, either like a particle or like a wave, and that the choice between the two types of behavior is determined by the nature of the observation being made. If the measurement one is making consists in detecting the quantum entity, then it will behave like a particle and leave a spot on the detection screen, just as a small projectile would. But if one chooses an interferometric experiment, then the quantum entity will behave like a wave and give rise to the typical interference pattern characteristic for waves. When referring to this picture, one usually speaks of Bohr's complementarity principle, thereby stressing the dual structure assumed for the quantum entity. This aspect of the Copenhagen interpretation has profound consequences for the general nature of reality. The complementarity principle introduces the necessity of a far-reaching subjective interpretation for quantum theory. If the nature of the behavior of a quantum entity (wave or particle) depends on the choice of the experiment that one decides to perform, then the nature of reality as a whole depends explicitly on the act of observation of this reality. As a consequence it makes no sense to speak about a reality existing independent of the observer.

This dramatic aspect of the Copenhagen interpretation is best illustrated by the delayed-choice experiments proposed by John Archibald Wheeler, where the experimental choice made at one moment can modify the past. Wheeler's reasoning is based on an experimental apparatus as shown in figure 14.1, where a source emits extremely low-intensity photons, one at a time, with a long time interval between one photon and the next. The light beam is incident on a semitransparent mirror A and divides into two beams, a northern beam n, which is again reflected by the totally reflecting mirror N and sent toward the photomultiplier D_1, and a southern beam s, which is reflected by the totally reflecting mirror S, and sent toward the photomultiplier D_2. We know that the outcome of the experiment will be that every photon will be detected either by D_1 or by D_2. Following the Copenhagen complementarity interpretation, this experimental situation pushes the photons to behave like a particle, that will either be detected in the northern detector D_2 or in the southern detector D_1. It is rather easy to introduce in the experimental setup an additional element that, following the Copenhagen interpretation, pushes the photons to behave like a wave. Wheeler pro-

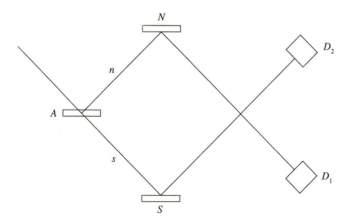

Fig. 14.1. The delayed-choice experimental setup as proposed by John Archibald Wheeler. A source emits extremely low intensity photons that are incident on a semitransparent mirror A. The beam divides into two, a northern beam n, which is again reflected by the totally reflecting mirror N and sent toward the photomultiplier D_1, and a southern beam s, which is reflected by the totally reflecting mirror S, and sent toward the photomultiplier D_2.

poses the following: we introduce a second semitransparent mirror B as shown on figure 14.2, and the thickness of B is calculated as a function of the wavelength of the light, such that the superposition of the northern beam and the southern beam generates a wave of zero intensity.

This means that nothing shall be detected in D_2 and all the light goes to D_1. This experimental setup pushes the photons of the beam into a total wave behavior: indeed, each photon interferes with itself in region B such that it is detected with certainty in D_1. So, we have two experimental setups, the one shown in figure 14.1 and the one shown in figure 14.2, that only differ by the insertion of a semitransparent mirror B. Wheeler proposes the semitransparent mirror B to be inserted or excluded at the last moment, when the photon has already left the source and interacted with the mirror N. Following the Copenhagen interpretation and this experimental proposal of Wheeler, the wave behavior or particle behavior of a quantum entity in the past could be decided upon by an experimental choice that is made in the present. We are dealing here with an inversion of the cause-effect relationship, that gives rise to a total upset of the temporal order of phenomena.

To indicate more drastically the profound subjective nature of the worldview that follows from a consistent application of the Copenhagen interpretation, Wheeler proposes an astronomical version of his delayed-choice experiment. He considers the observation on earth of the light coming from a distant star. The light reaches the Earth by two paths due to the presence of a gravitational lens, formed by a very massive

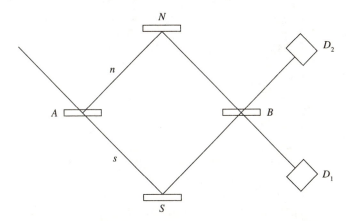

Fig. 14.2. The delayed-choice experimental setup as proposed by John Archibald Wheeler, where a second semitransparent mirror is introduced. Following the Copenhagen interpretation, in this experimental situation the photons shall behave like a wave.

galaxy between the Earth and the distant star. Wheeler observes that one may apply the scheme of figures 14.1 and 14.2, where instead of the semitransparent mirror A there is now the gravitational lens. The distant star may be billions of light years away, and by insertion or not of the semitransparent mirror, we can force the next photon that arrives to have traveled toward the Earth under a wave nature or under a particle nature. This means according to Wheeler that we can influence the past even on time scales comparable with the age of the universe.

Not all physicists believing in the correctness of the Copenhagen interpretation will go as far as Wheeler proposes. The general conclusion of Wheeler's example remains, however, valid. The Copenhagen interpretation makes it quite impossible to avoid the introduction of an essential effect on the nature and behavior of the quantum entity due to the choice of the type of measurement that one wants to perform on it. The determination of the nature and the behavior of a quantum entity independently of the specification of the measurement that one is going to execute is considered to be impossible in the Copenhagen interpretation.

2.3 *The Creation-Discovery View: Neither de Broglie nor Bohr*

The creation-discovery view that we want to bring forward is different from both of the previously mentioned interpretations, the de Broglie–Bohm interpretation and the Copenhagen interpretation. It is a realistic interpretation of quantum theory, in the sense that it considers the quantum entity as existing in the outside world, independent of us observing it, and with an existence and behavior that is also independent of the kind of observation to be made. In this sense it is strictly different from the Copenhagen interpretation, where the mere concept of quantum entity existing independently of a measurement process is declared to be meaningless. The creation-discovery view is, however, not like the de Broglie–Bohm theory, where it is attempted to picture quantum entities as point particles moving and changing in our three-dimensional Euclidean space, and where detection is considered just to be an observation that does not change the state of the quantum entity. In the creation-discovery view it is taken for granted that measurements, in general, *do change the state of the entity* under consideration. In this way the view incorporates two aspects, an aspect of "discovery" referring to the properties that the entity already had before the measurement started (this aspect is independent of the measurement being made), and an aspect of "creation," referring to the new properties that are created during the act of measurement (this aspect depends on the measurement being made).

The fact that there is a real effect of change during the act of a measurement is a rather plausible hypothesis. It is actually quite amazing, and probably due to the enormous prestige of the Copenhagen interpretation, that such a theory has never yet been tried out. In Brussels we are developing this kind of theory.[5] I must state immediately however that to be able to explain quantum theory by a *classical theory* with *explicit effect of change of state* during the act of measurement, we are obliged to drop some old and deep prescientific preconceptions about the nature of reality.

3. The New Methodology, the New "Physical" Formalism, and the Quantum Machine

The fact that it took so long to come to the kind of view that we propose is largely due to the way in which quantum mechanics was born as a physical theory. Indeed, the development of quantum mechanics proceeded in a rather haphazard manner, with the introduction of many ill-defined and poorly understood new concepts.

3.1 *Von Neumann and the Mathematical Quantum Generalizations*

During its first years (1890–1925, with Max Planck, Albert Einstein, Louis de Broglie, Hendrik Lorentz, Niels Bohr, Arnold Sommerfeld, and Hendrik Kramers), quantum mechanics (commonly referred to as the "old quantum theory"), did not even possess a coherent mathematical basis. In 1925 Werner Heisenberg[6] and Erwin Schrödinger[7] produced the first two versions of the new quantum mechanics, which then were unified by Paul Dirac[8] and John von Neumann[9] to form what is now known as the orthodox version of quantum mechanics. The mathematical formalism was elaborate and sophisticated, but the significance of the basic concepts remained quite vague and unclear. The predictive success of the theory was, however, so remarkable that it immediately was accepted as constituting a fundamental contribution to physics. Yet the problems surrounding its conceptual basis led to a broad and prolonged debate in which all the leading physicists of the time participated (Einstein, Bohr, Heisenberg, Schrödinger, Pauli, Dirac, von Neumann, etc.).

The orthodox quantum mechanics of von Neumann[10] is still dominant in the classroom, although a number of variant formalisms have since been developed with the aim of clarifying the basic conceptual shortcomings of the orthodox theory. In the 1960s and 1970s, new formalisms were being investigated by many research groups. In Geneva, the school of Josef Maria Jauch was developing an axiomatic formulation of quantum mechanics,[11] and Constantin Piron gave the proof of a fundamental representation theorem for the axiomatic structure.[12] Günther Ludwig's group in Marburg developed the convex ensemble theory, and in Massachusetts the group of Charles Randall and David Foulis was elaborating an operational approach.[13] Peter Mittelstaedt and his group in Cologne studied the logical aspects of the quantum formalism, while other workers (Jordan, Segal, Mackey, Varadarajan, Emch)[14] focused their attention on the algebraic structures, and Richard Feynman developed the path integral formulation. There appeared also theories of phase-space quantization, of geometrical quantization, and quantization by transformation of algebras.

These different formalisms all contained attempts to clarify the conceptual labyrinth of the orthodox theory, but none succeeded in resolving the fundamental difficulties. This was because they all followed the same methodology: first develop a mathematical structure, then pass to its physical interpretation. This is still the procedure followed in the most recent and authoritative theoretical developments in particle physics and unification theory, such as quantum chromodynamics and string theory. But from 1980 on, within the group of physicists involved in the study of quantum structures, there arose a growing feeling that a change of methodology was indispensable, that one should start from the physics of the problem, and only proceed to the construction of a theory after having clearly identified all basic concepts. Very fortunately, this change in attitude to theory coincided with the appearance of an abundance of new experimental results concerning many subtle aspects of microphysics, which previously could only have been conjectured upon. We here have in mind the experiments in neutron interferometry, in quantum optics, on isolated atoms, and others. The new insights as to the nature of physical reality, resulting in part from the new experimental data and in part from the new methodological approach, have made it possible to clarify some of the old quantum paradoxes and thereby to open the way to a reformulation of quantum mechanics on an adequate physical basis.

3.2 *The Quantum Machine: A Macroscopic Spin 1/2 Model*

In Brussels we have now decided to work explicitly along this new methodological approach, starting from the physics of the problem, and only proceeding to the construction of a theory after having clearly identified all basic concepts.[15] At the same time we have introduced a very simple example of a *quantum machine* that we shall use in the next section to put forward our explanation of quantum mechanics.

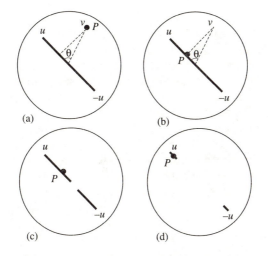

Fig. 14.3. A representation of the quantum machine. In (a) the physical entity P is in state p_v in the point v, and the elastic corresponding to the experiment e_u is installed between the two diametrically opposed points u and $-u$. In (b) the particle P falls orthogonally onto the elastic and sticks to it. In (c) the elastic breaks and the particle P is pulled toward the point u, such that (d) it arrives at the point u, and the experiment e_u gets the outcome o_1^u.

An entity S is in all generality described by the collection Σ of its possible states. A state p, at the instant t, describes the physical reality of the entity S at the time t.

The quantum machine (denoted qm in the following) that we want to introduce consists of a physical entity S_{qm} that is a point particle P that can move on the surface of a sphere, denoted *surf*, with center O and radius 1. The unit vector v giving the location of the particle on *surf* represents the state p_v of the particle (see figure 14.3a). Hence the collection of all possible states of the entity S_{qm} that we consider is given by $\Sigma_{qm} = \{p_v \,|\, v \in surf\}$. No mathematical structure is a priori assigned to this collection of states, contrary to what is done in quantum mechanics (a Hilbert space structure) or in classical mechanics (a phase space structure).

One considers further that experiments e are carried out on the entity S, and let $\mathscr{E}$ denote the collection of relevant experiments. For our quantum machine we introduce the following experiments.

For each point $u \in surf$, we introduce the experiment e_u. We consider the diametrically opposite point $-u$, and install an elastic band of length 2, such that it is fixed with one of its end points in u and the other end point in $-u$. Once the elastic is installed, the particle P falls from its original place v orthogonally onto the elastic, and sticks on it (figure 14.3b). Then the elastic breaks and the particle P, attached to one of the two pieces of the elastic (figure 14.3c), moves to one of the two end points u or $-u$ (figure 14.3d). Depending on whether the particle P arrives in u (as in figure 14.3) or in $-u$, we give the outcome o_1^u or o_2^u to e_u. Hence for the quantum machine we have $\mathscr{E}_{qm} = \{e_u \mid u \in surf\}$. Again, no a priori mathematical structure is imposed upon $\mathscr{E}$. The only assumption made is that when the entity S is in a state p and when an experiment e is carried out, a result x is obtained with some probability. Consequently, the state p will have changed into a new state q. We make the hypothesis that the elastic band breaks uniformly, which means that the probability that the particle, being in state p_v, arrives in u, is given by the length of L_1 (which is $1 + \cos\theta$) divided by the total length of the elastic (which is 2). The probability that the particle in state p_v arrives in $-u$ is the length of L_2 (which is $1 - \cos\theta$) divided by the total length of the elastic. If we denote these probabilities respectively by $P(o_1^u, p_v)$ and $P(o_2^u, p_v)$ we have:

$$P(o_1^u, p_v) = \frac{1 + \cos\theta}{2} = \cos^2\frac{\theta}{2}, \tag{1}$$

$$P(o_2^u, p_v) = \frac{1 - \cos\theta}{2} = \sin^2\frac{\theta}{2}. \tag{2}$$

In figure 14.4 we represent the experimental process connected to e_u in the plane where it takes place, and we can easily calculate the probabilities corresponding to the two possible outcomes. In order to do so, we remark that the particle P arrives in u when the elastic breaks in a point of the interval L_1, and arrives in $-u$ when it breaks in a point of the interval L_2 (see figure 14.4). The probabilities that we find in this way are exactly the quantum probabilities for measuring the spin of a spin 1/2 quantum entity; this means that we can describe this macroscopic machine using the ordinary quantum formalism with a two-dimensional complex Hilbert space as the carrier for the set of states of the entity.

Such an approach provides us with a very general description of the measuring process. If no measurement is made, an entity S is at all times in a well-defined state $p(t)$, and this state changes in function of time. Its dynamical evolution is, in the case of quantum mechanics, described by the Schrödinger equation, and in the case of classical mechanics, it is described by Newton's law. This "physical" formalism has already led

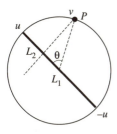

Fig. 14.4. A representation of the experimental process in the plane where it takes place. The elastic of length 2, corresponding to the experiment e_u, is installed between u and $-u$. The probability, $P(o_1^u, p_v)$, that the particle P ends up in point u is given by the length of the piece of elastic L_1 divided by the total length of the elastic. The probability, $P(o_2^u, p_v)$, that the particle P ends up in point $-u$ is given by the length of the piece of elastic L_2 divided by the total length of the elastic.

to a number of concrete and far-reaching results, of which we shall explain some in the following. The most important achievement however, in my opinion, consists in an explanation of the structure of quantum mechanics, and in identifying the reason why it appears in nature.

4. Explaining the Structure of Quantum and Getting Out of Space-Time

Already from the advent of quantum mechanics it was known that the structure of quantum theory is very different from the structure of the existing classical theories. This structural difference has been expressed and studied in different mathematical categories, and we mention here some of the most important ones: (1) if one considers the collection of properties (experimental propositions) of a physical entity, then it has the structure of a Boolean lattice for the case of a classical entity, while it is non-Boolean for the case of a quantum entity,[16] (2) for the probability model, it can be shown that for a classical entity it is Kolmogorovian,[17] while for a quantum entity it is not,[18] (3) if the collection of observables is considered, a classical entity gives rise to a commutative algebra, while a quantum entity does not.[19]

The presence of these deep structural differences between classical theories and quantum theory has contributed strongly to the earlier existing belief that classical theories describe the ordinary "understandable" part of reality, while quantum theory confronts us with a part of reality (the microworld) that is impossible to understand. Therefore there still now exists the strong paradigm that *quantum mechanics cannot be understood*. The example of our macroscopic machine with a quantum structure challenges this paradigm, because obviously the functioning of this machine can be understood. We want to show now that the main aspects of the quantum structures can indeed be explained in this way and identify the reason why they appear in nature. We shall focus here on the explanation in the category of the probability models, and refer to earlier work (Aerts and Van Bogaert, 1992; Aerts, Durt and Van Bogaert, 1993; Aerts, Durt, Grib, Van Bogaert and Zapatrin, 1993; Aerts, 1994; Aerts and Durt, 1994) for an analysis of this explanation in other categories.

4.1 *What Is Quantum Probability?*

The original development of probability theory aimed at a formalization of the description of a probability that appears as the consequence of *a lack of knowledge*. The probability structure appearing in situations of lack of knowledge was axiomatized by Kolmogorov and such a probability model is now called Kolmogorovian. Since the quantum probability model is not Kolmogorovian, it has now generally been accepted that the quantum probabilities are *not* associated with a *lack of knowledge*. Sometimes this conclusion is formulated by stating that the quantum probabilities are *ontological* probabilities, as if they were present in reality itself. In the approach that we follow in Brussels, and that we have named the *hidden-measurement approach*, we show that the quantum probabilities can be explained as being due to a *lack of knowledge*, and we prove that what distinguishes quantum probabilities from classical Kolmogorovian probabilities is the *nature of this lack of knowledge*. Let us go back to the quantum machine to illustrate what we mean.

If we consider again our quantum machine (figures 14.3 and 14.4), and look for the origin of the probabilities as they appear in this example, we can remark that the probability is entirely due to a *lack of knowledge* about the measurement process—namely the lack of knowledge of where exactly the elastic breaks during a measurement. More specifically, we can identify two main aspects of the experiment e_u as it appears in the quantum machine.

- The experiment e_u effects a real change on the state p_v of the entity S. Indeed, the state p_v changes into one of the states p_u or p_{-u} by the experiment e_u.
- The probabilities appearing are due to a *lack of knowledge* about a deeper reality of the individual measurement process itself, namely where the elastic breaks.

These two effects give rise to quantumlike structures, and the lack of knowledge about the deeper reality of the individual measurement process comes from "hidden measurements" that operate deterministically in this deeper reality;[20] and that is the origin of the name that we gave to this approach.

One might think that our "hidden-measurement" approach is in fact a "hidden-variable" theory. In a certain sense this is true. If our explanation for the quantum structures is the correct one, quantum mechanics is compatible with a deterministic universe on the deepest level. There is no need to introduce the idea of an ontological probability. Why then does there exist the generally held conviction that hidden-variable theories cannot substitute quantum mechanics? The reason is that those physicists who are interested in trying out hidden-variable theories, are not at all interested in the kind of theory that we propose here. They want the hidden variables to be hidden variables of the state of the entity under study, so that the probability is associated to a lack of knowledge about the deeper reality of this entity; as we have mentioned already, this gives rise to a Kolmogorovian probability theory. This kind of "state" hidden variables is indeed impossible for quantum mechanics for structural reasons, with exception of course of the de Broglie–Bohm theory: there in addition to the state hidden variables a new spooky entity of "quantum potential" is introduced to express the action of the measurement as a change in these state hidden variables.

If one wants to interpret our hidden measurements as hidden variables, then they are hidden variables of the measurement apparatus and not of the entity under study. In this sense they are highly contextual, since each experiment introduces a different set of hidden variables. They differ from the variables of a classical hidden-variable theory, because they do not provide an "additional deeper" description of the reality of the physical entity. Their presence, as variables of the experimental apparatus, has a well-defined philosophical meaning, and expresses the fact that we, human beings, want to construct a model of reality independent of our experience of this reality. The reason is that we look for "properties" or "relations between properties," and these are defined by our ability to make predictions independent of our experience. We want to model the structure of the world, independent of us observing and experimenting with this world. Since we do not control these variables in the experimental apparatus, we do not allow them in our model of reality, and the probability introduced by them cannot be eliminated from a predictive theoretical model. In the macroscopic world, because of the availability of many experiments with negligible fluctuations, we find an "almost" deterministic model.

We must now try to understand what is the consequence of our explanation for the quantum structure for the nature of reality. Since some of the less mathematically oriented readers may have had some difficulties in understanding our explanation of quantum mechanics by means of the quantum machine, we want to give another more metaphorical and less technical example of the creation-discovery view.

4.2 *Is Cracking Walnuts a Quantum Action?*

Consider the following experiment: "we take a walnut out of a basket, and break it open in order to eat it." Let us look closely at the way we crack the nut. We don't use a nutcracker, but simply take the nut between the palms of our two hands, press as hard as we can, and see what happens. Everyone who has tried this knows that different things can happen. A first possibility to envisage is that the nut is mildewed. If after cracking the shell the walnut turns out to be mildewed, then we don't eat it.

Assume for a moment that the only property of the nut that plays a role in our eating it or not is the property of being mildewed or not. Assume now that there are N walnuts in the basket. Then, for a given nut k (we have $1 \leq k \leq N$), there are always two possible results for our experiment: E_1, we crack the nut and eat it (and then following our hypothesis it was not mildewed); E_2, we crack the nut and don't eat it (and then it was mildewed). Suppose that M of the N nuts in the basket are mildewed. Then the probability that our experiment for a nut k yields the result E_1 is given by the ratio $\frac{(N-M)}{N}$, and that it yields the result E_2, by $\frac{N}{M}$. These probabilities are introduced by our lack of knowledge of the complete physical reality for the nut. Indeed, the nut k is either mildewed or not before we proceed to break it open. Had we known about the mildew without having to crack the nut, then we could have eliminated the probability statement, which is simply the expression of our lack of knowledge about the deeper unknown reality of the nut. We could have selected the nuts for eating by removing from the basket all the mildewed ones. The classical probability calculus is based, again, upon a priori assumptions as to the nature of existing probabilities.

Everyone who has had any experience in cracking walnuts knows that other things can happen. Sometimes, we crush the nut upon cracking the shell. We then have to make an assessment of the damage incurred, and decide whether or not it is worthwhile to try and separate out the nut from the fragments of the shell. If not, we don't eat the nut. Taking into account this more realistic situation, we have to drop our hypothesis that the only factor determining our eating the nut is the mildew, existing before the cracking. Now there are two factors: the mildew, and the state of the nut "after" the act of cracking. Again we have two possible results for our experiment: E_1, we don't eat the nut (then it was mildewed or is crushed upon cracking); and E_2, we eat it (then it was not mildewed and cleanly cracked). For a given nut k these two possible results will occur with a certain probability. We perceive immediately that this sort of probability depends on the way we crack the nut, and is thus of a different nature from the one only related to the presence of mildewed nuts. Before cracking the nuts, there is no way of separating out those which will be cleanly cracked and those which will be crushed. This distinction cannot be made because it is partly created by the cracking experiment itself. This is a nice example of how aspects of physical reality can be created by the measurement itself, namely, the cracking open of the walnuts.

We can state now easily our general creation-discovery view for the case of the nuts. The mildewed nature of the nut is a property that the nut has before and independent of the fact that we break it. When we break the nut and find out that it is mildewed, then this finding is a "discovery." These discoveries, related to outcomes of experiments, obey a classical probability calculus, expressing our lack of knowledge about something that was already there before we made the experiment. The crushed or cleanly cracked nature of the nut is not a discovery of the experiment of cracking. It is a creation. Indeed, depending on how we perform the experiment, and on all other circumstantial factors during the experiment, some nuts will come out crushed, while others will be cleanly cracked.

The mathematical structure of the probability model necessary to describe the probabilities for cleanly cracked or crushed nuts is quite different from that needed for mildewed or nonmildewed ones. More specifically:

- The probability structure corresponding to the indeterminism resulting from a lack of knowledge of an existing physical reality is a classical Kolmogorov probability model.
- The probability structure corresponding to the indeterminism resulting from the fact that during a measurement new elements of physical reality, which thus did not exist before the measurement, are created is a quantumlike probability model.[21]

Quantum probabilities can thus be taken as resulting from a lack of knowledge of the interaction between the measuring apparatus and the quantum entity during the measuring experiment. This interaction creates new elements of physical reality that did not exist before the measurement. This is the explanation we propose to account for quantum probabilities.

4.3 *Space and Walnuts*

Let us now assume that we have removed all the mildewed walnuts from the basket. We thus have the situation where none of the nuts is mildewed. In the physicist's jargon we say that the individual nuts are in a pure state, relative to the property of being mildewed or not. In the original situation when there were still mildewed nuts present, an individual nut was in a mixed state, mildewed and not mildewed, with weighting factors $\frac{M}{N}$ and $\frac{(N-M)}{N}$. In the new situation with the basket containing only nonmildewed nuts, we consider an event m: we take a nonmildewed walnut, and carry out the measurement consisting in cracking the nut. We have here the two possible results: E_1, the nut is cleanly cracked and we eat it; E_2, the nut is crushed and we don't eat it. The result depends on what takes place during the cracking experiment. We therefore here introduce the concept of potentiality. For the case of mildewed or nonmildewed nuts we could assert for each nut that, previously to the experiment, the nut was mildewed or not. For the case of cleanly cracked or crushed nuts, we cannot relate the outcome of the cracking to any anterior property of the walnut. What we can assert, however, is that each walnut is potentially cleanly cracked (and will then be eaten), or potentially crushed (and then will not be eaten).

Nobody will have any difficulty in understanding the walnut example. What we propose is that one should try to understand quantum reality in a similar manner. The only difference is that for the measurements in quantum mechanics that introduce a probability of the second type (i.e., with the creation of a new element of physical reality during the measurement), we find it difficult to visualize just what this creation is. This is the case for instance for detection experiments of a quantum entity. Intuitively, we associate the detection process with the determination of a spatial position which already exists. But now, we must learn to accept that the detection of a quantum entity involves, at least partially, the creation of the position of the particle during the detection process. Walnuts are potentially cleanly cracked or crushed, and likewise quantum entities are potentially within a given region of space or potentially outside it. The experiment consisting in finding or not finding a quantum entity in a given region takes place only after setting up in the laboratory the measuring apparatus used for the detection, and it requires the interaction of the quantum entity with that measuring apparatus. Consequently, the quantum entity is potentially present and potentially not present in the region of space considered.

It will be observed that this description of quantum measurements makes it necessary to reconsider our concept of space. If a quantum entity in a superposition state between two separated regions of space is only potentially present in both of these regions of space, then space is no longer the setting for the whole of physical reality. Space, as we intuitively understand it, is in fact a structure within which classical relations between macroscopic physical entities are established. These macroscopic entities are always present in space, because space is essentially the structure in which we situate these entities. This need not be, and is not the case for quantum entities. In its normal state, a quantum entity does not exist in space, it is only by means of a detection experiment that it is, as it were, pulled into space. The action of being pulled into space introduces a probability of the second type (the type associated with cracking the walnuts open), since the position of the quantum entity is partially created during the detection process.

4.4 *Quantum Entities, Neither Particles nor Waves*

Let us consider one photon in Wheeler's delayed-choice experiment. In the creation-discovery view we accept that the photon while it travels between the source and the detector is not inside space. It remains one entity traveling through reality and the two paths n and s are regions of space where the photon can be detected more easily than in other regions of space when a detection experiment is carried out. The detection experiment is considered to contain explicitly a creation element and pulls the photon inside space. If no detection experiment is carried out, and no physical apparatuses related to this detection experiment are put into place, the photon is not traveling on one of the two paths n or s. We can understand now how the "subjectif" part of the Copenhagen interpretation disappears. In the creation-discovery view the choice of the measurement, whether we choose to detect or to make an interference experiment (in the case of Wheeler's experiment this amounts to whether we choose the setting of figure 14.1 or the setting of figure 14.2), does not influence the intrinsic nature of the photon. In both choices the photon is traveling outside space, and the effect of an experiment appears only when the measurement related to the experiment starts. If a detection measurement is chosen, the photon starts to get pulled into a place in space where it localizes. If an interference experiment is chosen, the photon remains outside space, not localized, and interacts from there with the macroscopic material apparatuses and the fields, and this interaction gives rise to the interference pattern.

5. What about the Quantum Paradoxes in the Creation-Discovery View?

We have analyzed in foregoing sections in which way the creation-discovery view resolves the problems that are connected to the de Broglie theory and the Copenhagen interpretation. We would like to say now some words about the quantum paradoxes. Our main conclusion in relation with the quantum paradoxes is the following: some are due to intrinsic structural shortcomings of the orthodox theory and others find their origin in the nature of reality, and are due to the prescientific preconception about space that we come to explain. In this way we can state that the generalized quantum theories together with the creation-discovery view resolve all quantum paradoxes. We have no time to go into all the delicate aspects of the paradoxes, and refer therefore to the literature. We shall, however, present a sufficiently detailed analyses, such that it becomes clear in which way the paradoxes are solved in the generalized quantum theories and the creation-discovery view.

5.1 *The Measurement Problem: Is Schrödinger's Cat Dead or Alive or Neither?*

If one tries to apply orthodox quantum mechanics to describe a system containing both a quantum entity and the macroscopic measuring apparatus, one is led to very strange predictions. It was Schrödinger who discussed this problem in detail, so let us consider the matter from the point of view of his cat.[22] Schrödinger imagined the following thought experiment. He considered a room containing a radioactive source and a detector to detect the radioactive particles emitted. In the room there is also a flask of poison and a living cat. The detector is switched on for a length of time such that there is exactly a probability 1/2 of detecting a radioactive particle emitted by the source. Upon detecting a particle, the detector triggers a mechanism which breaks the flask, liberating the poison and killing the cat. If no particle is detected, nothing happens, and the cat stays alive. We can know the result of the experiment only when we go into the room to see what has happened. If we apply the orthodox quantum formalism to describe the experiment (cat included), then, until the moment that we open the door, the state of the cat, which we denote by p_{cat}, is a superposition of the two states "the cat is dead," written p_{dead}, and "the cat is alive," written p_{live}. Thus, $p_{cat} = (p_{dead} + p_{live})/\sqrt{2}$.

The superposition is suppressed, giving a change in the quantum mechanical state, only at the instant when we go into the room to see what has taken place. We first want to remark that if we interpret the state as described by the orthodox quantum mechanical wave function as a mathematical object giving exclusively *our knowledge* of the system, then there would be no problem with Schrödinger's cat. Indeed, from the point of view of our knowledge of the state, we can assume that before opening the door of the room the cat is already dead or is still alive, and that the quantum mechanical change of state simply corresponds to the change in our knowledge of the state. This *knowledge picture* would also resolve another problem. According to the orthodox quantum formalism, the superposition state $p_{cat} = (p_{dead} + p_{live})/\sqrt{2}$ is instantaneously transformed, at the instant when one opens the door, into one of the two component states p_{dead} or p_{live}. This sudden change of the state, which in the quantum mechanical jargon is called *the collapse of the wave function*, thus has a very natural explanation in the *knowledge picture*. Indeed, if the wave function describes our knowledge of the situation, then the acquisition of new information, as for instance by opening a door, can give rise to an arbitrarily sudden change of our knowledge and hence also of the wave function.

The *knowledge picture* cannot however be correct, because it is a hidden-variable theory. Indeed, the quantum mechanical wave function does not describe the physical reality itself, which exists independently of our knowledge of it, but describes only our knowledge of the physical reality. It would then follow, if the *knowledge picture* is correct, that there must exist an underlying level of reality that is not described by a quantum mechanical wave function. For the cat experiment, this underlying level describes the condition of the cat, dead or alive, independently of the knowledge of this condition we acquire by entering the room. The *knowledge picture* therefore leads directly to a *hidden-variable theory*, where hidden-variables describe the underlying level of reality. As we mentioned already, it can be shown that a probabilistic theory, in which a lack of knowledge of an underlying level of reality is the origin of the probabilistic description (a hidden-variable theory), always satisfies Kolmogorov's axioms. Now, the quantum mechanical theory does not satisfy these axioms, so that the *knowledge picture* is necessarily erroneous. One also has direct experimental evidence, in connection with the Bell inequalities, that confirms that any state-type hidden variable hypothesis is wrong.

Hence, the quantum mechanical wave function represents not our knowledge of the system, but its real physical state, independently of whether the latter is known or not. In that case, however, Schrödinger's cat presents us with a problem. Is it really possible that, before the door of the room is opened, the cat could be in a superposition state, neither living nor dead, and that this state, as a result of opening the door, is transformed into a dead or live state? It does seem quite impossible that the real world could react in this manner to our observation of it. A physical reality such that its states can come into being simply because we observe it is so greatly in contradiction with all our real experience that we can hardly take this idea seriously. Yet it does seem to be an unescapable consequence of orthodox quantum mechanics as applied to a global physical situation, with macroscopic components.

5.2 *Classical and Quantum Components of a General Description: The Resolution of Schrödinger's Cat Paradox*

In the new physical general description, that we have proposed already earlier,[23] it is perfectly possible and even very natural to make a distinction between different types of experiments. One will thus introduce the concept of a *classical experiment*: this is an experiment such that, for each state p of the entity S, there is a well-determined result x. For a classical experiment, the result is fully predictable even before the experiment is carried out. A collection $\mathscr{C}$ of relevant experiments will generally comprise both classical and nonclassical ones. It is possible to prove a theorem stating that the classical part of the description of an entity can always be separated out.[24] The collection of all possible states for an entity can then be expressed as the union of a collection of classical mixtures, such that each classical mixture is determined by a set of nonclassical microstates. When we formulate within this general framework the axioms of quantum mechanics, it can be shown that the set of states in a classical mixture can be represented by a Hilbert space. The collection of all the states of the entity is then described by an infinite collection of Hilbert spaces, one for each classical mixture. Orthodox quantum mechanics is in this formulation the limiting case for which no classical measurement appears, corresponding effectively to the existence of a single Hilbert space. Classical mechanics is the other limiting case, which is such that only classical measurements are present, and for which the formulation corresponds to a phase space description. The general case for an arbitrary entity is neither purely quantum nor purely classical, and can only be described by a collection of different Hilbert spaces. When one considers the measuring process within this general formulation, there is no Schrödinger cat paradox. Opening the door is a classical operation that does not change the state of the cat, and the state can thus also be described within the general formulation, and the quantum collapse occurs when the radioactive particle is detected by the detector, which is a nonclassical process, also within the general description.

The general formalism provides more than the resolution of the Schrödinger cat paradox. It makes it possible to consider quantum mechanics and classical mechanics as two particular cases of a more general theory. This general theory is quantumlike, but introduces no paradoxes for the measuring process because one can treat, within the same formalism, the measuring apparatus as a classical entity, and the entity to be measured as a quantum entity. The paradoxes associated with measurements result from the structural limitations of the orthodox quantum formalism.[25]

5.3 *The Einstein-Podolsky-Rosen Paradox: Further Limitations of the Orthodox Formalism*

The general existence of superposition states that lies at the root of the Schrödinger cat paradox was exploited by Einstein, Podolsky, and Rosen to construct a far subtler paradoxical situation. They consider the case of two disjoint entities S_1 and S_2, and the composite entity S, which these two entities constitute. They show that it is always possible to bring the composite entity S in a state in such a manner that a measurement on one of the component entities determines the state of the other component entity. For separated entities, this is a quantum mechanical prediction that contradicts the very concept of separateness. Indeed, for separated entities the state of one of the entities can a priori not be affected by how one acts upon the other entity, and this is confirmed by all experiments that one can carry out on separated entities.

Here again, we can resolve the paradox by considering the situation in the framework of the new general formalism. There, one can show that a composite entity S, made up of two separated entities S_1 and S_2, never satisfies the axioms of orthodox quantum mechanics, even if allowance is made for classical experiments as was done in the case of the measurement paradox.[26] Two of the axioms of orthodox quantum mechanics (*weak modularity* and the *covering law*) are never satisfied for the case of an entity S made up of two separated quantum entities S_1 and S_2. This failure of orthodox quantum mechanics is structurally much more far-reaching than that relating to the measuring problem. There one could propose a solution in which the unique Hilbert space of orthodox quantum mechanics is replaced by a collection of Hilbert spaces, and one remains more or less within the framework of the Hilbert space formalism (this is the way that superselection rules were described even within one Hilbert space). The impossibility of describing separated entities in orthodox quantum mechanics is rooted in the vector structure of the Hilbert space itself. The two unsatisfied axioms are those associated with the vector structure of the Hilbert space, and to dispense with these axioms, as is required if we wish to describe separated entities, we must therefore construct a totally new mathematical structure for the space of states.[27]

5.4 *Classical, Quantum, and Intermediate*

To abandon the vector space structure for the collection Σ of all possible states for an entity is a radical mathematical operation, but recent developments have confirmed its necessity. The possibility of accommodating within one general formalism both quantum and classical entities

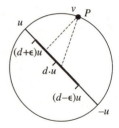

Fig. 14.5. A representation of the experiment $e^\epsilon_{u,d}$. The elastic breaks uniformly between the points $(d-\epsilon)u$ and $(d+\epsilon)u$, and is unbreakable in other points.

resolved the measurement paradox. If the quantum structure can be explained by the presence of a lack of knowledge on the measurement process, as it is the case in our "hidden-measurement" approach, we can go a step further, and wonder what types of structure arise when we consider the original models, with a lack of knowledge on the measurement process, and introduce a variation of the magnitude of this lack of knowledge. We have studied the quantum machine under varying "lack of knowledge," parameterizing this variation by a number $\epsilon \in [0, 1]$, such that $\epsilon = 1$ corresponds to the situation of maximal lack of knowledge, giving rise to a quantum structure, and $\epsilon = 0$ corresponds to the situation of zero lack of knowledge, generating a classical structure, and other values of ϵ correspond to intermediate situations, giving rise to a structure that is neither quantum nor classical.[28] We have called this model the ϵ-model, and we want to expose it shortly here.

We start from the quantum machine, but introduce now different types of elastic. An ϵ, d-elastic consists of three different parts: one lower part where it is unbreakable, a middle part where it breaks uniformly, and an upper part where it is again unbreakable. By means of the two parameters $\epsilon \in [0, 1]$ and $d \in [-1 + \epsilon, 1 - \epsilon]$, we fix the sizes of the three parts in the following way. Suppose that we have installed the ϵ, d-elastic between the points $-u$ and u of the sphere. Then the elastic is unbreakable in the lower part from $-u$ to $(d - \epsilon) \cdot u$, it breaks uniformly in the part from $(d - \epsilon) \cdot u$ to $(d + \epsilon) \cdot u$, and it is again unbreakable in the upper part from $(d + \epsilon) \cdot u$ to u (see figure 14.5). An e_u experiment performed by means of an ϵ, d-elastic shall be denoted by $e^\epsilon_{u,d}$. We have the following cases:

(1) $v \cdot u \leq d - \epsilon$. The particle sticks to the lower part of the ϵ, d-elastic, and any breaking of the elastic pulls it down to the point $-u$. We have $P^\epsilon(o^u_1, p_v) = 0$ and $P^\epsilon(o^u_2, p_v) = 1$.

(2) $d - \epsilon < v \cdot u < d + \epsilon$. The particle falls onto the breakable part of the ϵ, d-elastic. We can easily calculate the transition probabilities

and find:

$$P^\epsilon(o_1^u, p_v) = \frac{1}{2\epsilon}(v \cdot u - d + \epsilon), \tag{3}$$

$$P^\epsilon(o_2^u, p_v) = \frac{1}{2\epsilon}(d + \epsilon - v \cdot u). \tag{4}$$

(3) $d + \epsilon \leq v \cdot u$. The particle falls onto the upper part of the ϵ, d-elastic, and any breaking of the elastic pulls it upward, such that it arrives in u. We have $P^\epsilon(o_1^u, p_v) = 1$ and $P^\epsilon(o_2^u, p_v) = 0$.

Recent investigations of intermediate systems, neither quantum nor classical (hence for example the ϵ-example for the case $0 < \epsilon < 1$), has revealed that here again the same two axioms, weak modularity and the covering law, cannot be satisfied.[29] A new theory dispensing with these two axioms would allow for the description not only of structures that are quantum, classical, and mixed quantum-classical, but also of intermediate structures, which are neither quantum nor classical. This is then a theory for the mesoscopic region of reality, and we can now understand why such a theory could not be built within the orthodox theories, quantum or classical.

5.5 Why Orthodox Quantum Mechanics Is So Successful

As our ϵ version of the quantum machine shows, there are different quantumlike theories possible, all giving rise to quantumlike probabilities, that, however, differ numerically from the probabilities of orthodox quantum mechanics. These intermediate theories may allow us to generate models for the mesoscopic entities, and our group in Brussels is now investigating this possibility. The current state of affairs is the following: quantum mechanics and classical mechanics are both extremal theories, corresponding relatively to a situation with maximum lack of knowledge and a situation with zero lack of knowledge on the interaction between measuring apparatus and the physical entity under study. Most real physical situations will, however, correspond to a situation with lack of knowledge on the interaction between the measuring apparatus that is neither maximal nor zero, and as a consequence the theory describing this situation shall have a structure that is neither quantum nor classical. It shall be quantumlike, in the sense that the states are changed by the measurements, and there is a probability involved as in quantum mechanics, but the numerical value of this probability shall be different from the numerical value of the orthodox quantum mechanical probabilities. If this is the case, *why does orthodox quantum mechanics have so much success, also in its numerical predictions?* In this section we want to give a possible answer to this question.

We first want to consider again the quantum machine, and more specifically the ϵ-version of this quantum machine. Suppose that we consider a fixed angle α, such that $u \cdot v = \cos \alpha$. We know that for a given ϵ and d, the probability for the point particle P to arrive in u, and hence the probability for the state p_v to be transformed into the state p_u, is given by $P^\epsilon(o_1^u, p_v) = \frac{1}{2\epsilon}(\cos \alpha - d + \epsilon)$ (see (3)). We can remark that for some values of ϵ and d, this probability is smaller than the quantum probability $P_q(o_1^u, p_v) = \frac{1}{2}(1 + \cos \alpha)$, and for some values of ϵ and d this probability is larger than the quantum probability. More specifically, when

$$\frac{d}{1 - \epsilon} < \cos \alpha, \tag{5}$$

then $P^\epsilon(o_1^u, p_v) < P_q(o_1^u, p_v)$, and we are in a "superquantum situation" (the situation is stronger than quantum, and with stronger than quantum we mean that $\left|\frac{dP_q}{d\alpha}\right| \leq \left|\frac{dP_\epsilon}{d\alpha}\right|$), while when

$$\frac{d}{1 - \epsilon} > \cos \alpha, \tag{6}$$

then $P_q(o_1^u, p_v) > P_q(o_1^u, p_v)$, and we are in a "subquantum situation" (the situation is weaker than quantum, hence more close to classical). In figure 14.6 we give the example of two possible situations, one "superquantum" and the second "subquantum."

Suppose now for a moment that we perform $e_{u,d}^\epsilon$ experiments, and we choose "at random" one of the $e_{u,d}^\epsilon$ experiments, without knowing the values of ϵ and d. The experiment that we have chosen can then be classical, subquantum, quantum, or superquantum. We can ask ourselves then what would in this case be the probability for the point P to arrive in u. Let us call this probability $P_\Delta(p_u, p_v)$. Taking into account the well-known Bertrand paradox from probability theory, we know that there is no unique answer to this question. Indeed, the probability for the point P to arrive at u when an at-random choice is made for ϵ and d, depends on the way in which this choice is defined. What we do know, however, is that this probability $P_\Delta(p_u, p_v)$ is independent of ϵ and d, and shall, in our case of the quantum machine, only depend on v and u, the states before and after the measurement. We have made such a calculation for the case of the ϵ, d quantum machine, choosing the couple ϵ, d at random in the triangle defined by the lines $\epsilon = -d + 1$, $\epsilon = d + 1$, and

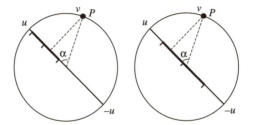

Fig. 14.6. An example of two possible ϵ, d situations. The first is a "superquantum" situation, the probability for the point P to arrive in u is bigger than the quantum probability, while the second is a "subquantum" situation, the probability for the point P to arrive in u is smaller than the quantum probability.

$\epsilon = 0$, as shown in figure 14.7. We find in this case:

$$P_\Delta(p_u, p_v) = \frac{1}{2}(1 + \cos \alpha) - \frac{1}{2}(1 + \cos \alpha)^2 \log \cos \frac{\alpha}{2}$$
$$+ \frac{1}{2}(1 - \cos \alpha)^2 \log \sin \frac{\alpha}{2}. \tag{7}$$

This probability $P_\Delta(p_u, p_v)$ resembles the orthodox quantum probability, but has some additional terms, which makes it numerically different. As we mentioned already, because of the Bertrand paradox, we can certainly invent a way of choosing ϵ and d at random in such a manner that $P_\Delta(p_u, p_v)$ equals the orthodox quantum probability. If we do this, is this then not just a way of cheating? Not really, because there is an element of the dimension of the example (we shall come back to this in the next section) that plays a role in this possibility of finding many different answers for P_Δ. The idea that we want to bring forward is however very fascinating.

Since, for two given states p_v and p_u, the hidden-measurement models of our general theories can lead to situations where the "lack of knowledge" on the interaction between the measurement apparatus and the entity (let us characterize the nature of this lack of knowledge by ϵ) is such that the resulting "transition probability" $P^\epsilon(p_u, p_v)$ is "superquantum" or "subquantum," could it not be so that an at-random choice between all these possible lack-of-knowledge situations (hence an at-random choice between all possible ϵ) gives rise to the orthodox quantum transition probability $P_q(p_u, p_v)$ between the states p_v and p_u?

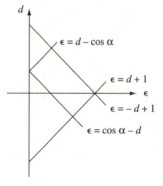

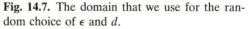

Fig. 14.7. The domain that we use for the random choice of ϵ and d.

If the answer would be "yes" to the foregoing question, this would mean that the probabilities of orthodox quantum mechanics can be interpreted as the probabilities corresponding to a first-order nonclassical theory—a nonclassical theory where we don't know anything about the amount of lack of knowledge on the interaction between the measuring apparatus and the entity. We shall see in the next section that we have reason to believe that we have found here a completely new explanation for quantum mechanics, and why its numerical values give good results in all regions of microphysics.

5.6 *Orthodox Quantum Mechanics as a First-Order Nonclassical Quantumlike Theory*

Suppose that we consider the situation of an entity S, and two possible states p_u and p_v corresponding to this entity. We also consider now all possible measurements that can be performed on this entity S, with the only restriction being that for each measurement considered it must be possible that when the entity is in state p_v, it can be changed by the measurement into state p_u. Among these measurements there shall be deterministic classical measurements, there shall be quantum measurements, but there also shall be superquantum measurements and subquantum measurements. All these different measurements are considered. We suppose now that we cannot distinguish between these measurements, and hence the actual "huge" measurement that we perform, and that we denote $\Delta(u, v)$, is an at-random choice between all these possible measurements. We shall call this measurement the "universal" measurement connecting p_v and p_u. We remark that, if we believe that there is "one" reality, then there is also only "one" universal measurement $\Delta(u, v)$ connecting p_v and p_u. We wonder now what is the probability $P_\Delta(p_u, p_v)$ that by performing the universal measurement $\Delta(u, v)$, the state p_v is changed into the state p_u.

There is a famous theorem in quantum mechanics that makes it possible for us to show that the universal transition probability $P_\Delta(p_u, p_v)$ corresponding to a universal measurement $\Delta(u, v)$ connecting states p_u and p_v is the quantum transition probability $P_q(p_u, p_v)$ connecting these two states p_v and p_u. This theorem is Gleason's theorem.

Gleason's theorem proves that for a given vector u of a Hilbert space $\mathcal{H}$, of dimension at least 3, there exists only one probability measure μ_u on the set of closed subspaces of this Hilbert space, with value 1 on the ray generated by u, and this is exactly the probability measure used to calculate the quantum transition probability from any state to this ray generated by u. To understand more clearly in which way Gleason's theorem can be used to determine the universal transition probability between states, let us consider the situation of a three-dimensional real Hilbert space.[30] The only positive function $w(p_v)$ that is defined on the rays p_v of a three-dimensional real Hilbert space R^3, and that has value 1 for a given ray p_u, and that is such that

$$w(p_x) + w(p_y) + w(p_z) = 1 \tag{8}$$

if the three rays p_x, p_y, p_z are mutually orthogonal, is given by

$$w(p_v) = |\langle u, v \rangle|^2. \tag{9}$$

Let us consider now two states p_u and p_v, and a measurement e (which is not a priori taken to be a quantum measurement) that has three eigenstates p_u, p_y, and p_z, which means that it transforms any state into one of these three states after the measurement. The probability $P_e(p_u, p_v)$, that the measurement e transforms the state p_v into the state p_u is given by a positive function $f(v, u, x, y)$ that can depend on the four vectors $v, u, x,$ and y. In the same way we have $P_e(p_x, p_v) = f(v, x, y, u)$, $P_e(p_y, p_v) = f(v, y, u, x)$, and $f(v, u, x, y) + f(v, x, y, u) + f(v, y, u, x) = 1$. This is true, independent of the nature of the measurement e. If e is a quantum measurement, then $f(v, u, x, y) = |\langle v, u \rangle|^2$, and the dependence on x and y disappears, because the quantum transition probability only depends on the state before the measurement and the eigenstate of the measurement that is actualized, but not on the other eigenstates of the measurement. Gleason's theorem states that "if the transition probability depends only on the state before the measurement and on the eigenstate of the measurement that is actualized after the measurement, then this transition probability is equal to the quantum transition probability." But this Gleason property (dependence of the transition probability only on the state before the measurement and the eigenstate that is actualized after the measurement) is exactly a property that is satisfied by what we have called the "universal" measurements. Indeed, the transition probability of a universal measurement, by definition of this measurement, only depends on the state before the measurement and the actualized state after the measurement. Hence Gleason's theorem shows that the transition probabilities connected with universal measurements are quantum mechanical transition probabilities. We go a step further and want to interpret now the quantum measurements as if they are universal measurements. This means that quantum mechanics is the theory that describes the probabilistics of possible outcomes for measurements that are mixtures of all imaginable types of measurements. Quantum mechanics is then the first-order nonclassical theory. It describes the statistics that goes along with an at-random choice between any arbitrary type of manipulation that changes the state p_v of the system under study into the state p_u, in such a way that we don't know anything of the mechanism of this change of state. The only information we have is that "possibly" the state before the measurement, namely p_v, is changed into a state after the measurement, namely p_u. If this is a correct explanation for quantum statistics, it explains its success in so many regions of reality, also concerning its numerical statistical predictions.

6. Entities, to Be or Not to Be?

Let us now finally come to the basic question of this chapter: what about the concept of entity? As we mentioned already in the introduction, the creation-discovery view chooses in a certain sense for the conservation of the notion of entity. But we should now also finally say what we mean by the concept of entity in general. An entity is a collection of properties that have a certain state of permanence to be clustered together, and a property is a state of prediction toward a certain experiment. A property, as elements of the collection of properties that defines an entity, can be actual, which means that the corresponding outcome can be predicted with certainty, or potential, which means that the outcome cannot be predicted with certainty, but that the actuality of this property is available. This seems at first sight to be a very abstract notion, but it is not. Let us give some examples. Suppose that we consider an entity that is a point particle with a certain mass in space, moving with a certain velocity. Then the cluster of properties that defines the particle is made up of position properties and momentum properties. In a certain state, some of the position and momentum properties shall be actual (those where the particle is, and those that correspond to the particle's momentum) and others potential (those where the particle is not, but where it can come, and those that do not correspond to the particle's momentum, but could do so). Clusters of properties that have enough permanence are entities. With this definition for an entity, quantum objects are entities in the creation-discovery view. Indeed, careful experiments have meanwhile shown that even a quantum object that is in a pure superposition state—for example, of two states corresponding to localization in spatially separated regions of space (e.g., a photon in the delayed-choice experiment of Wheeler)—behaves as an entity. This is most obviously demonstrated by Helmut Rauch in his famous neutron interferrometer experiments. There Rauch manages to manipulate the neutron in the superposition state, and the result is that it still behaves as a neutron.[31] So we can conclude by stating that in our creation-discovery view we retain the concept of entity as a cluster of properties that are more or less permanently joined, and we drop the preconception that such clusters of properties are in space and carry a definite impact.

Notes

Supported by the IUAP-III no. 9. I want to thank my friend and colleague George Severne and my students Thomas Durt, Bob Coecke, Sven Aerts, Frank Valckenborgh, and Bart D'Hooghe for the stimulating discussions we had about the subject of this article.

1. For example, C. W. Misner, K. S. Thorne, and J. A. Wheeler, *Gravitation* (San Francisco: W. H. Freeman, 1973).
2. L. de Broglie, "Sur la possibilité de relier les phénomènes d'interférence et de diffraction à la théorie des quanta de lumière," *Comptes Rendus* 183 (1926): 447–448.
3. D. Bohm and J. P. Vigier, "Model of the Causal Interpretation of Quantum Theory in Terms of a Fluid with Irregular Fluctuations," *Physical Review* 96 (1954): 208–216.
4. See D. Bohm, *Wholeness and the Implicate Order* (London: Routledge, 1980), chaps. 4.6 and 4.7.
5. D. Aerts, "The One and the Many," (Ph.D. diss. Free University of Brussels, 1981); D. Aerts, "Classical Theories and Non Classical Theories as a Special Case of a More General Theory," *Journal of Mathematical Physics* 24 (1983): 2441–2453; D. Aerts, "How Do We Have to Change Quantum Mechanics in Order to Describe Separated Systems," in S. Diner, D. Fargue, G. Lochak, and F. Selleri, eds., *The Wave-Particle Dualism* (Dordrecht: Reidel, 1984); D. Aerts, "A Possible Explanation for the Probabilities of Quantum Mechanics and Example of a Macroscopical System That Violates Bell Inequalities," in P. Mittelstaedt and E. W. Stachow, eds., *Recent Developments in Quantum Logic* (Mannheim: Bibliographisches Institut, 1985); D. Aerts, "A Possible Explanation for the Probabilities of Quantum Mechanics", *Journal of Mathematical Physics* 27 (1986): 202–210; D. Aerts, "The Origin of the Non-Classical Character of the Quantum Probability Model," in A. Blanquiere, S. Diner, and G. Lochak, eds., *Information, Complexity, and Control in Quantum Physics* (Berlin: Springer, 1987); D. Aerts, "An Attempt to Imagine Parts

of the Reality of the Micro-world," in J. Mizerski, A. Posievnik, J. Pykacz, and M. Zukowski, eds., *Problems in Quantum Physics II* (Singapore: World Scientific, 1990); D. Aerts and J. Reignier, "The Spin of a Quantum Entity and Problems of Non-Locality," in P. Lahti and P. Mittelstaedt, eds., *Symposium on the Foundations of Modern Physics, 1990* (Singapore: World Scientific, 1990); D. Aerts and J. Reignier, "On the Problem of Non-Locality in Quantum Mechanics," *Helvetica Physica Acta* 64 (1991): 527–547; D. Aerts, "A Macroscopic Classical Laboratory Situation with Only Macroscopic Classical Entities Giving Rise to a Quantum Mechanical Probability Model," in L. Accardi, ed., *Quantum Probability and Related Topics,* vol. 6 (Singapore: World Scientific, 1991); D. Aerts, T. Durt, and B. Van Bogaert, "A Physical Example of Quantum Fuzzy Sets, and the Classical Limit," *Tatra Montains Math. Publ.* 1 (1992): 5–15; D. Aerts and B. Van Bogaert, "A Mechanical Classical Laboratory Situation with a Quantum Logic Structure," *International Journal of Theoretical Physics* 31 (1992): 1839–1848; D. Aerts, T. Durt, and B. Van Bogaert, "Quantum Probability, the Classical Limit and Non-Locality," in T. Hyvonen, ed., *Symposium on the Foundations of Modern Physics* (Singapore: World Scientific, 1993); D. Aerts, T. Durt, A. A. Grib, B. Van Bogaert, and R. R. Zapatrin, "Quantum Structures in Macroscopic Reality," *International Journal of Theoretical Physics* 32 (1993): 489–498; D. Aerts, "Quantum Structures Due to Fluctuations of the Measurement Situation," *International Journal of Theoretical Physics* 32 (1993): 2207–2220; D. Aerts, "Quantum Structures, Separated Physical Entities and Probability," *Foundations of Physics* 24 (1994): 1227–1258; D. Aerts and T. Durt, "Quantum, Classical and Intermediate: An illustrative example," *Foundations of Physics* 24 (1994): 1353–1368; D. Aerts, and T. Durt, "Quantum, Classical and Intermediate: A Measurement Model," in T. Havyonen, ed., *Symposium on the Foundations of Physics* (Helsinki, 1994); D. Aerts and S. Aerts, "Applications of Quantum Statistics in Psychological Studies of Decision Processes," *Foundations of Science* 1 (1995): 85–97; D. Aerts, "Quantum Structures: An Attempt to Explain the Origin of Their Appearance in Nature," *International Journal of Theoretical Physics* 34 (1995): 1165–1186; D. Aerts and S. Aerts, "Interactive Probability Models: From Quantum to Kolmogorovian" (preprint, CLEA, VUB, Brussels, 1995); D. Aerts, B. Coecke, and F. Valckenborgh, "A Mechanistic Macroscopic Physical Entity with a Three Dimensional Hilbert Space Quantum Description," *Helvetica Physica Acta* 70 (1997): 793–814. B. Coecke, "A Hidden Measurement Representation for Quantum Entities Described by Finite Dimensional Complex Hilbert Spaces" (preprint TENA, Free University of Brussels, 1995); B. Coecke, "A Hidden Measurement Model for Pure and Mixed States of Quantum Physics in Euclidean Space," *International Journal of Theoretical Physics* 34 (1995): 1313–1320; B. Coecke, "Generalization of the Proof on the Existence of Hidden Measurements to Experiments with an Infinite Set of Outcomes" (preprint TENA, Free University of Brussels, 1995); F. Valckenborgh, "Closure Structures and the Theorem of Decomposition in Classical Components," in Proceedings of the 5th Winterschool on Measure Theory, Liptowski, 1995.

6. W. Heisenberg, "Über quantentheoretische Umdeutung kinematischer und mechanischer Beziehungen," *Zeitschrift für Physik* 33 (1925): 879–893.

7. E. Schrödinger, "Quantisiering als Eigenwertproblem," *Annalen der Physik* 79 (1926): 361–376.

8. P.A.M. Dirac, *The Principles of Quantum Mechanics* (Oxford: Oxford University Press, 1978; orginally published in 1930).

9. J. von Neumann, *Mathematische Grundlagen der Quantenmechanik* (Berlin: Springer, 1932).

10. The von Neumann theory constitutes the orthodox mathematical model of quantum mechanics (von Neumann, *Mathematische Grundlagen*). The state of a quantum entity is described by a unit vector in a separable complex Hilbert space; an experiment is described by a self-adjoint operator on this Hilbert space, with as eigenvalues the possible results of the experiment. As the result of an experiment, a state will be transformed into the eigenstate of the self-adjoint operator corresponding to a certain experimental result, with a probability given by the square of the scalar product of the state vector and of the eigenstate unit vector. It follows that, if the state of the quantum entity is not an eigenstate of an operator associated with a given experiment, then the experiment can yield any possible result, with a probability determined by the scalar product of the state and eigenstate vectors as indicated earlier. The dynamical evolution of the state of a quantum entity is determined by the Schrödinger equation.

11. J. M. Jauch, *Foundations of Quantum Mechanics* (Reading, Mass.: Addison-Wesley, 1968).

12. C. Piron, *Foundations of Quantum Physics* (Reading, Mass.: W. A. Benjamin, 1976).

13. C. Randall and D. Foulis, "The Operational Approach to Quantum Mechanics," in C. A. Hooker, ed., *Physical Theory as Logico-Operational Structure* (Dordrecht: Reidel, 1979); C. Randall and D. Foulis, "A Mathematical Language for Quantum Physics," in C. Gruber et al., eds., *Les fondements de la mécanique quantique* (Lausanne, 1983).

14. I. E. Segal, "Postulates for General Quantum Mechanics," *Annals of Mathematics* 48 (1947): 930–948; G. G. Emch, *Mathematical and Conceptual Foundations of 20th Century Physics* (Amsterdam: North-Holland, 1984).

15. Aerts, "The One and the Many"; Aerts, "Quantum Structures."

16. G. Birkhoff and J. von Neumann, "The Logic of Quantum Mechanics," *Annals of Mathematics* 37 (1936): 823–843; Jauch, *Foundations of Quantum Mechanics;* Piron, *Foundations of Quantum Physics.*

17. The axioms formulated by Kolmogorov in 1933 relate to the classical probability calculus as introduced for the first time by Simon Laplace. Quantum probabilities do not satisfy these axioms. John von Neumann was the first to prove a "no go" theorem for hidden-variable theories (von Neumann, *Mathematische Grundlagen*). Many further developments were however required before it was definitively proved that it is impossible to reproduce quantum probabilities from a hidden-variable theory. See I. Pitowski, *Quantum Probability—Quantum Logic* (Berlin: Springer, 1989).

18. Randall and Foulis, "The Operational Approach"; Randall and Foulis, "A Mathematical Language"; S. P. Gudder, *Quantum Probability* (Boston: Academic Press, 1988); L. Accardi and A. Fedullo, "On the Statistical Meaning of the Complex Numbers in Quantum Mechanics," *Lettere al Nuovo Cimento* 34 (1982): 161–172; Pitowski, *Quantum Probabability.*

19. Segal, "Postulates for General Quantum Mechanics"; Emch, *Mathematical and Conceptual Foundations.*

20. Aerts, "A Possible Explanation"; Aerts, "The Origin of the Non-Classical Character"; Aerts, "A Macroscopic Classical Laboratory Situation."

21. We should point out that the non-Kolmogorovian nature of the probability model corresponding to situations of creation cannot be shown for the case of a single experiment, as considered. At least three different experiments with two outcomes of the creation type are necessary to prove in a formal way that a description within a Kolmogorovian model is not possible. We refer to Aerts 1986 for the details of such a proof for the quantum spin 1/2 model. The fact that we need at least three experiments does not however suppress the fact that the physical origin of the non-Kolmogorovian behavior is clearly due to the presence of explicit creation aspects. See, for example, Aerts and Aerts, "Applications of Quantum Statistics."

22. E. Schrödinger, "Die gegenwärtige Situation in der Quantenmechanik I–III," *Die Naturwissenschaften* 23 (1935): 807–812, 823–828, 844–849.

23. Aerts, "Quantum Structures."

24. Aerts, "The One and the Many"; Aerts, "Classical Theories and Non Classical Theories"; Valckenborgh, "Closure Structures."

25. This decomposition theorem of a general description into a direct product of irreducible descriptions, where each irreducible description corresponds with one Hilbert space, had been shown already within the mathematical generalizations to quantum mechanics (Jauch, *Foundations of Quantum Mechanics*; Piron, *Foundations of Quantum Physics*). The aim was then to give an explanation for the existence of superselection rules. The decomposition was later generalized for the physical formalisms (see n. 24).

26. Aerts, "The One and the Many"; D. Aerts, "Description of Many Physical Entities without the Paradoxes Encountered in Quantum Mechanics," *Foundations of Physics* 12 (1982): 1131–1170; Aerts, "Quantum Structures."

27. Aerts, "How Do We Have to Change Quantum Mechanics"; D. Aerts, "The Physical Origin of the Einstein Podolsky Rosen Paradox," in G. Tarozzi and A. van der Merwe, eds., *Open Questions in Quantum Physics* (Dordrecht: Reidel, 1985); Aerts, "A Possible Explanation."

28. Aerts, Durt, and Van Bogaert, "A Physical Example of Quantum Fuzzy Sets"; Aerts, Durt, and Van Bogaert, "Quantum Probability"; Aerts and Durt, "Quantum, Classical and Intermediate: An Illustrative Example"; Aerts and Durt, "Quantum, Classical and Intermediate: A Measurement Model"; Aerts, "Quantum Structures."

29. See n. 28.

30. Gleason's theorem is only valid for a Hilbert space of dimension at least three. The essential part of the demonstration consists of proving the result for a three-dimensional real Hilbert space. Indeed, the three-dimensional real

Hilbert space case contains already all the aspects that make Gleason's theorem such a powerful result. This is also the reason that we want to illustrate our "interpretation" of Gleason's result for the case of a three-dimensional real Hilbert space.

31. Aerts and Reignier, "The Spin of a Quantum Entity"; Aerts and Reignier, "On the Problem of Non-Locality."

15

Dynamical Reduction Theories as a Natural Basis for a Realistic Worldview

Gian Carlo Ghirardi

With reference to recently proposed theoretical models accounting for reduction in terms of a unified dynamics governing all physical processes, we analyze the problem of working out a worldview accommodating our knowledge about natural phenomena. We stress the relevant conceptual differences between the considered models and standard quantum mechanics. In spite of the fact that both theories describe systems within a genuine Hilbert space framework, the peculiar features of the spontaneous reduction models limit drastically the states that are dynamically stable. This fact by itself allows one to work out an interpretation of the formalism that makes it possible to give a satisfactory description of the world in terms of the values taken by an appropriately defined mass density function in ordinary configuration space. A topology based on this function and which is radically different from the one characterizing the Hilbert space is introduced and in terms of it the idea of similarity of macroscopic situations is precisely defined. Finally, the formalism and the interpretation are shown to yield a natural criterion for establishing the psychophysical parallelism. The conclusion is that, within the considered theories and at the nonrelativistic level, one can satisfy all sensible requirements for a completely satisfactory macroobjective description of reality.

1. Introduction

A quite natural question that all scientists who are concerned about the meaning and the value of science have to face is whether one can elaborate a worldview that can accommodate our knowledge about natural phenomena. Such a program has been appropriately denoted by A. Shimony as *closing the circle*.[1] As is well known, this desideratum does not raise particular problems within classical physics for various reasons, which have been lucidly pointed out by J. Bell:

> Of course it is true that also in classical mechanics any isolation of a particular system from the world as a whole involves approximations, but at least one can envisage an accurate theory of the universe, to which the restricted account is an approximation,

and moreover

> even a human observer is no trouble (in principle) in classical theory—it can be included in the system (in a schematic way) by postulating a psycho-physical parallelism—i.e. by supposing his experience to be correlated with some function of the coordinates.[2]

The situation is quite different in quantum theory, due to the specific peculiarities of the formalism. To elucidate this point we recall that in the early days of wave mechanics E. Schrödinger tried to interpret the square modulus of the wave function as describing the charge density:

> we regard $|\Psi|^2$ as representing in the general case the density of the electricity, which is "associated" with one of the particles of classical mechanics, or which "originates in it," or which "corresponds to it in wave mechanics": the integral of $|\Psi|^2$ taken over all those coordinates of the system, which in classical mechanics fix the position of the rest of the particles, multiplied by a certain constant, the classical "charge" of the first particle. The resultant density of charge at any point of space is then represented by the sum of such integrals taken over all the particles.[3]

The deep motivation for such an attempt derives from his firm conviction that one cannot give up a space-time description of natural processes:

> From the philosophical standpoint, I would consider a conclusive decision in this sense as equivalent to a complete surrender. For we cannot really alter our manner of thinking in space and time, and what we cannot comprehend within it we cannot understand at all.[4]

However, just as a consequence of his evolution equation, wave packets diffuse, and nevertheless, no matter how far a wave function has extended, the reaction of a detector to a particle remains spotty. Thus Schrödinger abandoned his original idea: how could one reconcile the fact that the "particle" is spread out in large regions while exhibiting always pointlike behavior in interactions? Stated differently, how can one avoid the conflict between the real (i.e., diffused) status of the particle and the localized outcomes of experiments?

The enormous successes of quantum theory have compelled us to radically change our conceptions about natural phenomena and, at least for what concerns the microscopic world, to accept that electrons can *enjoy the cloudiness of waves*[5]—that is, that their "real status" may not correspond to what we find when we try to detect them. As appropriately pointed out by E. Castellani in the introduction of this book, among others, even the now mentioned features have led various philosophers to consider entities like microphysical particles as "borderline cases" of physical objects. However, things are quite different when we come to consider macroscopic phenomena. The theory allows also macroscopic objects to be in linear superpositions of macroscopically different states.[6] This renders problematic the attribution of definite properties to them, giving rise to a discrepancy between the real status of a macrosystem and what we perceive about it, a fact that is unacceptable from a realistic standpoint.

Various solutions to this puzzling situation have been proposed; for our purposes it turns out not to be relevant to discuss their specific features and/or to comment on their pros and cons. What has to be pointed out is that now everybody agrees that one needs a reinterpretation or a modification of the formalism (such as breaking the von Neumann chain, introducing hidden variables, limiting measurability, or modifying the evolution law) that does not appreciably alter quantum predictions for microsystems but implies or makes legitimate the statement that macroobjects have definite macroproperties.

All these attempts attribute to positions a privileged role in the description of the macroscopic world.[7] This is quite natural since the definiteness, the particularity of the world of our experience, derives from our perceiving physical objects in definite places, which is also why the prescriptions for establishing the psychophysical correspondence usually involve positions.

This chapter is devoted to show how, within the context of the recently introduced nonrelativistic models of spontaneous dynamical reduction,[8] one can give a consistent description of the universe satisfying all previous requirements in terms of mass density in ordinary space.

In section 2 we briefly sketch a dynamical reduction model of the type presented in 1986,[9] in which, however, the mass plays a privileged role. Handling this model is sufficient to grasp completely the relevant arguments of the chapter. For those interested in a more technical and detailed presentation we have confined in the appendix the sketchy presentation of the most elaborate nonrelativistic model of dynamical reduction worked out till now.[10] We also discuss how the considered models, while yielding a solution to the measurement problem, require us to consider a question that has been raised by Shimony, and by Albert and Loewer[11] and which will be referred to as *the problem of the tails* of the wave function. This alleged difficulty will be shown to find a natural solution when one adopts the point of view proposed in this chapter.

Sections 3 and 4, which in a sense represent the core of the chapter, deal with a reinterpretation of the wave function allowing one to describe the macroscopic world in terms of an objective mass density in ordinary space.

Section 5 is devoted to sketch a possible way of establishing the psychophysical correspondence and to prove its consistency.

2. A Dynamical Reduction Model

As already mentioned, models have recently been developed that, by considering nonlinear and stochastic modifications of Schrödinger's dynamics, imply, without entailing any violation of established experimental facts, wave packet reduction with fixed pointer positions in measurement processes and, more generally, forbid the persistence of linear superpositions of macroscopically distinguishable states.[12]

Here we discuss a slightly modified version (in which mass has a privileged role) of the first and simplest model of this kind, quantum mechanics with spontaneous localization (QMSL). The model is based on the assumption that, besides the standard evolution, physical systems are subjected to spontaneous localizations occurring at random times and affecting their elementary constituents. Such processes, which we will call "hittings," are formally described in the following way.

We start by considering a system of one particle of mass m. When it suffers a hitting its wave function changes according to

$$\Psi(\mathbf{r}) \to \Psi_{\mathbf{x}}(\mathbf{r}) = \Phi_{\mathbf{x}}(\mathbf{r})/\|\Phi_{\mathbf{x}}\|,$$

$$\Phi_{\mathbf{x}}(\mathbf{r}) = (\alpha/\pi)^{3/4} e^{-\frac{\alpha}{2}(\mathbf{r}-\mathbf{x})^2} \Psi(\mathbf{r}). \qquad (2.1)$$

The probability density of the process occurring at point $\mathbf{x}$ is given by $\|\Phi_{\mathbf{x}}\|^2$.

To investigate the effect of a hitting process, we consider an initial wave function $\Psi(\mathbf{r})$ of the type

$$\Psi(\mathbf{r}) = k[\Psi_{-\mathbf{a}}(\mathbf{r}) + \Psi_{\mathbf{a}}(\mathbf{r})] = k[e^{-\frac{\delta}{2}(\mathbf{r}+\mathbf{a})^2} + e^{-\frac{\delta}{2}(\mathbf{r}-\mathbf{a})^2}], \qquad (2.2)$$

where $\delta \gg \alpha$, $a \gg 1/\sqrt{\alpha}$ and k is a normalization constant. The state $\Psi(\mathbf{r})$ is the superposition of two gaussian functions $\Psi_{-\mathbf{a}}(\mathbf{r})$ and $\Psi_{\mathbf{a}}(\mathbf{r})$ well peaked (with respect to the length $1/\sqrt{\alpha}$) around the positions $-\mathbf{a}$ and $\mathbf{a}$, respectively. If the hitting (see figure 15.1) occurs at a point $\mathbf{x}$

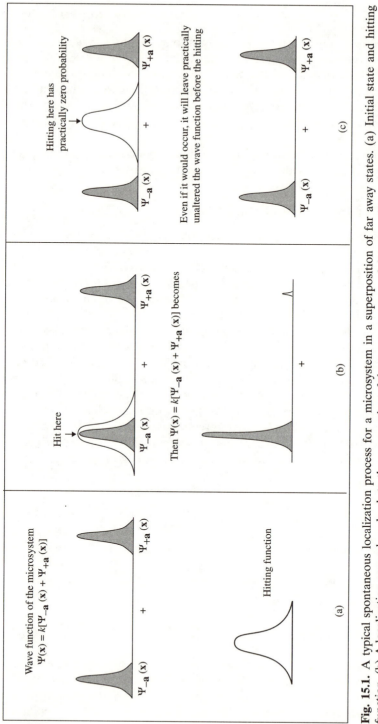

Fig. 15.1. A typical spontaneous localization process for a microsystem in a superposition of far away states. (a) Initial state and hitting function. (b) A localization around $-\mathbf{a}$ leads to the suppression of the $+\mathbf{a}$ component of the state vector. (c) A very improbable localization has practically no effect.

around $-\mathbf{a}$ (**a**), the term $\Psi_{\mathbf{a}}(\mathbf{r})$ ($\Psi_{-\mathbf{a}}(\mathbf{r})$) of the superposition turns out to be exponentially suppressed with respect to the other so that we can state that the particle has been localized around the point $-\mathbf{a}$ (**a**). If the hitting occurs, for example, at $\mathbf{x} = 0$, then, as one can easily see, the wave function after normalization is practically unchanged, so that no localization of the particle occurs. However, the probability of hittings of this type is extremely small and practically all hittings lead to localizations.

For what concerns the temporal features of the processes we assume that hittings occur at randomly distributed times with a mean frequency λ_m, which depends on the mass of the considered particle. We choose $\lambda_m = \frac{m}{m_0}\lambda$, where m is the mass of the particle, m_0 is the nucleon mass, and λ is of the order of $10^{-16}\,\mathrm{sec}^{-1}$. The localization parameter $1/\sqrt{\alpha}$ is assumed to take the value 10^{-5} cm.

We consider now a system of N particles. We suppose that the hittings on the various constituents occur independently from each other and that when the i-th particle of the system suffers a localization the wave function changes just in the same way as in the case of a single particle

$$\Psi(\mathbf{r}_1, \ldots, \mathbf{r}_N) \rightarrow \Psi_{\mathbf{x}}(\mathbf{r}_1, \ldots, \mathbf{r}_N) = \Phi_{\mathbf{x}}(\mathbf{r}_1, \ldots, \mathbf{r}_N)/\|\Phi_{\mathbf{x}}\|,$$

$$\Phi_{\mathbf{x}}(\mathbf{r}_1, \ldots, \mathbf{r}_N) = (\alpha/\pi)^{3/4} e^{-\frac{\alpha}{2}(\mathbf{r}_i - \mathbf{x})^2} \Psi(\mathbf{r}_1, \ldots, \mathbf{r}_N). \tag{2.3}$$

Let a macroscopic system be in a state $|\Psi\rangle = |\Psi_1\rangle + |\Psi_2\rangle$ which is a superposition of two states $|\Psi_1\rangle$ and $|\Psi_2\rangle$ in which a certain number of particles are in different positions. When one of these particles suffers a hitting that localizes it in the position corresponding to the state $|\Psi_1\rangle$ ($|\Psi_2\rangle$), the other term of the superposition is exponentially suppressed. Therefore, the macroscopic system jumps either to the state $|\Psi_1\rangle$ or to $|\Psi_2\rangle$ every time one of the particles differently located in the two states suffers a hitting. This implies that the frequency leading to the suppression of the coherence between the two states increases proportionally to the masses that are displaced (see figure 15.2).

The fact that the wave function after a hitting process is still a superposition of two terms, even though one of them is exponentially suppressed with respect to the other, has been considered as a difficulty of the theory by Shimony[13] and by Albert and Loewer[14] on the basis of the probabilistic interpretation of the wave function. In particular, Shimony claims that, within a dynamical reduction model, *one should not tolerate tails in wave functions which are so broad that their different parts can be discriminated by the senses, even if very low probability amplitude is assigned to them.*[15] In the next section we will show how, when the state vector is reinterpreted according to the lines we are going to propose, the problem of the tails will find a natural solution.

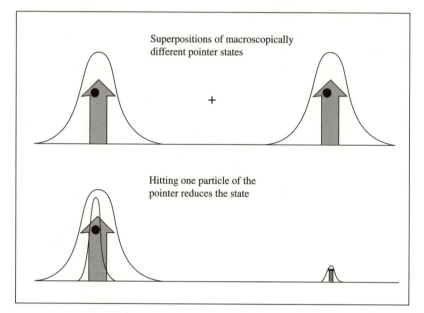

Fig. 15.2. The amplification mechanism: any localization of one of the constituents of a macroscopic pointer amounts to a localization of the whole pointer.

Once one has grasped the important fact that the dynamics implies that any individual physical system is associated at any time to a definite state vector, to investigate the relevant physical features of the reduction mechanism one can study the evolution equation for the statistical operator. As can be easily seen, one has:[16]

$$\frac{d\rho(t)}{dt} = -\frac{i}{\hbar}[H, \rho(t)] + \sum_k \sum_{i=1}^{n^{(k)}} \frac{m^{(k)}}{m_0} \lambda \{T_i[\rho(t)] - \rho(t)\}, \qquad (2.4)$$

where the index (k) denotes the type of particle, $n^{(k)}$ is the total number of particles of such a type, and

$$T_i[\rho] = \left(\frac{\alpha}{\pi}\right)^{3/2} \int d\mathbf{x}\, e^{-\frac{\alpha}{2}(\hat{\mathbf{r}}_i - \mathbf{x})^2} \rho\, e^{-\frac{\alpha}{2}(\hat{\mathbf{r}}_i - \mathbf{x})^2}. \qquad (2.5)$$

The process considered here does not respect the symmetry properties of the wave function in the case of identical constituents. The appropriate generalization satisfying such a requirement, the continuous spontaneous localization model (CSL), which has been presented and discussed in various papers,[17] is briefly sketched in the appendix. Here, to allow the reader who is not familiar with stochastic differential calculus to grasp the relevant physical aspects of the general formalism, we present a reduction model of the "hitting type," relating decoherence to mass differences and taking into account the requirements imposed by the identity of the particles.

We consider the mass density operators

$$M(\mathbf{r}) = \sum_k m^{(k)} N^{(k)}(\mathbf{r}), \tag{2.6}$$

where $m^{(k)}$ is the mass of the particles of type k and $N^{(k)}(\mathbf{r})$ are the number density operators defined by equation (A.11) of the appendix, giving the average density of particles of type k in a volume of about 10^{-15} cm^3 around the point $\mathbf{r}$. The hitting processes affect the state vector in the following way:

$$|\Psi\rangle \to |\Psi_{\mathbf{r}, z}\rangle = \frac{|\Phi_{\mathbf{r}, z}\rangle}{\||\Phi_{\mathbf{r}, z}\rangle\|},$$

$$|\Phi_{\mathbf{r}, z}\rangle = \left(\frac{\beta(\mathbf{r})}{\pi}\right)^{1/4} e^{-\frac{1}{2}\beta(\mathbf{r})[M(\mathbf{r}) - z]^2} |\Psi\rangle. \tag{2.7}$$

Here $\beta(\mathbf{r})$ is a positive scalar function with dimensions reciprocal to a square of a mass density and z is a real c-number having the dimensions of a mass density. As usual, the probability density of a process taking place at z, for a given $\mathbf{r}$, is $\||\Phi_{\mathbf{r}, z}\rangle\|^2$. The considered processes occur independently with an appropriate frequency $\mu(\mathbf{r})$.

At this stage one can already grasp the physical implications of the modified dynamics; suppose one has a superposition of two states $|\Psi_{\mathbf{r}, z_1}\rangle$ and $|\Psi_{\mathbf{r}, z_2}\rangle$ which are "eigenstates" of $M(\mathbf{r})$ corresponding to the values z_1 and z_2. As before, only the hittings characterized either by z_1 or by z_2 have an appreciable probability of occurrence and lead to the suppression of the superposition.

One can show that, if one takes the limits

$$\beta(\mathbf{r}) \to 0, \quad \mu(\mathbf{r})\beta(\mathbf{r}) = \frac{\gamma}{m_0^2}, \tag{2.8}$$

one gets the model of the appendix.[18]

Once more, to evaluate the physical consequences of this model one can consider, in place of the evolution of the individual state vectors, the equation for the statistical operator:

$$\frac{d\rho(t)}{dt} = -\frac{i}{\hbar}[H, \rho(t)] + \frac{\gamma}{m_0^2} \int d\mathbf{r} \, M(\mathbf{r})\rho(t)M(\mathbf{r})$$

$$-\frac{\gamma}{2m_0^2}\left\{\int d\mathbf{r} \, M^2(\mathbf{r}), \rho(t)\right\}. \tag{2.9}$$

In the preceding equation m_0 is the nucleon mass and the parameter γ is assumed to take a value of the order of 10^{-30} cm^3 sec^{-1}. With this choice the decoherence is governed by the mass of the nucleons in ordinary matter, the contribution due to electrons being negligible.

One of the main motivations to relate reduction to the mass derives from the desire to make gravity responsible for the violation of linearity, as suggested by various authors[19] (a model with analogous characteristics has been presented by Ghirardi, Grassi, and Rimini).[20] Another important feature of this choice has been pointed out by Pearle and Squires.[21]

In what follows, for simplicity's sake, we will consider a version of the model obtained by disregarding the Hamiltonian term in equation (2.9) and discretizing the space. We divide the space into cells of volume $(\alpha/4\pi)^{-3/2}$ and we denote by M_i the mass operator corresponding to the i-th cell:

$$M_i = \sum_k m^{(k)}N_i^{(k)}, \tag{2.10}$$

where $N_i^{(k)}$ is the number operator counting the particles of type k in the i-th cell. The corresponding equation for the statistical operator reads:

$$\frac{d\rho(t)}{dt} = \frac{\gamma}{m_0^2}\left(\frac{\alpha}{4\pi}\right)^{3/2}\left[\sum_i M_i\rho(t)M_i - \frac{1}{2}\left\{\sum_i M_i^2, \rho(t)\right\}\right]. \tag{2.11}$$

In accordance with the relation discussed in the appendix, we will often use the QMSL frequency parameter λ in place of the expression $\gamma(\alpha/4\pi)^{3/2}$. If we denote by $|m_1, m_2, \ldots, m_i, \ldots\rangle$ the state with the indicated masses in the various cells, the solution of equation (2.11) reads, in the considered basis:

$$\langle m_1, m_2, \ldots | \rho(t) | m_1', m_2', \ldots \rangle$$

$$= e^{-\frac{\lambda}{2m_0^2}\sum_i(m_i - m_i')^2 t}\langle m_1, m_2, \ldots | \rho(0) | m_1', m_2', \ldots\rangle. \tag{2.12}$$

Equation (2.12) shows that linear superpositions of states with different masses in the various cells are dynamically reduced to one of the superposed states with a time rate depending on the expression $\frac{\lambda}{2m_0^2} \sum_i (m_i - m_i')^2$.

In the following, when we will study systems of identical particles, in place of the vectors $|m_1, m_2, \ldots, m_i, \ldots\rangle$ we will consider the vectors $|n_1, n_2, \ldots, n_i, \ldots\rangle$ where n_i is the occupation number of the i-th cell. Obviously, m_i is equal to mn_i, with m the mass of the particles. This notation will allow us to use in many computations (i.e., when the considered particles are nucleons so that $m = m_0$) the formulas and the figures of standard CSL. In such a case expressions of the type $\frac{\lambda}{m_0^2} \sum_i (m_i - m_i')^2$ become simply $\lambda \sum_i (n_i - n_i')^2$.

3. How to Describe the Macroscopic World within a Dynamical Reduction Context

In this section we will show how, by taking advantage of the specific features of the dynamical reduction mechanisms, one can give a description of the world in terms of the mean values $\mathcal{M}(\mathbf{r}, t)$, at different places and at different times, of appropriately defined mass density operators. The procedure will involve the following steps. First, we will show how, if one does not restrict the set of all possible states of the Hilbert space of "our universe," one unavoidably meets situations that cannot be consistently described in terms of the function $\mathcal{M}(\mathbf{r}, t)$. Fortunately, one can show that the universal dynamics of the reduction models does not permit the persistence for *more than a split second*[22] of the unacceptable states, thus allowing one to use the function $\mathcal{M}(\mathbf{r}, t)$ as the basic element for the description of the world. In terms of it one can then define an appropriate "topology" that is the natural candidate for establishing a satisfactory psychophysical correspondence.

3.1 *Mass Density Function*

Let us now consider a physical system S, which will constitute "our universe," and let us denote by $H^{(S)}$ the associated Hilbert space. Let $|\Psi(t)\rangle$ be the normalized state vector describing our individual system at time t; in terms of it we define an average mass density c-number function $\mathcal{M}(\mathbf{r}, t)$ in ordinary space as

$$\mathcal{M}(\mathbf{r}, t) = \langle \Psi(t) | M(\mathbf{r}) | \Psi(t) \rangle. \tag{3.1}$$

If one assumes, as one can consistently do within a nonrelativistic quantum framework, that the system S contains a fixed and finite number of particles, equation (3.1) establishes, for a given t, a mapping of $H^{(S)}$ into the space of positive and bounded functions of $\mathbf{r}$.

Obviously this map is many to one; in particular, to better focus on this point as well as for future purposes, it is useful to compare two state vectors[23] $|\Psi^{\oplus}\rangle$ and $|\Psi^{\otimes}\rangle$ defined as follows. Let us consider a very large number N of particles and two space regions A and B with spherical shape and radius R. The state $|\Psi^{\oplus}\rangle$ is the linear superposition, with equal amplitudes, of two states $|\Psi_N^A\rangle$ and $|\Psi_N^B\rangle$ in which the N particles are well localized with respect to the characteristic length (10^{-5} cm) of the model and uniformly distributed in regions A and B, respectively, in such a way that the density turns out to be of the order of $1\,\text{gr/cm}^3$. On the other hand, $|\Psi^{\otimes}\rangle$ is the tensor product of two states $|\Phi_{N/2}^A\rangle$ and $|\Phi_{N/2}^B\rangle$ corresponding to $N/2$ particles being uniformly distributed in region A and $N/2$ in region B, respectively:

$$|\Psi^{\oplus}\rangle = \frac{1}{\sqrt{2}}\{|\Psi_N^A\rangle + |\Psi_N^B\rangle\}, \quad |\Psi^{\otimes}\rangle = |\Phi_{N/2}^A\rangle \otimes |\Phi_{N/2}^B\rangle. \qquad (3.2)$$

It is trivially seen that the two considered N-particle states give rise to the same function $\mathcal{M}(\mathbf{r})$ and it is clear that if one attempts to give some meaning to it one has to be very careful in keeping in mind from which state $\mathcal{M}(\mathbf{r})$ originates.

In particular, in the case of $|\Psi^{\oplus}\rangle$ it is quite obvious that $\mathcal{M}(\mathbf{r})$ cannot be considered as an "objective" mass density function. To see this, let us suppose that one can use quantum mechanics to describe the gravitational interaction between massive bodies and let us consider the following *Gedanken* experiment: a test mass is sent through the middle point of the line joining the centers of regions A and B with its momentum orthogonal to it (see figures 15.3a, b). In the case of the state $|\Psi^{\otimes}\rangle$ for the system of the N particles, quantum mechanics predicts that the test particle will not be deflected. On the other hand, if the same test is performed when the state is $|\Psi_N^A\rangle$ $(|\Psi_N^B\rangle)$, quantum mechanics predicts an upward (downward) deviation of the test particle. Due to the linear nature of the theory this implies that if one would be able to prepare the state $|\Psi^{\oplus}\rangle$ the final state would be

$$|\phi\rangle = \frac{1}{\sqrt{2}}\{|\Psi_N^A\rangle \otimes |\phi^{UP}\rangle > +|\Psi_N^B\rangle \otimes |\phi^{DOWN}\rangle\}, \qquad (3.3)$$

with obvious meaning of the symbols. If one includes the test particle into the "universe" and considers the mass density operator in regions corresponding to the wave packets $|\phi^{UP}\rangle$ and $|\phi^{DOWN}\rangle$, one discovers once more that nowhere in the universe is there a density corresponding

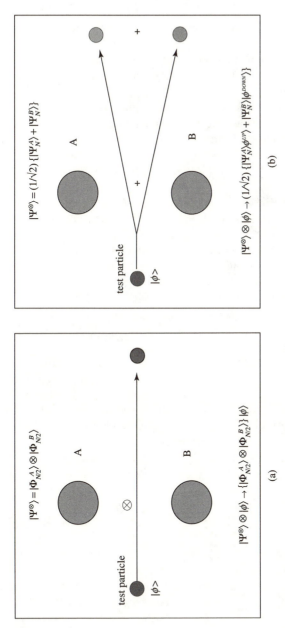

$|\Psi^{\otimes}\rangle = |\Phi^A_{N/2}\rangle \otimes |\Phi^B_{N/2}\rangle$

A

⊗

B

test particle
$|\phi\rangle$

$|\Psi^{\otimes}\rangle \otimes |\phi\rangle \rightarrow \{|\Phi^A_{N/2}\rangle \otimes |\Phi^B_{N/2}\rangle\}|\phi\rangle$

(a)

$|\Psi^{\otimes}\rangle = (1/\sqrt{2})\{|\Psi^A_N\rangle + |\Psi^B_N\rangle\}$

A

+

B

test particle
$|\phi\rangle$

+

$|\Psi^{\otimes}\rangle \otimes |\phi\rangle \rightarrow (1/\sqrt{2})\{|\Psi^A_N\rangle\phi^{UP}\rangle + |\Psi^B_N\rangle|\phi^{DOWN}\rangle\}$

(b)

Fig. 15.3. Objective and nonobjective mass density distribution $\mathcal{M}(\mathbf{r})$. The shading intensity is proportional to the value of $\mathcal{M}(\mathbf{r})$ in the shaded region. In case (a), corresponding to the factorised state $|\Psi^{\otimes}\rangle$, the mass density in regions A and B is objective and the test particle, interacting with $|\Psi^{\otimes}\rangle$ behaves in such a way as to give rise to the appropriate density along its natural trajectory. In case (b), corresponding to the superposition $|\Psi^{\oplus}\rangle$, the densities in A and B are nonobjective and the same holds for the density distribution generated by the interaction of the test particle with $|\Psi^{\oplus}\rangle$.

to the density of the test particle. In a sense, if one would insist in giving a meaning to the density function he would be led to conclude that the particle has been split by the interaction into two pieces of half its density.

This analysis shows that great attention should be paid in attributing an "objective" status to the function $\mathcal{M}(\mathbf{r})$. We will tackle this problem in the next subsection.

Before going on we consider also another quantity that will be useful in what follows. It is the mass density variance at $\mathbf{r}$ at time t defined by the following map from $H^{(S)}$ into $\mathcal{R}^3$:

$$\mathcal{V}(\mathbf{r}, t) = \langle \Psi(t) | [M(\mathbf{r}) - \langle \Psi(t) | M(\mathbf{r}) | \Psi(t) \rangle]^2 | \Psi(t) \rangle, \qquad (3.4)$$

$|\Psi(t)\rangle$ being a normalized state vector.

With these premises we have all the elements which are necessary to discuss the problems one meets when dealing with $\mathcal{M}(\mathbf{r}, t)$ and the way to overcome them. We will do this in the next subsection.

Before doing that, we consider it appropriate to simply mention the obvious fact that the states giving rise to puzzling, nonobjective density functions are those corresponding to superpositions of differently located macroscopic bodies, that is, the infamous states that are at the center of the long-debated problems about the meaning of quantum mechanics at the macrolevel.

For future purposes it is useful to introduce a mathematical criterion that consents to make clear the different status of the mass densities in the two previously considered cases (corresponding to the states $|\Psi^\oplus\rangle$ and $|\Psi^\otimes\rangle$ respectively). This is more easily expressed by resorting once more to a discretization of space in analogy with what has been done in section 2. Obviously, in place of the space functions $\mathcal{M}(\mathbf{r}, t)$ and $\mathcal{V}(\mathbf{r}, t)$ we will consider the mean value $\mathcal{M}_i(t)$ and the variance $\mathcal{V}_i(t)$ of the mass operator in the i-th cell. In correspondence of an arbitrary cell i we define the ratio:

$$\mathcal{R}_i^2 = \mathcal{V}_i / \mathcal{M}_i^2. \qquad (3.5)$$

We then state that the mass $\mathcal{M}_i$ is objective if $\mathcal{R}_i$ turns out to be much smaller than one. This criterion is clearly reminiscent of the probabilistic interpretation of the state vector in standard quantum mechanics. Actually, within such a theory equation (3.5) corresponds to the fact that the spread of the mass operator $\mathcal{M}_i$ is much smaller than its mean value. Even though in this chapter we take a completely different attitude with respect to the mean value $\mathcal{M}_i$, it turns out to be useful to adopt the preceding criterion also within the new context. In fact, as we will discuss in what follows, when one has a space region such that for all cells contained in it (3.5) holds, it behaves as if it would have the "classical" mass corresponding to $\mathcal{M}_i$.

With reference to the previous example we stress that in the case of $|\Psi^{\otimes}\rangle$ all cells within regions A and B are such that criterion (3.5) is very well satisfied. In the case of $|\Psi^{\oplus}\rangle$ for the same cells one has:

$$\mathcal{M}_i \cong \frac{n}{2}m_0, \qquad \mathcal{V}_i \cong \frac{n^2}{4}m_0^2, \tag{3.6}$$

where n is the number of particles per cell. There follows

$$\mathcal{R}_i \cong 1. \tag{3.7}$$

3.2 The Mass Density Function in the Context of Dynamical Reduction Models

In the previous subsection we have presented a meaningful example of the difficulties one meets when one keeps the standard quantum dynamics and tries to base a description of the world on the mass density function $\mathcal{M}(\mathbf{r})$. The unacceptable features find their origin in the fact that, when the macrostate is $|\Psi^{\oplus}\rangle$ while the density function takes the value of about $(1/2)\,\text{gr/cm}^3$ within regions A and B, if one performs a measurement of the density in the considered regions, or if a measurement-like process (such as the passage of the test particle in between A and B) occurs, things proceed in a way that is incompatible with the preceding density value. Actually one could state that no outcome emerges in the measurement. To understand fully the meaning of this statement, one could identify, for example, the final position of the test particle with a pointer reading; the pointer would then not point to the middle position (corresponding to equal densities in A and B) but would be split into "two pointers of half density" pointing upward and downward, respectively (compare with figure 15.3b).

If one attempts to take an analogous attitude with reference to dynamical reduction theories, one does not meet the same difficulties because they imply that linear superpositions of states corresponding to far-apart macroscopic systems are dynamically suppressed in extremely short times and measurements have outcomes.[24] Therefore, we can guess that, within the context of the dynamical reduction program, the description of the world in terms of the mass density function $\mathcal{M}(\mathbf{r})$ is a *good* description; moreover, it is such as to allow one to base on it a sensible psychophysical correspondence.

Obviously, some *fuzzy* situations can occur also in this context, when the mass density is not "objective"—that is, when (in the simplified discretized version) the criterion that expression (3.5) be much smaller than 1 is not satisfied. However, as we are going to show, this does not give rise to any difficulty for the program we are furthering.

In order to show this, we will examine, along the suggested lines, the status of the mass density function $\mathcal{M}(\mathbf{r})$ for the various possible states that are not forbidden by the reducing dynamics. We will discuss the cases of microsystems and macrosystems, and, with reference to the latter, we will identify two physically relevant classes of states that can occur. As we have done previously we will deal with a discretized space.

Microscopic Systems

For the sake of simplicity, let us consider a single nucleon. As is well known, the reducing dynamics does not forbid the persistence, for extremely long times, of linear superpositions of faraway states of the particle, typically states like

$$|\phi\rangle = \frac{1}{\sqrt{2}}(|0, 0, \ldots, 1_i, \ldots, 0_j, \ldots\rangle + |0, 0, \ldots, 0_i, \ldots, 1_j, \ldots\rangle), \quad (3.8)$$

where i and j are two distinct and far apart cells. Such microscopic states that are not eigenvectors of the operators M_i will be called "microscopically nondefinite," the term "nondefinite" making reference to the characteristic preferred basis of the model. As it is evident from (3.8), the mean values of M_i and M_j are $(1/2)m_0$ and the criterion $\mathcal{R}_i \ll 1$ is not satisfied in correspondence of both cells. A measurement of the mass in one of these two cells would give the *definite outcome* 0 or m_0 with equal probability and not $(1/2)m_0$, the value taken by the density function within the considered cells. This discrepancy, this *nonclassical* character of $\mathcal{M}_i$ and $\mathcal{M}_j$, cannot, however, be considered a difficulty for the theory with the proposed interpretation; it simply amounts to a recognition that we cannot legitimately apply our *classical* pictures to the microworld. On the contrary, we must allow *microsystems to enjoy the cloudiness of waves.*[25]

Macroscopic Systems

The theory allows the persistence of two general classes of states for macroscopic systems: those corresponding to a macroscopic number of microsystems in microscopically nondefinite states and states like those

of almost rigid bodies with a sharply defined (with respect to the characteristic length of the model) center of mass position. Due to the fact that
the center of mass wave function has, in general, noncompact support,
this second class obviously includes also the states that are of particular interest for our discussion, the states that, being brought out by the
reducing dynamics, have "tails."

Concerning states of the first class it is of extreme relevance to have
clear that they have a conceptual status that is very different from the
one of the superpositions of macroscopically distinguishable states like
$|\Psi^{\oplus}\rangle$ of subsection 3.1. This important difference has already been appropriately stressed by A. Leggett,[26] who, even though in a different
context, has introduced the mathematically precise concept of discon-
nectivity to distinguish states of this type from states like $|\Psi^{\oplus}\rangle$.

To be more specific let us consider a system of N nucleons and a
discretization of space in cells of linear dimensions 10^{-8} cm. We consider
again two macroscopic regions A and B, and we label with the indices
k_A and k_B the pairs of cells located symmetrically with respect to the
middle point of the line connecting the centers of A and B, respectively.
For $k_A \neq \tilde{k}_A$ the two cells are disjoint and the union of all cells k_A (k_B)
covers the region A (B). The indices k_A and k_B run from 1 to N, a very
large number; typically if A and B have volumes of the order of $1\,\text{dm}^3$,
N will be of the order of 10^{27}.

Let us denote by $|\phi_{k_A}\rangle$ and $|\phi_{k_B}\rangle$ the states of a particle whose wave
functions in coordinate representation are well localized within k_A and
k_B, respectively. As an example we could choose

$$\langle \mathbf{r}|\phi_{k_A}\rangle = \chi(k_A), \tag{3.9}$$

$\chi(k_A)$ being the characteristic function of cell k_A. We now consider the
following microscopically nondefinite state for the k-th particle

$$|\phi^k\rangle = \frac{1}{\sqrt{2}}(|\phi_{k_A}\rangle + |\phi_{k_B}\rangle) \tag{3.10}$$

and the factorized state of the N particles

$$|\phi\rangle = |\phi^1\rangle \otimes \cdots \otimes |\phi^k\rangle \otimes \cdots \otimes |\phi^N\rangle. \tag{3.11}$$

In spite of the fact that the state $|\phi\rangle$ is a direct product of microscop-
ically nondefinite states it is nevertheless "almost" an eigenstate of the
operators M_i (remember that the linear dimensions of the cell to which
the index i refers are of the order of 10^{-5} cm so that one such cell contains about $n \cong 10^9$ cells of the kind of k_A (k_B)). Taking into account

this fact, one can easily see that $|\phi\rangle$ gives rise to "objective" mass $\mathcal{M}_i$ in regions A and B respectively:[27]

$$\langle M_{i(A,B)}\rangle \cong \frac{1}{2}nm_0, \qquad \langle M_{i(A,B)}^2\rangle \cong \frac{1}{4}(n^2+n)m_0^2, \qquad (3.12)$$

hence

$$\mathcal{V}_{i(A,B)}^2 \cong \frac{1}{4}nm_0^2 \quad \text{and} \quad \mathcal{R}_{i(A,B)} \cong \frac{1}{\sqrt{n}} \ll 1. \qquad (3.13)$$

To clarify the physical implications of the state $|\phi\rangle$, from the point of view that interests us here, we can imagine to perform once more the *Gedanken* experiment with a test particle we have already considered in the previous subsection, assuming, for simplicity, that the interactions between the test particle and the considered N particles do not change the state of the latter.[28] By substituting equation (3.10) into equation (3.11), we see that $|\phi\rangle$ is a superposition of 2^N states in which each particle is well localized. In such a superposition all states have an equal amplitude $(1/\sqrt{2^N})$ and almost all states correspond to about $N/2$ particles being in regions A and B respectively. Therefore, in the language of dynamical reduction models, the probability of occurrence of a realization of the stochastic potential leading to the "actualization" of an almost completely undeflected trajectory for the test particle is extremely close to one.[29] This shows that the mass density function $\mathcal{M}(\mathbf{r})$ corresponding to the state $|\phi\rangle$ behaves in a "classical way," so that no trouble arises in this case. It has to be noted that, obviously, the mass $\mathcal{M}_i$ corresponding to the state (3.11) coincides with the one corresponding to the state $|\Psi^\otimes\rangle$ of subsection 3.1, in spite of the fact that both states are dynamically allowed and are quite different as physical states. However, as we have shown, the masses $\mathcal{M}_i$ in the two cases behave practically in the same way and give rise to no trouble contrary to what happens in the case of $|\Psi^\oplus\rangle$.

We come now to the consideration of the other type of allowed states of interest for us, macroscopic states that are "almost" eigenstates of the mass operators M_i but which however have tails. Let $|\Psi\rangle$ be the normalized state

$$|\Psi\rangle = \alpha|\Psi_N^A\rangle + \beta|\Psi_N^B\rangle, \qquad (3.14)$$

where $|\Psi_N^A\rangle$ and $|\Psi_N^B\rangle$ are the states defined in the previous subsection and $|\beta|^2$ is extremely close to zero. In region A we have

$$\mathcal{M}_{i(A)} \cong |\alpha|^2 nm_0, \qquad \mathcal{V}_{i(A)} \cong |\alpha|^2|\beta|^2 n^2 m_0^2 \quad \text{and} \quad \mathcal{R}_{i(A)} \cong |\beta|^2 \ll 1, \qquad (3.15)$$

so that the masses $\mathcal{M}_{i(A)}$ are objective and practically equal to those corresponding to the state $|\Psi_N^A\rangle$. In region B we have

$$\mathcal{M}_{i(B)} \cong |\beta|^2 n m_0, \qquad \mathcal{V}_{i(B)} \cong |\alpha|^2 |\beta|^2 n^2 m_0^2 \quad \text{and} \quad \mathcal{R}_{i(B)} \cong |\beta|^{-2} \gg 1, \tag{3.16}$$

hence the masses $\mathcal{M}_{i(B)}$ are not objective.

At this point it is appropriate to make a detailed analysis, within our dynamical scheme, to get a quantitative estimate of $\mathcal{R}_{i(A)}$ and of the total mass in region B. To this purpose (as it is evident from equations (3.15) and (3.16)), one has to evaluate explicitly the order of magnitude of the parameter $|\beta|^2$ implied by the reducing dynamics. In order to do this, to cover also the case of nonhomogeneous bodies, we consider again two far apart regions A and B, each containing K cells and a system of nucleons which at time $t = 0$ is in a (normalized) state of the type (the overall-phase factor being irrelevant)

$$|\Psi\rangle = \alpha(0)|n_{1(A)}, \ldots, n_{K(A)}, 0, \ldots, 0\rangle$$
$$+ \beta(0)e^{i\gamma(0)}|0, \ldots, 0, n_{1(B)}, \ldots, n_{K(B)}\rangle, \tag{3.17}$$

where $\alpha(0)$ and $\beta(0)$ are comparable positive numbers and $n_{i(A,B)}$ represents the occupation number in the i-th cell in regions A and B respectively.[30] We then study the ensemble of systems brought in by the reducing dynamics after a time interval of the order of, for example, 10^{-2} sec (the reason for this choice will become clear in what follows).

According to the CSL model discussed in the appendix, after such a time interval the normalized state corresponding to a definite realization of the stochastic potential would be of the type

$$|\Psi_B(t)\rangle = \alpha_B(t)|n_{1(A)}, \ldots, n_{K(A)}, 0, \ldots, 0\rangle$$
$$+ \beta_B(t)e^{i\gamma(0)}|0, \ldots, 0, n_{1(B)}, \ldots, n_{K(B)}\rangle, \tag{3.18}$$

with $\alpha_B(t)$ and $\beta_B(t)$ as positive numbers. The ensemble of systems corresponding to all possible realizations of the stochastic potential would be described by the statistical operator

$$\rho(t) = \int dB_1 \ldots dB_{2K} P_{cook}[B(t)]|\Psi_B(t)\rangle\langle\Psi_B(t)| \tag{3.19}$$

satisfying

$$\langle n_{1(A)}, \ldots, n_{K(A)}, 0, \ldots, 0|\rho(t)|0, \ldots, 0, n_{1(B)}, \ldots, n_{K(B)}\rangle$$

$$= e^{-\lambda t \sum_i^K n_i^2} \langle n_{1(A)}, \ldots, n_{K(A)}, 0, \ldots, 0|\rho(0)|0, \ldots, 0,$$

$$n_{1(B)}, \ldots, n_{K(B)}\rangle \quad (3.20)$$

with $\lambda t \cong 10^{-18}$. From (3.20) we see that the matrix elements of $\rho(t)$ between the considered states are exponentially damped by a factor that is proportional to $\sum_i^K n_i^2$.

In the following we consider only situations in which $\sum_i^K n_i^2$ turns out to be much greater than 10^{18}, so that in the considered time interval the linear superposition (3.17) is actually suppressed, that is, either $\alpha_B(t)$ or $\beta_B(t)$ of equation (3.18) becomes very small. The states at time t are then typical states with "tails," that is, states whose existence is considered as a drawback of the theory proposed by Shimony[31] and Albert and Loewer.[32] Equation (3.20) implies (taking into account equations (3.18) and (3.19)) that

$$\int dB_1 \ldots dB_{2K} P_{cook}[B(t)]\alpha_B(t)\beta_B(t) = \alpha_B(0)\beta_B(0)e^{-\lambda t \sum_i^K n_i^2}. \quad (3.21)$$

From (3.21), since $\alpha_B(t)$ and $\beta_B(t)$ are positive, one can easily deduce that the probability of occurrence of realizations of the stochastic potential that would lead to a value for the product $\alpha_B(t)\beta_B(t)$ much greater than $e^{-\lambda t \sum_i^K n_i^2}$ must be extremely small. Therefore, one can state that in practically all cases

$$\alpha_B(t)\beta_B(t) \cong e^{-\lambda t \sum_i^K n_i^2}. \quad (3.22)$$

If we assume that $\alpha_B(t) \cong 1$—that is, we consider an individual case for which the reduction leads to the state corresponding to the nucleons being in region A—$|\beta_B(t)|^2$ must be of the order of $e^{-2\lambda t \sum_i^K n_i^2}$. On the basis of this fact we can then estimate the value of $|\beta_B|^2$, for example, for a homogeneous sphere of normal density (so that $n_i = n \cong 10^9$ is the number of particles per cell) and of size $1\,\mathrm{dm}^3$ (so that $K \cong 10^{18}$ is the number of cells in regions A and B), getting a figure of the order of $e^{-10^{18}}$. Correspondingly, we have

$$\mathcal{R}_{i(A)} \cong e^{-10^{18}} \quad (3.23)$$

while for the total mass in region B we get the value

$$\mathcal{M}_B \cong e^{-10^{18}} 10^{17} m_0. \quad (3.24)$$

Equation (3.23) shows that the mass in region A is "objective" to an extremely high degree of accuracy and equation (3.24) shows that the total mass in region B is much smaller than the mass of a nucleon. If we consider a situation in which K or n is greater than those of the example we have discussed now, we find values for $\mathcal{R}_{i(A)}$ and $\mathcal{M}_B$ that are even smaller than those of equations (3.23) and (3.24).[33] This fact by itself (see also the analysis of the following subsection) shows that the states with "tails" allowed by CSL cannot give rise to difficulties for the proposed interpretation of the theory. If we would perform the usual *Gedanken* experiment with the test particle it would be deflected just as if in region A there would be the "classical" mass Knm_0.

Concluding, we have made plausible that in the context of the dynamical reduction program one can consistently describe the macroworld, at a given time, in terms of the mass density function $\mathcal{M}(\mathbf{r})$. Obviously, since with the elapsing of time the state of the world changes, a complete description requires the consideration of the motion picture of the density, that is, of $\mathcal{M}(\mathbf{r}, t)$ defined in equation (3.1). We will discuss in greater detail this crucial point in section 4.

3.3 *Defining an Appropriate Topology for the CSL Model*

Let us consider a system S of finite mass that will constitute our "universe" and its associated Hilbert space $H^{(S)}$. We denote by $\mathcal{U}^{(S)}$ the unit sphere in $H^{(S)}$ and we consider the nonlinear map $\mathcal{M}$ associating to the element $|\phi\rangle$ of $\mathcal{U}^{(S)}$ the element $\mathbf{m} = \{\mathcal{M}_i(|\phi\rangle)\}$ of l_2, $\mathcal{M}_i(|\phi\rangle)$ being the quantity $\langle\phi|M_i|\phi\rangle$.[34]

On $\mathcal{U}^{(S)}$ we define a topology by introducing a mapping $\Delta: \mathcal{U}^{(S)} \otimes \mathcal{U}^{(S)} \to \mathcal{R}^+$ according to:

$$\Delta(|\phi\rangle, |\psi\rangle) = d(\mathbf{m}, \mathbf{n}) = \sqrt{\sum_i (m_i - n_i)^2}, \qquad (3.25)$$

where $\mathbf{m} = \{\mathcal{M}_i(|\phi\rangle)\}$, $\mathbf{n} = \{\mathcal{M}_i(|\psi\rangle)\}$. Such a mapping is not a distance since, as it emerges clearly from the analysis of the previous subsection, it may happen that $\Delta(|\phi\rangle, |\psi\rangle) = 0$ even though $|\phi\rangle \neq |\psi\rangle$. However, Δ meets all other properties of a distance:

$$\Delta(|\phi\rangle, |\psi\rangle) = \Delta(|\psi\rangle, |\phi\rangle) \geq 0, \qquad (3.26)$$

and

$$\Delta(|\phi\rangle, |\psi\rangle) \leq \Delta(|\phi\rangle, |\chi\rangle) + \Delta(|\chi\rangle, |\psi\rangle), \qquad (3.27)$$

as one easily proves by taking into account the fact that d is a distance in l_2.

From now on we will limit our considerations to the proper subset $\mathscr{A}^{(S)}$ of $\mathscr{U}^{(S)}$ of those states which are allowed by the CSL dynamics. In the previous subsection we have already identified, even though in a rough way, the set $\mathscr{A}^{(S)}$. Obviously, one could be very precise about such a set by adopting, for example, the following criterion: be $|\phi\rangle \in \mathscr{U}^{(S)}$ and let us consider the ensemble $\mathscr{A}^{(S)}(|\phi\rangle)$ of states which have a nonnegligible (this obviously requires the definition of a threshold) probability of being brought in by the reducing dynamics after a time interval of the order of 10^{-2} sec for the given initial condition $|\phi\rangle$. The union of all subsets $\mathscr{A}^{(S)}(|\phi\rangle)$ for $|\phi\rangle$ running over $\mathscr{U}^{(S)}$ is then $\mathscr{A}^{(S)}$. For our purposes, however, it is not necessary to go through the cumbersome management of a very precise definition of the set $\mathscr{A}^{(S)}$, the consideration of the cases we have discussed in the previous subsection being sufficient to lead to the interesting conclusions.

For any element $|\phi\rangle$ of $\mathscr{A}^{(S)}$ we consider the set of states of $\mathscr{A}^{(S)}$ for which $\Delta(|\phi\rangle, |\psi\rangle) \gtrsim \epsilon$. Here the quantity ϵ has the dimensions of a mass and is chosen of the order of $10^9 \, m_0$, with m_0 the nucleon mass. From the properties of the map Δ it follows that:

i. $\{\Delta(|\phi\rangle, |\psi\rangle) \gtrsim \epsilon$ and $\Delta(|\phi\rangle, |\chi\rangle) \gtrsim \epsilon\}$ implies $\Delta(|\chi\rangle, |\psi\rangle) \gtrsim \epsilon$.

ii. $\{\Delta(|\phi\rangle, |\psi\rangle) \gg \epsilon$ and $\Delta(|\phi\rangle, |\chi\rangle) \gtrsim \epsilon\}$ implies $\Delta(|\chi\rangle, |\psi\rangle) \gg \epsilon$.

We have introduced the parameter ϵ in such a way that it turns out to be sensible to consider similar to each other states whose "distance" Δ is smaller than (or of the order of) ϵ. More specifically, when

$$\Delta(|\phi\rangle, |\psi\rangle) \gtrsim \epsilon \tag{3.28}$$

we will say that $|\phi\rangle$ and $|\psi\rangle$ are "physically equivalent."

To understand the meaning of this choice it is useful to compare it with the natural topology of $H^{(S)}$. We begin by pointing out the inappropriateness of the Hilbert space topology to describe the concept of similarity or difference of two macroscopic states. In fact, suppose our system S is an almost rigid body and let us consider the following three states: $|\phi^A\rangle$, $|\phi^B\rangle$, and $|\tilde{\phi}^A\rangle$. The state $|\phi^A\rangle$ corresponds to a definite internal state of S and to its center of mass being well localized around A, the state $|\phi^B\rangle$ is simply the translation of $|\phi^A\rangle$ so that it is well localized in a far region B, the state $|\tilde{\phi}^A\rangle$ differs from $|\phi^A\rangle$ simply by the fact that one or a small number of its "constituents" are in states that are orthogonal to the corresponding ones in $|\phi^A\rangle$.

It is obvious that, on any reasonable assumption about similarity or difference of the states of the universe, $|\tilde{\phi}^A\rangle$ must be considered very similar (identical) to $|\phi^A\rangle$ while $|\phi^B\rangle$ must be considered very different from $|\phi^A\rangle$. On the other hand, according to the Hilbert space topology

$$\left\|(|\phi^A\rangle - |\tilde{\phi}^A\rangle)\right\| = \left\|(|\phi^A\rangle - |\phi^B\rangle)\right\| = \sqrt{2}. \qquad (3.29)$$

This shows with striking evidence that the Hilbert space topology is totally inadequate for the description of the macroscopic world. As a consequence such topology is also quite inadequate to base on it any reasonable psychophysical correspondence.

We now discuss the "distorted" (with respect to the Hilbert space one) topology associated to the "distance" Δ. First of all we stress that the two states $|\phi^A\rangle$ and $|\tilde{\phi}^A\rangle$, which are maximally distant in the Hilbert space topology, turn out to be equivalent, that is, to satisfy condition (3.28) in the new topology. This represents an example showing how such a topology takes more appropriately into account the fact that, under any sensible assumption, the "universes" associated to the considered states are very similar.

Obviously, one problem arises. Criterion (3.28) leads to consider equivalent states that are quite different from a physical point of view, even at the macroscopic level. To clarify this statement we take into account two states $|\phi\rangle$ and $|\psi\rangle$ corresponding to an almost rigid body located, at $t = 0$, in the same position but with macroscopically different momenta, let us say $P = 0$ and P, respectively. Even though the two states are physically quite different, their distance at $t = 0$ is equal to zero. However, if one waits up to the time in which the state $|\psi\rangle$ has moved away from $|\phi\rangle$, the "distance" $\Delta(|\phi(t)\rangle, |\psi(t)\rangle)$ becomes large and the two states are no longer equivalent. We will discuss the now outlined problem in great details in the next section.

Before concluding this part it is important to analyze the case of two states $|\psi\rangle$ and $|\psi_T\rangle$ such that $|\psi\rangle$ corresponds to an almost rigid body with a center of mass wave function perfectly localized and $|\psi_T\rangle$ corresponds to the same body with a "tail" in a distant region. As we have already discussed, the CSL dynamics allows the existence of this latter type of state, however, it tends to depress more and more the tail in such a way as to make the mass in the distant region extremely close to zero (much less than one nucleon mass) in very short times. As a consequence, according to the topology that we propose the two states $|\psi\rangle$ and $|\psi_T\rangle$ turn out to be identical. This is quite natural. In fact, in the same way in which taking away a single particle from a macroscopic system would be accepted as being totally irrelevant from a macroscopic point of view, when one chooses, as we do, to describe reality in terms of mass density, one must consider equivalent situations in which their difference derives entirely from the location of a small fraction of the mass of a nucleon in the whole universe. We remark that $|\psi\rangle$ and $|\psi_T\rangle$ are extremely close to each other also in the standard Hilbert space topology.

4. Deepening the Proposed Interpretation

We consider it appropriate to devote this section to discuss in great generality the problem of giving an acceptable description of the world within a given theory. Usually one tries to do so by resorting to the notion of observable. As repeatedly remarked, such an approach meets, within standard quantum mechanics, serious difficulties since the formal structure of the theory allows only probabilistic statements about outcomes conditional under the measurement being performed. In brief, the theory deals with *what we find*, not with *what is*. This is why J. S. Bell has suggested to replace the notion of observable with the one of "beable," from the verb to be, to exist.[35] Obviously, the identification of the beables, of what is real, requires the selection of some formal ingredients of the theory with which we are dealing.

4.1 *The Case of the Pilot-Wave Theory*

To clarify our point, it turns out to be useful to analyze the de Broglie–Bohm Pilot-Wave theory. It describes the world in terms of the wave function and of the actual positions of the particles of our "universe," each of which follows a definite trajectory. Therefore, in such a theory it is quite natural to consider as the beables the positions (which are the local elements accounting for reality at a given instant) and the wave function (which is nonlocal and determines uniquely the evolution of the positions). It is important to stress that, within the theory under discussion, all other "observables" (in particular, e.g., the spin variables) turn out, in general, to be contextual. This simply means that *the truth value* of a statement about the outcome of the measurement of one such observable (which in turn is simply a statement about the future positions of some particles) may in general depend (even nonlocally) on the *overall* context. This obviously implies that the attribution of a value to the considered observable cannot be thought as corresponding, in general, to an "objective property" of the system.

Before coming to discuss the problem of the beables within CSL we would like to call attention to the fact that within the Pilot-Wave theory, one can construct, from the microscopic variables **r** macroscopic variables **R** including pointer positions, images on photographic plates, and so on.[36] Obviously this requires some fuzziness, but such a limitation is not relevant for a consistent account of reality. Thus, in this theory we are led to suppose that it is from the **r**, rather than from the wave function, that the observables we use to describe reality are constructed. The positions are also the natural candidates to be used in defining a psychophysical parallelism, if we want to go so far. An appropriate way to express the now discussed features of the theory derives from denoting, as J. S. Bell proposed, as "exposed variables" the positions of particles and as a "hidden variable" the wave function Ψ.

4.2 *The Case of CSL*

Let us now perform a corresponding analysis for the model theory considered in section 2. Since, as it should be clear from the discussion given there, the most relevant feature of the modified dynamics is that of suppressing linear superpositions corresponding to different mass distributions, one is actually led to identify as the local beables of the theory the mass density function $\mathcal{M}(\mathbf{r}, t)$ at a given time. Obviously, also within CSL just as for the Pilot-Wave, the wave function plays a fundamental role for the evolution so that it too acquires the status of a nonlocal beable.

It has to be remarked that in the interpretation we are proposing, even though the wave function is considered as one of the beables of the theory, the "exposed variables" are the values of the mass density function at different points. It is then natural to relate to them, as we have done in the previous section, the concept of similarity or difference between universes.

In doing so, one is led to consider equivalent, at a fixed time t, two "universes" that are almost identical in the exposed beables (i.e., they satisfy the condition (3.28)). Obviously the fact that the preceding condition holds at t by no means implies that the two universes will remain equivalent as time elapses.

It has to be stressed that the previously mentioned feature is not specific of the model and of the interpretation we are proposing, but it is quite general and occurs whenever one tries to make precise the idea of "similarity" of physical situations. In fact, within all theories we know, and independently of the variables we choose to use to define nearness, situations can occur for which *nearby states* at a given time can evolve in extremely short times in *distant states*.

To focus on this important fact, we can consider even classical mechanics with the assumption that both positions and momenta are the beables of the theory.[37] As it is obvious, even if such an attitude is taken there are at least two reasons for which nearby points in phase space can rapidly evolve into distant ones. First of all, one must take into account one of the most important conceptual achievements of recent times, that is, the discovery that many systems exhibit dynamical instability so that the distance between "trajectories" may grow exponentially with time. Second, even for a "dynamically standard" situation one can consider cases in which just the present conditions can give rise to completely different evolutions depending on some extremely small difference in the whole universe. Suppose in fact you consider two universes A, $\tilde{A}$ differing only in the *direction* of propagation of a single particle (such universes have to be considered as very close in any sensible objective interpretation). If the trajectory of the particle in $\tilde{A}$ is such that within a very short time it triggers, for example, the discharge of a Geiger counter, which in turn gives rise to some relevant macroscopic effect, while in A it does not, the evolved universes become soon quite different. An analogous argument obviously holds for standard quantum mechanics, the Pilot-Wave theory, and, as previously remarked, for CSL too.

It is appropriate to stress that, in a sense, the foregoing considerations favor taking a position about reality which can be described in the following terms. One chooses the sensible "beables" for its theory at a fixed time and one distinguishes similar or different universes on the basis of such a snapshot. Obviously, one must then also pay attention to the way in which the beables evolve, that is, to compare snapshots at different times.[38]

4.3 *The Role of Mass Density*

The previous analysis has shown that the proposed interpretation can be consistently taken. Obviously it gives an absolutely prominent role to the mass in accordance with the fact that mass is the handle by which the reduction mechanism induces macroobjectification.

Other features of natural phenomena, such as the effects related to the charge are, in a sense less fundamental since to become objective they need mass as a support. To clarify this point we consider two states $|C_0\rangle$ and $|C_c\rangle$ of a condenser corresponding to its plates being neutral or having been charged by displacing 10^{12} *electrons* from one plate to the other. We remark that the electric field within the plates is zero or about 10^8 V/m in the two states, respectively. In spite of this, due to the smallness of the electron mass the linear superposition $1/\sqrt{2}(|C_0\rangle + |C_c\rangle)$ could persist for more than ten years. Suppose now we send through the plates of the condenser a test particle of normal density and of radius 10^{-3} cm carrying a charge corresponding to 10^4 electrons. The final state is the entangled state

$$|\Psi(t)\rangle = \frac{1}{\sqrt{2}}\big[|C_0\rangle|undeflected\rangle + |C_c\rangle|deflected\rangle\big]. \qquad (4.1)$$

It is easily seen that the macroscopic force acting on the particle when the condenser is in the state $|C_c\rangle$ leads to a displacement of the order of its radius in about 10^{-5} sec and that within the same time the reducing effect of the dynamics suppresses one of the two terms of the superposition.

This example is quite enlightening since it shows that superpositions of charge distributions generating macroscopically different fields are not suppressed unless they induce displacements of masses.

It goes without saying that any attempt to relate reduction to charge is doomed to fail since it will not suppress superpositions of macroscopically different but electrically neutral mass distributions.

We hope to have made clear, with this perhaps tedious analysis, the real significance of treating the mass function as the "exposed beables" allowing one to describe reality.

4.4 *The Stochastic Nature of the Evolution*

So far we have discussed the description of the world allowed by the CSL theory in terms of the values taken by the mass density function $\mathcal{M}(\mathbf{r})$, which have been recognized to constitute the exposed beables of the theory. According to equation (3.1) it is the wave function associated with the system that determines $\mathcal{M}(\mathbf{r})$. It is useful to analyze the evolution of the beables. As we discuss in the appendix, the dynamical evolution equation for the wave function is *fundamentally* stochastic, being governed by the stochastic processes $w(\mathbf{r}, t)$. The "cooked" probability of occurrence of such processes, based on the analogous of the prescription following equation (2.7), depends on the wave function that describes the system and this fact is of crucial importance for getting the "right" (i.e., the quantum) probabilities of measurement outcomes. Therefore, in the CSL theory, the wave function has both a descriptive (since it determines $\mathcal{M}(\mathbf{r})$) and a probabilistic (since it enters in the prescription for the cooking of the probability of occurrence of the stochastic processes) role.

As we prove in appendix A.2, also the "tails" of the wave function have a precise role. In fact, suppose our "universe" is described at $t = 0$ by a normalized state

$$|\Psi(0)\rangle = \alpha(0)|a\rangle + \beta(0)|b\rangle, \qquad (4.2)$$

with $|\beta(0)|^2$ being extremely small. The "reality" of the universe at $t = 0$ is "determined" by the state $|a\rangle$, as we have explicitly shown in section 3. However, one cannot ignore the (extremely small) probability $|\beta(0)|^2$ that a realization of the stochastic potential occurs that, after a sufficiently long time, leads to a normalized state

$$|\Psi(t)\rangle = \alpha(t)|\tilde{a}\rangle + \beta(t)|\tilde{b}\rangle, \qquad (4.3)$$

with $|\alpha(t)|^2$ extremely small and with $|\tilde{a}\rangle$ and $|\tilde{b}\rangle$ two of the most probable states at time t for the initial conditions $|a\rangle$ and $|b\rangle$, respectively. Then, the "reality" at time t is that associated to the state $|\tilde{b}\rangle$ which has its origin in the negligible component $|b\rangle$ at time $t = 0$. Thus, some "memory" of a situation which at time zero did not correspond to the "reality" of the world remains at time t. Obviously, if such an extremely improbable case would occur one would be tempted (wrongly) to retrodict that "reality" at $t = 0$ was the one associated to $|b\rangle$ and not the one associated to $|a\rangle$. However, we stress that such peculiar events, which we could denote as the "reversal of the universe," have absolutely negligible probabilities. As made plausible by the estimate for the values of $\beta(t)$ given in section 3, the "risk to be wrong" in retrodicting from the present to the past "status of the world" is even smaller than the probability that observing now a table standing on the floor, and knowing that it has been kept isolated, we can infer that it was standing there even one hour ago, in spite of the fact that thermodynamically a very peculiar situation corresponding to its "levitation" at that time could in principle have occurred.

5. The Psychophysical Parallelism within CSL

The most characteristic and appealing feature of CSL and of its interpretation we have proposed in this chapter consists in the fact that it allows one to give a satisfactory account of reality, to take a realistic view about the world, to talk about it as if it is really there even when it is not observed. However, one cannot avoid raising the problem of including also conscious observers into the picture: *what is interesting if not experienced?*[39] Thus one is led to consider the problem of the psychophysical parallelism within the considered theory.

The previous analysis has already given clear indications about the way to reach this goal. In this section we will first of all clarify how to relate the "different states of the universe" as characterized by the formalism according to the lines of the previous sections to the specificity of conscious perceptions. Second, we will show how the theory itself supports the proposed correspondence. To get this we will perform a very sketchy analysis, from the point of view of CSL, of the current ideas about the physical processes leading to perceptions.

5.1 *External World and Internal Perceptions*

As we have seen, the CSL dynamics leads naturally to consider as the exposed beables accounting for "reality" the values taken by the mass density function $\mathcal{M}(\mathbf{r}, t)$, at different points and at different times. We have also discussed, in the case of macroscopic objects, how the dynamical evolution forces the mass density to be "objective" at almost all times in the regions where such objects are. According to the foregoing picture, "reality," "what is out there," is identified with a precise mass distribution in real space. The reality of a massive macroobject in front of us corresponds to the fact that in the region it occupies there is the objective mass density that characterizes it.

On the other hand, it is a fundamental feature of our perceptions that they correspond to objects having precise locations and extensions. The problem of establishing a map between reality and perceptions is then naturally solved by correlating our perceptions to mass density distributions.[40]

After these simple remarks, we can come to discuss how one can account, within CSL, for the emergence of perceptions.

5.2 *Describing the Perceptive Process*

To clarify how CSL is able to "describe" the occurrence of definite perceptions of conscious beings, it turns out to be quite appropriate to start by discussing a criticism that has been put forward recently.[41] The idea is quite simple but raises a problem that deserves a detailed investigation. One considers the following process: a neutral microsystem with spin is sent through a Stern-Gerlach apparatus. The spin state is such that the system ends up in the superposition of being deflected, with equal probabilities, upward or downward, respectively. The two "potential trajectories" cross a fluorescent screen in two macroscopically far apart regions A and B, respectively. The particle-screen interaction is such as to lead to the excitation of a small number, say of about ten atoms, which subsequently undergo a transition to the ground state accompanied by the emission of photons.

The argument proposed in a paper by Albert and Vaidman and in other works by Albert goes as follows.[42] Since only a few atoms of the screen are excited during the process, since their excitations involve displacements that are much smaller than the characteristic localization length of the model, since photons are not spontaneously localized, there is no way for the CSL mechanism to suppress one of the two states. Thus, the superposition of states corresponding to ten photons emerging from the different space regions A and B will persist for extremely long times. On the other hand, since the visual perception threshold is quite low (about seven photons) there is no doubt that the naked eye of a human observer is sufficient to detect whether the luminous spot on the screen is at A or at B. This raises an interesting question: in the considered situation how can it happen that a definite perception about the location of the spot on the screen emerges? Are we compelled to accept that, at least in some circumstances, also within CSL the conscious observer and his perceptions play a peculiar role analogous to the one they have in Wigner's views?

That this is not the case has been discussed in great detail in a recent article.[43] If one takes, as one must, the preceding remark seriously, one is compelled to consider the actual systems that enter into play and to analyze the implications of the CSL dynamics for them. We simply sketch here the argument of Aicardi et al.[44] One takes into account what we know about the transmission of nervous signals from, let us say, the retina to the higher visual cortex. Such a transmission requires, among other changes, the displacement of ions along the axons involved in the process. A very rough estimate of the mass associated to these ions and of the displacements they have to make to flow through the ion channels that open at Ranvier's nodes to transmit the electric pulse can be given. This estimate makes perfectly plausible that the conditions that are sufficient, according to CSL, for the suppression of one of the two superposed states (nervous signals) within the visual perception time (which is of the order of 10^{-2} sec) are satisfied.

We do not want to be misunderstood; this analysis does by no means attribute a special role to the conscious observer or to his perception. The observer's brain is simply the only system among those which enter into play in which a superposition of two states involving different locations of a large number of particles occurs. As such it is the only place where the reduction can and actually must take place according to the theory. If, in place of the eye of a human observer, one puts in front of the photon beams a spark chamber or any device leading to the displacement of a macroscopic pointer or producing ink spots on a computer output, reduction will take place. In the considered example, the human nervous system is simply a physical system, a specific assembly of particles, which has the same function as any one of these devices, if no such device is interacting with the photons before the human being does.[45] In section 3, in order to study the states allowed by the CSL theory and the "size of tails," we have considered a time interval of the order of 10^{-2} sec. The reason for such a choice should now be clear: it is the time interval corresponding to the perception time, the time in which our brain, acting as a physical system, must (and actually does) suppress the linear superpositions of states corresponding to different stimuli in order that the observer has a definite perception. Analogous considerations have been taken into account in choosing the parameter ϵ characterizing the similarity or difference of physical situations.

The preceding analysis should be sufficient by itself to clarify our position. Our perceptions are triggered by our sensory apparatuses. In many cases, such as for the auditory or tactile perceptions, the stimulus itself cannot be ambiguous, that is, it cannot correspond to superpositions of different perceptions, since it requires macroscopic displacements (of the hearing membrane or of the skin). In other cases, as in the one just discussed, the nervous signal can be triggered by a microscopic system, which can very well be in a superposition of states capable to induce different perceptions. But in all cases the nervous transmission of the signal involves a "macroscopic" (on the appropriate scale) displacement of mass in the brain. And, as repeatedly stressed, the CSL dynamics does not tolerate nonobjective macroscopic mass distributions lasting for a time of the order of the perception time.

In a sense the analysis shows that the problem of the psychophysical correspondence admits a simple solution quite similar to the one that is usually assumed to hold for classical theories. Reality and perceptions involve the same and fundamental type of "exposed beables": the values of the mass density function. On one hand "reality," as previously discussed, is related to mass density and macroscopic situations to "objective" mass density distributions. On the other, definite perceptions are related to objective macroscopic mass density distributions within the brain.

We do not want to spend further time in elaborating on this point. Our aim here is not that of performing a detailed technical analysis of any conceivable situation but simply that of making plausible that the problem of the psychophysical parallelism admits, within CSL, a solution that is quite analogous to the one of the unproblematic classical case. Thus CSL can be claimed, according to J. S. Bell's definition, to be an "exact theory" in the precise and limited sense that *it neither needs nor is embarrassed by an observer.*[46]

6. Concluding Remarks

We consider the previous analysis sufficient to give a clear idea of the reasons and the formal aspects that allow to close the circle within the dynamical reduction program. We cannot, however, conclude our analysis, without stressing the crucial role of some stimulating remarks by

J. S. Bell for the elaboration of the ideas of this chapter. He has been the first one to call attention to the fact that spontaneous reduction models, unlike standard quantum mechanics, allow one to take a density rather than a probabilistic interpretation of the modulus square of the wave function.[47] Therefore, the line of thought of this chapter represents, in a sense, an implementation of his suggestions. But there are important differences between the present attitude and the one he was inclined to take. He repeatedly insisted that the density he was referring to was not a mass or a charge density but the density of stuff of which the world is made. Moreover, he stressed strongly (probably since he was worried about the consequences of adopting an interpretation à la Schrödinger) that the density function had to be taken seriously only in the $3N$-dimensional configuration space and not in the real three-dimensional space.[48] We hope to have shown that, within CSL, a quite satisfactory interpretation can be obtained along the (more traditional) lines we have presented in this chapter.

Appendix

A.1 *The CSL Model*

The model is based on a linear stochastic evolution equation for the state vector. The evolution does not preserve the norm but only the average value of the square norm. The equation, in the Stratonovich version, is:

$$\frac{d|\Psi_w(t)\rangle}{dt} = \left[-\frac{i}{\hbar}H + \sum_i A_i w_i(t) - \gamma \sum_i A_i^2 \right]|\Psi_w(t)\rangle. \qquad (A.1)$$

In equation (A.1), the quantities A_i are commuting self-adjoint operators, while the quantities $w_i(t)$ are c-number stochastic processes with probability of occurrence satisfying

$$P_{cook}[w(t)] = P_{raw}[w(t)] \big\| |\Psi_w(t)\rangle \big\|^2. \qquad (A.2)$$

In equation (A.2) $P_{raw}[w(t)]$ is equal to

$$P_{raw}[w(t)] = \frac{1}{\mathcal{N}} e^{-\frac{1}{2\gamma} \sum_i \int_0^t d\tau w_i^2(\tau)}, \qquad (A.3)$$

$\mathcal{N}$ being a normalization factor, that is, to the probability density of a white noise process satisfying

$$\langle\langle w_i(t)\rangle\rangle = 0, \qquad \langle w_i(t)w_j(t')\rangle\rangle = \gamma\delta_{ij}\delta(t - t'). \qquad (A.4)$$

To clarify the physical meaning of the model, let us assume, for the moment, that the operators A_i have a purely discrete spectrum and let us denote by M_σ their common eigenmanifolds and by P_σ the associated projection operators.

Then we make the following precise assumption: if a homogeneous ensemble (pure case) at the initial time $t = 0$ is associated to the state vector $|\Psi(0)\rangle$, then the ensemble at time t is the union of homogeneous ensembles associated with the normalized vectors $|\Psi_w(t)\rangle / \||\Psi_w(t)\rangle\|$, where $|\Psi_w(t)\rangle$ is the solution of equation (A.1) with the assigned initial conditions and for the specific stochastic process $w(\tau)$ that occurred in the interval $(0, t)$. The probability density for such a subensemble is that given by equation (A.2).

One can prove that the map from the initial ensemble to the final ensemble obeys the forward time translation semigroup composition law.[49] It is also easy to prove that the evolution, at the ensemble level, is governed by the dynamical equation for the statistical operator

$$\frac{d\rho(t)}{dt} = -\frac{i}{\hbar}[H, \rho(t)] + \gamma \sum_i A_i \rho(t) A_i - \frac{\gamma}{2} \left\{ \sum_i A_i^2, \rho(t) \right\}, \qquad (A.5)$$

from which one immediately sees that, if one disregards the Hamiltonian evolution, the off-diagonal elements $P_\sigma \rho(t) P_\tau$ ($\sigma \neq \tau$) of the statistical operator are exponentially damped.

For our concerns, the relevant feature of the dynamical process (A.1) with the prescription (A.2) is that it drives the state vector of each individual member of the ensemble into one of the common eigenmanifolds of the operators A_i, with the appropriate probability. To clarify this, we consider a simplified case in which only one operator A appears in equation (A.1).[50] The solution of this equation corresponding to the particular initial condition (involving only two eigenmanifolds of A with eigenvalues α, β)

$$|\Psi(0)\rangle = P_\alpha |\Psi(0)\rangle + P_\beta |\Psi(0)\rangle, \qquad (A.6)$$

when the Hamiltonian is disregarded, is:[51]

$$|\Psi_B(t)\rangle = e^{\alpha B(t) - \alpha^2 \gamma t} P_\alpha |\Psi(0)\rangle + e^{\beta B(t) - \beta^2 \gamma t} P_\beta |\Psi(0)\rangle. \qquad (A.7)$$

Here $B(t)$ is the Brownian process

$$B(t) = \int_0^t d\tau \, w(t). \qquad (A.8)$$

Taking into account equation (A.7) and the cooking prescription, one gets the cooked probability density for the value $B(t)$ of the Brownian process at time t:

$$P_{cook}[B(t)] = \left\| P_\alpha | \Psi(0) \rangle \right\|^2 \frac{1}{\sqrt{2\pi\gamma t}} e^{-\frac{1}{2\gamma t}(B(t) - 2\alpha\gamma t)^2}$$

$$+ \left\| P_\beta | \Psi(0) \rangle \right\|^2 \frac{1}{\sqrt{2\pi\gamma t}} e^{-\frac{1}{2\gamma t}(B(t) - 2\beta\gamma t)^2}. \qquad (A.9)$$

From (A.9) it is evident that for $t \to \infty$, the process $B(t)$ can assume only values belonging to an interval of width $\sqrt{\gamma t}$ around either the value $2\alpha\gamma t$ or the value $2\beta\gamma t$.[52] The corresponding probabilities are $\left\| P_\alpha | \Psi(0) \rangle \right\|^2$ and $\left\| P_\beta | \Psi(0) \rangle \right\|^2$, respectively. The occurrence of a value "near" to $2\alpha\gamma t$ for the random variable $B(t)$ leads, according to equation (A.7), to a state vector that, for $t \to \infty$, lies in the eigenmanifold corresponding to the eigenvalue α of A. In fact, one gets:

$$\frac{\left\| P_\beta | \Psi_B(t) \rangle \right\|^2}{\left\| P_\alpha | \Psi_B(t) \rangle \right\|^2} \cong e^{-2\gamma t(\alpha-\beta)^2} \frac{\left\| P_\beta | \Psi_B(0) \rangle \right\|^2}{\left\| P_\alpha | \Psi_B(0) \rangle \right\|^2} \xrightarrow{t \to \infty} 0. \qquad (A.10)$$

Analogously, when the random variable $B(t)$ takes a value "near" to $2\beta\gamma t$, for $t \to \infty$, the state vector is driven into the eigenmanifold corresponding to the eigenvalue β of A.

It is then clear that the model establishes a one-to-one correspondence between the "outcome" (the final "preferred" eigenmanifold into which an individual state vector is driven) and the specific value (among the only ones having an appreciable probability) taken by $B(t)$ for $t \to \infty$, a correspondence irrespective of what $|\Psi(0)\rangle$ is.[53] In the general case of several operators A_i, a similar conclusion holds for the "outcomes" α_i of A_i and the corresponding Brownian processes $B_i(t)$.

This concludes the exposition of the general structure of the CSL model. Obviously, to give a physical content to the theory one must choose the so-called preferred basis, that is, the eigenmanifolds on which reduction takes place or, equivalently, the set of commuting operators A_i. In accordance with the remarks of section 2 about the privileged role we assign to the mass density, we identify the discrete index i and the operators A_i of the preceding formulas with the continuous and discrete indices $(\mathbf{r}, k)$ and the operators:

$$M^{(k)}(\mathbf{r}) = m^{(k)} N^{(k)}(\mathbf{r}) \qquad (A.11)$$

with

$$N^{(k)}(\mathbf{r}) = \left(\frac{\alpha}{2\pi} \right)^{\frac{3}{2}} \sum_s \int d\mathbf{q} \, e^{-\frac{\alpha}{2}(\mathbf{q}-\mathbf{r})^2} a_k^+(\mathbf{q}, s) a_k(\mathbf{q}, s). \qquad (A.12)$$

Here $a_k^+(\mathbf{q}, s)$ and $a_k(\mathbf{q}, s)$ are the creation and annihilation operators of a particle of type k and mass $m^{(k)}$ (e.g., $k =$ electron, proton,...) at point $\mathbf{q}$ with spin component s, satisfying the canonical commutation or anticommutation relations. Correspondingly one has a continuous family of stochastic Gaussian processes satisfying:

$$\langle\langle w_k(\mathbf{r}, t)\rangle\rangle = 0,$$
$$\langle\langle w_k(\mathbf{r}, t)w_j(\mathbf{r}', t')\rangle\rangle = [\gamma/m_0^2]\delta_{kj}\delta(\mathbf{r} - \mathbf{r}')\delta(t - t'). \quad \text{(A.13)}$$

The parameter α is assumed to take the same value ($10^{10}\,\text{cm}^{-2}$) as in the case of QMSL, while γ is related to the frequency $\lambda = 10^{-16}\,\text{sec}^{-1}$ of that model according to $\gamma = \lambda(4\pi/\alpha)^{3/2}$ and m_0 is the nucleon mass.

Due to the choice of the parameters the considered dynamics has the following nice features:

- In the case of microscopic systems the non-Hamiltonian terms have negligible effects.
- On the contrary, in the macroscopic case the reduction mechanism is extremely effective in suppressing linear superpositions of states in which a macroscopic number of particles is displaced by more than the characteristic localization length.

The amplification process in going from the microscopic to the macroscopic case and the preferred role assigned to the mass density make it clear how such models overcome the difficulties of quantum measurement theory. In fact, in measurement processes one usually assumes that different eigenstates of the measured microobservable trigger (through the system-apparatus interaction) different displacements of a macroscopic pointer from its "ready" position. The unique dynamical principle leads, in extremely short times, to the dynamical suppression, with the appropriate probability, of one of the terms in the superposition, that is, to the emergence of an outcome.

A.2 *"Outcomes" in Dynamical Reduction Models*

For the following analysis and with reference to the fundamental issue of the objectification of properties, that is, of the dynamical emergence of outcomes, it is important to deepen the discussion of the formal and physical aspects of the theory and to mention some peculiar situations that may occur. To this purpose we confine our considerations to the case in which only two *macroscopically different* outcomes α and β, which we identify with the eigenvalues of an operator A, can occur. We consider the CSL model with A taking the place of the operators A_i of equation (A.1), and we assume that the initial state vector has the form (A.6). As repeatedly remarked, when one disregards the Hamiltonian evolution, the CSL dynamics, for $t \to \infty$, drives the state vector either within the eigenmanifold M_α or within M_β. However, it is important to stress that for any finite time t, no one among the states evolved from the considered initial state vector can be exactly an eigenstate of A. As discussed in (A.1), after a characteristic reduction time Δt (defined through $e^{-2\gamma\Delta t(\alpha-\beta)^2} \ll 1$), for all values of the Brownian process $B(\Delta t)$ that have an appreciable probability of occurrence (i.e., those for which $B(\Delta t) \approx 2\alpha\gamma\,\Delta t$ or $B(\Delta t) \approx 2\beta\gamma\Delta t$), the normalized state vector describing an individual system will have a negligible component on one of the two eigenmanifolds. Since one wants outcomes to emerge in the characteristic reduction time Δt, one is compelled to attribute to the system e.g. the "definite outcome α" also when $\|P_\alpha|\Psi\rangle\|^2/\||\Psi\rangle\|^2$ is extremely close but not exactly equal to one.[54] This corresponds to the same attitude taken in this chapter when we have stipulated that an object is localized in a certain region even though its wave function has a (vanishingly) small tail in another region.

We come now to the discussion of some peculiar aspects of the theory.[55] The first one derives from the fact that, in principle, also in the macroscopic case it could happen that even for a time larger than Δt no outcome has emerged. In fact, if one considers for the Brownian process $B(\Delta t)$ the value $(\alpha + \beta)\gamma\,\Delta t$ whose probability density, although very small, is not zero, one can easily show that equation (A.7) leads to a state vector that coincides, apart from a normalization factor, with the initial one. In other words, no reduction has taken place and no outcome has been obtained. Since, as already remarked, the probability of such a peculiar event is extremely small, its occurrence cannot be considered as a drawback of the theory.

Another peculiar situation can occur, namely the "reversal" of a macroscopic outcome. To see this, suppose one has a normalized state vector $|\Psi\rangle$ which "almost" belongs to the eigenmanifold M_α, that is, for which $\|P_\alpha|\Psi\rangle\|^2$ is extremely close but not identical to 1. One can then state that the definite outcome α has occurred. According to the theory, however, there is a very small probability $\|P_\beta|\Psi\rangle\|^2$ that in the far future the Brownian process $B(t)$ takes a value such that in the state vector (A.7) the norm of the second term becomes overwhelmingly large with respect to that of the first. Correspondingly, one would be led to conclude that the outcome β has emerged. This shows that, though with extremely low probabilities, even definite macroscopic outcomes can (spontaneously) change.[56] The analysis we have just performed and the conclusions we have reached hold for QMSL too.

Notes

1. A. Shimony, "Search for a Worldview Which Can Accommodate Our Knowledge of Microphysics," in J. T. Cushing and E. McMullin, eds., *Philosophical Consequences of Quantum Theory: Reflections on Bell's Theorem* (Notre Dame, Ind.: University of Notre Dame Press, 1989).

2. J. S. Bell, "Quantum Mechanics for Cosmologists," in C. Isham, R. Penrose, and D. Sciama, eds., *Quantum Gravity 2* (Oxford: Clarendon Press, 1981).

3. E. Schrödinger, "Quantisiering als Eigenwertproblem (Vierte Mitteilung)," *Annalen der Physik* 81 (1926): 109–139, English translation in *Collected Papers on Wave Mechanics*, 2d ed. (New York: Chelsea, 1978), 102.

4. E. Schrödinger, "Quantisiering als Eigenwertproblem (Zweite Mitteilung)," *Annalen der Physik* 79 (1926): 489–527, English translation in *Collected Papers on Wave Mechanics*, 13.

5. J. S. Bell, "Six Possible Worlds of Quantum Mechanics," in S. Allén, ed., *Proceedings of the Nobel Symposium 65: Possible Worlds in Arts and Sciences* (Stockholm: Nobel Foundation, 1986).

6. In particular, as is well known, this situation occurs in measurement-like processes in which, after the system-apparatus interaction is over, one has, in general, a linear superposition of macroscopically distinguishable apparatus states.

7. Obviously, a satisfactory description of the macroworld also requires taking into account how positions change with time.

8. G. C. Ghirardi, A. Rimini, and T. Weber, "Unified Dynamics for Microscopic and Macroscopic Systems," *Physical Review D* 34 (1986): 470–491; P. Pearle, "Combining Stochastic Dynamical State-Vector Reduction with Spontaneous Localization," *Physical Review A* 39 (1989): 227–239; G. C. Ghirardi, P. Pearle, and A. Rimini, "Markov Processes in Hilbert Space and Continuous Spontaneous Localization of Systems of Identical Particles," *Physical Review A* 42 (1990): 78–89; G. C. Ghirardi and A. Rimini, "Old and New Ideas in the Theory of Quantum Measurement," in A. Miller, ed., *Sixty-Two Years of Uncertainty* (New York: Plenum, 1990); G. C. Ghirardi, R. Grassi, and F. Benatti, "Describing the Macroscopic World: Closing the Circle within the Dynamical Reduction Program," *Foundations of Physics* 25 (1995): 5–38.

9. Ghirardi, Rimini, and Weber, "Unified Dynamics."

10. Pearle, "Combining Stochastic Dynamical State-Vector Reduction"; Ghirardi, Pearle, and Rimini, "Markov Processes"; Ghirardi and Rimini, "Old and New Ideas"; Ghirardi, Grassi, and Benatti, "Describing the Macroscopic Word."

11. A. Shimony, "Desiderata for a Modified Quantum Dynamics," in A. Fine, M. Forbes, and L. Wessels, eds., *PSA 1990*, vol. 2 (East Lansing, Mich.: Philosophy of Science Association, 1991); D. Z. Albert and B. Loewer, "Wanted Dead or Alive: Two Attempts to Solve Schrödinger's Paradox," in A. Fine, M. Forbes, and L. Wessels, eds., *PSA 1990*, vol. 1 (East Lansing, Mich.: Philosophy of Science Association, 1990).

12. Cf. n. 10.

13. Shimony,"Desiderata for a Modified Quantum Dynamics."

14. Albert and Loewer, "Wanted Dead or Alive."
15. Shimony, "Desiderata for a Modified Quantum Dynamics," 53.
16. Ghirardi, Rimini, and Weber, "Unified dynamics."
17. Cf. n. 10.
18. Ibid.
19. F. Karolyhazy, "Gravitation and Quantum Mechanics of Macroscopic Objects," *Il Nuovo Cimento A* 42 (1966): 390–402, and references therein; A. B. Komar, "Qualitative Features of Quantized Gravitation," *International Journal of Theoretical Physics* 2 (1969): 157–160; R. Penrose, "Gravity and State-Vector Reduction," in R. Penrose and C. J. Isham, eds., *Quantum Concepts in Space and Time* (Oxford: Clarendon Press, 1986); J. Ellis, S. Mohanty, and D. V. Nanopoulos, "Quantum Gravity and the Collapse of the Wavefunction," *Physics Letters B* 221 (1989): 113–119; L. Diosi, "Models for Universal Reduction of Macroscopic Quantum Fluctuations," *Physical Review A* 40 (1989): 1165–1174; A. Frenkel, "Spontaneous Localizations of the Wave Function and Classical Behavior," *Foundations of Physics* 20 (1990): 159–188.
20. G. C. Ghirardi, R. Grassi, and A. Rimini, "Continuous-Spontaneous-Reduction Model Involving Gravity," *Physical Review A* 42 (1990): 1057–1064.
21. P. Pearle and E. Squires, "Bound State Excitation, Nucleon Decay Experiments, and Models of Wave Function Collapse," *Physical Review Letters* 73 (1994): 1–5.
22. J. S. Bell, "Are There Quantum Jumps?," in C.E.W. Kilmister, ed., *Schrödinger-Centenary Celebration of a Polymath* (Cambridge: Cambridge University Press, 1987).
23. In what follows we will often drop, for simplicity and since we will be mainly interested in investigating the mathematical features of the map (3.1), the time parameter from the state vectors.
24. Ghirardi, Rimini, and Weber, "Unified Dynamics"; Pearle, "Combining Stochastic Dynamical State-Vector Reduction"; Ghirardi, Pearle, and Rimini, "Markov Processes"; Ghirardi and Rimini, "Old and New Ideas"; G. C. Ghirardi, R. Grassi, and P. Pearle, "Relativistic Dynamical Reduction Models: General Framework and Examples," *Foundations of Physics* 20 (1990): 1271–1316; G. C. Ghirardi, R. Grassi, and P. Pearle, "Relativistic Dynamical Reduction Models," in P. Lahti and P. Mittelstaedt, eds., *Symposium on the Foundations of Modern Physics, 1990* (Singapore: World Scientific, 1990); G. C. Ghirardi and P. Pearle, "Dynamical Reduction Theories: Changing Quantum Theory so the Statevector Represents Reality," in A. Fine, M. Forbes, and L. Wessels, eds., *PSA 1990,* vol. 2 (East Lansing, Mich.: Philosophy of Science Association, 1991); F. Benatti, G. C. Ghirardi, A. Rimini, and T. Weber, "Quantum Mechanics with Spontaneous Localization and the Quantum Theory of Measurement," *Il Nuovo Cimento* 100 B (1987): 27–41.
25. Bell, "Six Possible Worlds."
26. A. J. Leggett, "Macroscopic Quantum Systems and the Quantum Theory of Measurement," *Progress of Theoretical Physics Supplements* 69 (1980): 80–100.
27. In making the computations we have identified the operators M_i with the sum of the projectors (multiplied by the nucleon mass m_0) of the various particles on the i-th cell.

28. At any rate possible changes in such a state would be symmetrical with respect to the middle plane so that the subsequent considerations would still hold true.

29. It could be useful to remark that if one would analyze the same experiment in terms of the linear quantum dynamics, the test particle would end up in the linear superposition of an extremely large number of states. However, since such states correspond to trajectories which are very near and almost undeflected, the evaluation of the mass density associated with the final state vector would show that in the "middle" region there would practically be the total mass of the test particle. Therefore, this represents a case in which even without any reduction process the mass density referring to the test particle would correspond to a precise outcome of the measurement.

30. We disregard the cells that are not contained in regions A and B since they are irrelevant for the following discussion.

31. Shimony,"Desiderata for a Modified Quantum Dynamics."

32. Albert and Loewer, "Wanted Dead or Alive."

33. Note that this holds also for objects like a galaxy or a neutron star.

34. To be rigorous one should consider the map $\mathcal{M}$ from the unit sphere of $H^{(S)}$ into the space L^2 of the square integrable functions of $\mathbf{r}$. However we can deal, without any loss of generality, with the discretized version of the model.

35. J. S. Bell, "Beables for Quantum Field Theory," CERN-TH. 4035/84 (1984).

36. Bell, "Quantum Mechanics for Cosmologists."

37. Obviously, within classical mechanics any function of these variables can be considered as a beable, but since all information about the system can be derived from the positions and the momenta, consideration of such variables is sufficient.

38. From this point of view, one could state that also the classical world would be most appropriately described in terms of positions at fixed times.

39. Bell, "Quantum Mechanics for Cosmologists."

40. Obviously, our perceptions are much richer than those (corresponding to position and shape) we have listed here. In the next subsection we will make clear how also more complex perceptions (such as color etc.) can be naturally included in the picture we are presenting.

41. Shimony, "Desiderata for a Modified Quantum Dynamics"; Albert and Loewer, "Wanted Dead or Alive"; D. Z. Albert and L. Vaidman, "On a Proposed Postulate of State-Reduction," *Physics Letters A* 139 (1989): 1–4; D. Z. Albert, "On the Collapse of the Wave Function," in A. Miller, ed., *Sixty-Two Years of Uncertainty* (New York: Plenum, 1990); D. Z. Albert, *Quantum Mechanics and Experience* (Cambridge, Mass.: Harvard University Press, 1992).

42. Albert and Vaidman, "On a Proposed Postulate of State-Reduction"; Albert, "On the Collapse of the Wave Function"; Albert, *Quantum Mechanics and Experience*.

43. F. Aicardi, A. Borsellino, G. C. Ghirardi, and R. Grassi, "Dynamical Models for State-Vector Reduction: Do They Ensure That Measurements Have Outcomes?," *Foundations of Physics Letters* 4 (1991): 109–116.

44. Ibid.

45. We consider appropriate a specification. The above analysis could be taken as indicating that we adopt a very naive and oversimplified attitude about the deep problem of the brain-mind correspondence. We do not claim and we do not pretend that CSL yields a physicalistic explanation of consciousness. We simply point out that, for what we know about the purely physical aspects of the perceptual process, the conditions guaranteeing that superpositions of different perceptions cannot occur are satisfied.
46. Bell, "Quantum Mechanics for Cosmologists."
47. J. S. Bell, "Against 'Measurement'," in A. Miller, ed., *Sixty-Two Years of Uncertainty* (New York: Plenum, 1990).
48. Of course, the interpretation we have adopted is not à la Schrödinger; while, for example, for a particle, the wave function, and consequently the mass density, is extended in space, the detection of the particle remains spotty. This fact, as we have already discussed, simply implies that the mass density function associated to the particle is not "objective."
49. Cf. n. 10.
50. Ibid.
51. In equation (A.7) and following we have changed the notation for the state vector from the one labeled by the white noise symbol w as in equation (A.1) to the one labeled by the Brownian motion symbol B, to stress the fact that, under our assumptions, the state at time t does not depend on the specific sample function $w(\tau)$ in the interval $(0, t)$ but only on its integral equation (A.8).
52. Note that, even though the spread $\sqrt{\gamma t}$ tends to ∞ for $t \to \infty$, its ratio to the distance $2(\alpha - \beta)\gamma t$ between the two considered peaks of the distribution tends to zero.
53. Obviously $|\Psi(0)\rangle$ enters in a crucial way in determining the probability of occurrence of the Brownian processes $B(t)$.
54. Ghirardi, Grassi, and Pearle, "Relativistic Dynamical Reduction Models"; Ghirardi and Pearle, "Dynamical Reduction Theories." It is useful to remark that, also within standard quantum mechanics with the reduction postulate, since outcomes are usually related to positions of macroscopic pointers and no wave function can have, in general, compact support in configuration space, one is unavoidably led to adopt an analogous criterion for the attribution of "outcomes."
55. G. C. Ghirardi, R. Grassi, J. Butterfield, and G. N. Fleming, "Parameter Dependence and Outcome Dependence in Dynamical Models of State Vector Reduction," *Foundations of Physics* 23 (1993): 341–364; J. Butterfield, G. N. Fleming, G. C. Ghirardi, and R. Grassi, "Parameter Dependence in Dynamical Models for Statevector Reduction," *International Journal of Theoretical Physics* 32 (1993): 2287–2304.
56. To avoid misunderstandings we consider it appropriate to stress that, when one is dealing with an entangled state, this "reversal," if it takes place, preserves the correlations implied by the state vector.

16

Microphysical Objects and Experimental Evidence

Giulio Peruzzi

What is the physicist's conception of a "particle"? What is the experimental evidence supporting the particle picture resulting from the quantum field theory approach to the physical world? The main purpose of this chapter is a tentative answer to such questions.

We shall not explicitly take into account the philosophical questions that are analyzed in the literature on foundations and philosophy of physics. Our central concern here is to understand the physicist's point of view, through an excursion on the (not so well defined) borderline between the theory and the experimental evidence.

1. What Is a Particle? The Framework of Relativistic Quantum Field Theory

The present theoretical framework of particle physics is relativistic quantum field theory. Our aim in this first section is to provide a brief survey of some of the basic aspects of this subject. The emphasis is on the so-called standard model of particle physics, at present the most satisfactory account for the particle properties.

From an experimental point of view, particles are detectable packets of energy and momentum. This is true for particles as elementary constituents of matter but also for the so-called quasi particles (phonons, plasmons, etc.), which are collective excitations of a many-particle system. Our treatment in the following is devoted to particles, but most of the ideas and results in this subject can be applied also to quasi particles.

When atomic nuclei and forces within them are probed by high energy collisions, many new particles appear that are unstable and decay into more familiar particles (electrons, photons, protons, etc.). Because of the high number of these new particles, it is necessary to have some basic principle to order and classify this "particle zoology."

The contemporary theory of matter and forces, known as *relativistic quantum field theory*, is the result of combining relativity theory and quantum mechanics. The fundamental assumptions of relativistic quantum field theory are Poincaré invariance, microcausality, and the structure of the energy momentum spectrum (i.e., the principle of positivity of the energy).[1] As required by Poincaré invariance and the general principle of causality (no effect can precede its cause), measurements separated by a spacelike interval cannot influence one another: local observables exist and local observables relative to spacelike separated regions commute. Microcausality is this quantum-relativistic version of the causality principle.

The theoretical classification of particles is, first of all, based on the assumption that physical laws are invariant under Poincaré transformations: a particle is defined as a state of a quantum field that transforms under elements of the Poincaré group according to a definite irreducible representation. As showed by Wigner,[2] this result implies that a particle has a definite *mass* and *spin* (mass and spin are the labels that determine the irreducible representation up to unitary equivalence) and that each particle has an associated antiparticle of the same mass and spin. The assumptions of microcausality and of the principle of positivity of energy in (local) quantum field theory imply a connection between spin and statistics—formally stated in the "spin-statistics theorem": particles with integer spin are *bosons* and particles with half-integer spin are *fermions*.

A further step in the classification is based on the inclusion of the interactions among particles.[3] Interactions are required to be invariant under space-time symmetry (Poincaré group) and, in addition, under *internal symmetry groups* related with space-time-independent transformations of particles states. The invariance of the interactions under internal symmetry groups gives rise to further quantum numbers labeling particle states (besides mass and spin), such as electric charge and isospin.

At our present level of knowledge the fundamental constituents of matter—that is, the most elementary interacting units in the theory—are *leptons* and *quarks*, which are all fermions (spin equal to one half).[4]

The number of these "fundamental" particles has grown up to the present value of twenty-four particles. For every particle there is an associated antiparticle of the same mass and spin but opposite charge and color.[5]

Strong and electroweak interactions among "fundamental" particles can be described in terms of a unique *gauge principle*, which summarizes the idea that the origin of all basic interactions is found in invariance properties (local symmetries). Poincaré invariance and gauge symmetries specify the nature and structure of the weak, electromagnetic, and strong interactions.

All basic physical processes correspond to the exchange of a quantum of the gauge field[6]—or to the exchange of a particle, since particles and quanta of field are, in some sense, equivalent concepts in relativistic quantum field theory[7]—and the number of different gauge fields is specified in each case by the gauge symmetry group. For the electroweak interaction the gauge group is $SU(2) \times U(1)$. In virtue of that gauge group there are four gauge fields and the basic physical processes correspond to the exchange of a photon (electromagnetic coupling), to the exchange of a charged $W^{\pm}$ (standard weak coupling) and to the exchange of a neutral Z^0 (neutral current weak coupling). The gauge group for the strong interaction is $SU(3)$ that implies the "existence" of eight gluons each carrying color charge.[8]

The gauge theory of strong and electroweak interactions—usually referred to as $SU(3) \times SU(2) \times U(1)$—is the so-called standard model (SM), today's dominant theoretical framework of particle physics. But the theory in its present stage poses a number of outstanding problems. First of all, SM is compatible with all known experimental data, in particular with the scattering experimental data, but this compatibility must be carefully analyzed. The electroweak ($SU(2) \times U(1)$) sector of SM has been stringently tested by quantum electrodynamics (QED) and the W and Z properties. The situation is different in probing the strong interaction ($SU(3)$) sector of SM. Starting from the strong interaction theory (QCD), physicists are far from being able to reproduce all the experimental results on hadron properties and their interactions. This interaction is characterized by an effective running coupling constant ($\alpha_s(Q^2)$) that goes to zero (the interaction strength goes to zero) when the energy scale goes to infinity, that is, at short distances (this property of the coupling constant is called *asymptotic freedom*). We can therefore use QCD in a perturbative way at large momentum transfers (or at small distances) having good agreement with experimental results, for example, in deep inelastic lepton-hadron scattering. But at larger distances, where quarks get confined into hadrons, QCD is not computable in a satisfactory way, leaving us with only qualitative agreement. As a matter of fact, at large distances (small momentum transfers) we have a model (actually a set of models, mainly based on lattice theories) and we perform computer (model-dependent) simulations. In some cases these simulations need a computer power that is still not available for more precise estimations. Furthermore, the use of a computer is actually an extra experiment: we by no means have the possibility to control all the flow of the computer procedures, which actually become a new kind of experiment. Here the connections between theory and experiment require a new careful analysis.

But there are also other questions that cannot be answered in a satisfactory way within the standard model. For example, Why are there so many undetermined free parameters (twenty-one or more of them)? Why is the gauge structure a product of three gauge groups? Why are there three generations of leptons and quarks?[9] Where does gravity come in?

In spite of the presence of unsolved fundamental problems, at the present (high) level of accuracy the standard model describes very well the results of laboratory experiments—while, on the contrary, many of the most recent theoretical activities are still highly speculative.

2. Testing Microphysical Objects: Scattering Techniques

Rutherford's experiment of the scattering of alpha particles by matter is the prototype of scattering techniques in testing microphysical objects. By impinging a beam of alpha particles on a thin layer of matter, from the angular distribution of the scattered particles Rutherford drew the picture of atoms having small massive positive charged nuclei at their centers.

The same rather simple idea is fundamental still today in high-energy physics, even if the technological (and theoretical) complexity has grown up more and more and the interrelations between theoretical models and data are less direct. The scheme is the following. Let us take a *beam* of particles and fire it at a *target* (that at present is usually another beam of particles moving against the other). In the interactions within the target some of the beam particles are scattered and possibly additional particles are created. The particles emerging from the collision region are then registered in specific *detectors*. This is more or less what happens in our ordinary way of seeing (macroscopic) objects via the scattering of light photons which are detected by our eyes. The main difference rests on the very poor sensitivity of eyes: we cannot see in regions of the electromagnetic spectrum beyond the visible and we cannot use interactions besides the electromagnetic one.

The main reasons for using high energy to investigate microphysical objects and their interactions are the increasing of space resolution with energy[10] and the possibility of new particle production.[11] High-energy physics is based on two fundamental characteristics of microphysical objects: their behavior like waves in specified experimental situations and the fact that they are ruled by the laws of relativity theory. At the same time, the experiments in high-energy physics are a way of testing these specific quantum-relativistic characters of microobjects.

The experimental results in high-energy physics provide us with precision tests of SM and also with the possibility to detect new physical phenomena, some deviations heralding new physics beyond SM. In order to understand better the connection between experimental technique and theoretical analysis we have therefore chosen to include some details of accelerator machines, which are the main experimental tools in high-energy physics. Particle accelerators can be divided into *fixed-target* and *colliding-beam* (or colliders) machines. In fixed-target accelerators, particles are accelerated to the highest operating energy and then the beam is extracted from the machine and directed on a stationary target. The main disadvantage of fixed-target machines concerns the fact that one has to work at higher center-of-mass energies (which is a measure of the energy available to create new particles and to improve space resolution). If E_L is the energy of the beam particle in the laboratory, then it can be proved that, at high energies, the E_{CM} (the total center-of-mass energy) increases only as $E_L^{1/2}$ and most of the beam energy is unavailable for particle production. In a colliding-beam accelerator, two beams of particles traveling in almost opposite directions are made to collide at a small or zero crossing angle. If the colliding particles have the same mass (that is the case in most of the current available colliders) and if they collide at zero crossing angle with the same energy E_L, the E_{CM} is equal to $2E_L$. This increases linearly with the energy E_L of the accelerated particles improving significantly on the fixed-target result.

Without entering into more technical aspects,[12] let us just recall that beam, target, and detectors are the main components in the experimental apparatus to test the fundamental microphysical objects. Beside energy, the other fundamental parameter is luminosity ($\mathcal{L}$), that is, the product of the incident beam flux (particles per second) times the mean target density (particles per unit area); low luminosity implies low event rate. In a fixed-target experiment (lower energies available in the center-of-mass frame) we can use a solid or liquid in which the density of nucleons is very high (high $\mathcal{L}$) in comparison with a target made by a colliding beam (higher energies available in the center-of-mass frame) in which particles are more like a low-density gas (low $\mathcal{L}$). The progress of collider technology has partly solved these practical problems by pinching the beams down to very small size at the crossing points and also by bringing the beam around repeatedly in circular colliders to get multiple crossings. However many technical difficulties (often related to some apparently fundamental limitations) hinder the manipulation of bunches of particles. Nevertheless colliders are now established and privileged tools of high-energy physics.

We can register the production of predetermined species of particles emerging from the target at given angles and with given momenta only if we are able to select and separate these particles from all other possible (particle-like) events. In order to obtain this result, we have preliminarily to set out: (1) the beam of specified energy and intensity containing a specified kind of particles; (2) the target particles; (3) the detectors, their configuration in space and electronic circuits (data acquisition system). This latter choice is crucial in the selection and separation of predetermined events and it is essentially based on a *trigger* that permits to measure only the events that satisfy specific preconditions. Further selection cuts are then made to refine the sample of events. In the case of rare events, the data acquisition system must be very sophisticated and very severe in the event selection. In the early search of Z^0 at CERN, for example, less than one event was retained out of every 10^9 interactions, and in the first communication about the discovery of the W-particle, as reported by Pais, "Rubbia explained how six out of one billion recorded events had been singled out as bearing the indubitable 'signature' of the W-particle. This production rate, six in a billion, agreed well with theoretical expectations, as did the crude determination of the W-mass."[13]

This is actually a "crucial" point. Here we have a paradigmatic example of the so called "theory ladenness" of observation. The complexity of the observational data is now so high that it is impossible to understand anything if we have no theory guiding us in the selection and in the ordering of the phenomena (even if this order should not be considered at all as the final order of events). That is, *until we know what we are looking for, we have little chance of finding it*. This situation, always present in the physical analysis of the experimental data, acquires indeed new nuances nowadays, leading to a view in which fundamental physics seems to have evolved in an irreversible way from an inductive to a deductive procedure. The only reliable way for "new physics," that is, for unexpected new effects, is to measure very accurately particular pieces of precisely predictable standard physics looking for any discrepancy. Every significant discrepancy could guide us to something new. But SM has a large—too large—number of free parameters, which can be adjusted to fit new effects. The central problem is not so much theory ladenness of observations but rather the question of how precisely fundamental physics can predict observational theory-laden results. We shall consider this problem in the last part of the chapter.

3. What Is Measured in a Typical Scattering Experiment?

A typical scattering experiment measures *cross sections*, a very central notion in the conceptualization of microphysical objects. Cross sections represent the effective area of interaction between beam and target particles. Moreover, as we shall see, cross sections are also understood as measures of probability of different kind of events. We shall briefly review the role of cross sections in the description of scattering processes, starting with the classical theory of scattering.

If the particles are viewed as hard elastic spheres—as in the early conception of an atom—each particle presents to the oncoming particle a target area equal to $\sigma = \pi d^2$ (where d is the distance between the centers of two colliding particles). This target area is the cross section for this particular collision problem. Suppose that we are dealing with a target containing many particles. The total target area is then just the sum of the cross sections of the individual particles.[14] In other words, a sheet of material of area A and thickness dx (containing $\rho A\,dx$ particles, where ρ is the density of particles) presents an effective target area equal to $\rho A\sigma\,dx$. The probability that an oncoming particle makes a collision is equal to the fraction of the effective (total) target area with respect to the total area A—that is, $\rho A\sigma\,dx/A = \rho\sigma\,dx$—and we can write:

$$dP = \rho\sigma\,dx. \qquad (1)$$

This result can be obtained also computing the probability that a spherical particle, in the short time dt, collides with another in the target. Let us consider a cylinder with a base of area πd^2 and height equal to the distance $dx = v\,dt$ traveled by the particle during this time. The probability of collision is then equal to the probability that the center of another particle lies in this cylindrical region, that is

$$dP = \rho\pi d^2 v\,dt. \qquad (2)$$

Writing $\pi d^2 = \sigma$ we have the same result as equation (1).[15]

Equation (1) is the basic relation connecting the scattering cross section (a geometrical concept) with the probability of collision. On the ground of equation (1) we can think the cross section either as an effective area of interaction between beam and target particles, or as a measure of probability of different kinds of colliding events. This ontological ambiguity can be found also in the generalization of the cross-section concept from the classical to the quantum theory. This can explain, in part, why some aspects of the classical description of objects can be found also in the quantum mechanical framework. A more careful analysis of the cross-section notion in classical and quantum mechanics will permit us to understand what in summary is the influence of classical analogies in the physicist's conceptualization of microobjects.

Cross sections are usually expressed as functions of scattering angles. For example, in the case of hard, perfectly elastic, identical spheres ($d = 2a$, where a is the radius of the particles) we obtain

$$\sigma_{tot}(\theta) = 4\pi a^2 \cos^2 \frac{\theta}{2},$$

where $\sigma_{tot}(\theta)$ is the *total cross section* for scattering through an angle θ or greater (if θ is equal to zero we arrive again to the above result $\sigma_{tot} = \pi d^2 = 4\pi a^2$), that is, the effective area for producing collisions with deflections larger than θ. Another cross section is the *differential cross section*, $\sigma(\theta)$, which is such that $\sigma(\theta)\,d\theta$ is the cross section for producing deflections that lie between θ and $\theta + d\theta$.[16]

The assumption that the particles behave as hard elastic spheres is too restrictive. A more general theory of scattering considers potentials smooth with respect to the sharp potential between (idealized) hard elastic spheres.[17] For example, the Coulomb force between charged particles (corresponding to the potential $V = e^2/r$) is so soft that the idealization of hard spheres is no more a good approximation. In spite of that, (micro) physical objects are always thought of as some sort of spherical objects (pointlike or not) that interact each other within an effective area shaped by the specific potentials of the interacting forces and by the specific symmetries related to those forces. So, already in the classical theory, there are fundamental cases in which the geometrical shape of the objects is not as central as the space-time behavior of the particular force field generated by the particular object; the observational properties by which an entity is defined as a physical object are no more only related to the localization (permanent or not, sharp or fuzzy) in the space-time but they are enriched in connection with observational properties of the force field generated by the object. Hence the progressive identification of the objects with specific force fields, which becomes even more fundamental in the quantum theory, is already present in the classical theory.

To calculate cross sections for an arbitrary spherically symmetrical law of force, we can note, first of all, that the particle orbit (or trajectory) in this case lies always in a plane. We can define, in analogy with the case of hard spheres, a collision parameter b as the distance between the original direction of motion of one (beam) particle and the center of force (target particle), either thought as pointlike particles.[18] The net deflection in the orbit of the incoming particle can be denoted by θ (the angle between the original and final direction of motion).

Now we have two different possibilities to use this scheme. First of all, if we know the law of force we can solve (by exact or approximate methods) the equations of motion finding the function that connects the deflection θ to the collision parameter b. The (differential) cross section will be the area of the ring $2\pi b\, db$ within which the particle must be if it will suffer a scattering between θ and $\theta + d\theta$. The total cross section is just the integral of the differential cross section and it is equal to $\sigma_{tot} = \pi b^2(\theta)$ for scattering into angles greater or equal to θ (if we put $\theta = 0$, we have the total cross section through all possible values up to $\theta = \pi$). This method was used, for example, by Rutherford to interpret the results of α-particles scattering experiments from which he deduced the current model of an atom. Conversely, if we do not know the law of force (and in general that is the case), we can use the information obtained by experimental measures on cross sections to investigate the law of force[19] and the size of the objects.

The scattering processes have to be analyzed in a rather different way in quantum theory. In the classical theory we have always in principle the possibility of describing the motion of the particles with *complete* accuracy by means of classical orbits; the specification of the orbit permits to calculate the detailed transfers of momentum to the particle at every (space-time) point. This possibility breaks down in quantum theory where we must use wave packets whose *average* coordinates give the classical orbits: the scattering processes have to be described by wave functions that are solutions of Schrödinger's equations, and particles cannot have simultaneously well defined position and momentum. The cross section is therefore better interpreted as a measure of the relative probabilities of different events rather than as an effective area of interaction, even if this latter interpretation maintains a role in the representation of physical microobjects.

If we use a space-time description of the scattering processes (via the wave picture, where the intensity of the scattered wave yields the probability that the particles have been scattered through a given angle), we cannot have a detailed description of how the particle obtains its momentum. If we choose, on the contrary, to specify the momentum of the particle, we have to use the momentum representation of the wave function in the description of scattering processes (the probability of scattering through a given angle is obtained by finding the probability of the corresponding momentum transfer). Therefore the deflection is something that is caused by the force fields implied in the scattering processes, but we cannot know exactly where the momentum was transferred within the region of a wave packet: all parts of the potentials covered by a wave packet are simultaneously involved in the scattering processes. Both the methods give the same result for the probability of scattering through a given angle in a given scattering process, what changes is the description of the intermediate mechanism.

Furthermore, if we are in the context of relativistic quantum theory (i.e., quantum field theory), we have the violation of the law of conservation of the number of particles: particles can be created or annihilated.[20] In such a context the geometrical interpretation of cross sections is apparently even more inadequate.

From the experimental point of view, however, we have always to analyze the results of measurement on scattering processes, even if these results are subject to further interpretative constraints in comparison with the classical case. It turns out that, without the possibility to associate classical trajectories linking the incoming with the outcoming macroscopic tracks of the scattering, we can not describe what happened point by point in the space-time during the scattering processes but only outside a region of space-time with an irreducible finite size, a sort of black box.

The scattering results in the quantum theory show, in spite of these conceptual difficulties, some analogies with those in the classical theory. First of all, using specific detectors (very large *multicomponent detectors*[21] are the devices commonly used in modern particle physics) we have the detection of the elementary particles outcoming from the scattering process—actually every scattering process can be analyzed only referring to an ensemble of single scattering occurring in the different parts of multicomponent detectors. Such a detection needs a resolution in space and time sufficient in order to determine which are the particles associated with a specific event, with the capability to identify each particle and measure its energy and momentum. All these data are then used in the measurements of the cross sections, which are functions of the center-of-mass energy and of the energies and momenta of the particles detected. From these measurements we obtain the informations about the properties of the force fields and we can understand if the particles are really elementary—pointlike—or, on the contrary, composite[22] and, in this latter case, what are the properties of the elementary constituents.

4. Main Results in the Scattering Picture of Quantum Objects

Given what we have said so far, we can proceed to analyze some of the main features of the image of microphysical objects resulting from today's experimental physics.

Let us start from the "size" of microphysical objects. The experimental evidence that leptons are pointlike and that, on the contrary, hadrons are not pointlike is today stringently tested. In this light it is useful to enter into some detail of the commonly accepted picture of hadrons. The most important *short-range* interactions with hadrons are due to the strong nuclear force, which, unlike the electromagnetic interaction, is as important for neutral particles as for charged particles. If we plot the experimental results of the total cross section[23] of processes of scattering of some hadrons (like pions) on protons or neutrons[24] as a function of the beam particle laboratory momentum, we can observe a clear dependence on energy. At low energies there are several peaks that are interpreted as associated with the formation and subsequent decay of specific entities called *nucleon resonances.*[25] For sufficiently high energies, however, above about a few GeV, the total cross section is slowly varying and of a much larger size than the elastic cross section (that is to say that the dominant term is the contribution due to inelastic processes). The values of the total cross section for scattering of hadrons on nucleons at high energies lie typically in the range of 10 to 50 mb (millibarn). This range corresponds to a geometrical cross section πr^2 where r is approximately equal to 1 fm (fermi), which is the approximate range of the strong interaction between hadrons. In some sense we can speak of protons (or neutrons) as nonpointlike objects with a radius equal to 1 fm.

This result about the size of nucleons is confirmed by the measurements of the values of the proton form factors in elastic electron-proton scattering.[26] From these measurements we obtain the average value for the radius of the proton equal to 0.85 ± 0.02 fm, according to the previous result obtained via hadrons scattering.

Moreover, if we perform measurement on (deep) inelastic lepton (electron or muon) proton scattering, we are led to the first clear evidence of individual pointlike constituents within the proton: the quarks.[27] Quarks and gluons can be almost directly "seen" in the so-called *jet events* observed at LEP. The simplest and dominant process consists of the formation, in electron-positron collision, of a Z^0 which decays into a quark-antiquark pair.[28] The quark pair is followed by a strong interaction process, called *fragmentation*, which converts the high energy and momentum of the quark pair into two jets of hadrons (mainly pions), which leave little doubt as to their origin. The data analysis depends sensitively on the definability of jet axes. It can be proved that the jet angular distributions reflect the angular distributions of the quark-antiquark pair. Part of the time the production of the quark-antiquark pair is accompanied by the radiation of one energetic gluon which gives a third jet.

These jets are closely related to the underlying quark and gluon interactions, and they are the closest thing to quark and gluon "tracks" that we can today imagine to see. These conclusions are strictly related to the analysis of jet events, which makes use again of the cross-section interpretative scheme.[29] In the case of two-jet events the interpretation of their angular distributions—as related to those of the quark-antiquark pair—is confirmed by comparing their angular distributions, in the reaction $e^+ + e^- \to$ photon $\to$ quark + anti-quark $\to$ two jets of hadrons, with those of muons (μ) produced in the reaction $e^+ + e^- \to$ photon $\to \mu^+ + \mu^-$. The mechanism for the two reactions is identical and we can obtain the angular distributions comparing the two differential cross sections, that is, the directly measured cross section for muons with the indirectly (via two-jet events) measured cross section for quarks. Moreover, the ratio between the total cross section for electron-positron annihilation to hadrons and the total cross section for muon production gives us new hints about the jet process and about the properties of quarks. The near constancy of this ratio follows from the dominance of two jets mechanism, while the value of the ratio directly confirms the existence of three color states, each with the same electric charge, for a given quark flavor.

Furthermore, the observed rate for production of three-jet events can be used to determine a value of the strong coupling constant and to confirm its running. In fact the probability that a quark will emit a gluon is determined by the value of the strong coupling constant, $\alpha_s(Q^2)$, in the same way that the probability that an electron will emit a photon is determined by the fine structure constant.

Let us stop here with the details and briefly summarize some general results. It is clear that microphysical objects are experimentally analyzed on the ground of the notion of cross section together with its geometrical and probabilistic interpretation. In conclusion, all the measurements of sets of invariants for the identification of microobjects are based on the measurements of cross sections in the different parts of the experimental device. All informations are deeply related to peculiar radiation fields with specific resolution power. Using this radiation in a way roughly similar to the light, we obtain the possibility of "seeing" microobjects and their properties.

The first problem comes out when we turn to the fact that such informations are actually carried through a succession of experimental devices that, at the end, give informations in the form of macroscopic space-time tracks. An analysis of this question, with reference to the background information useful to interpret the different steps, has been debated for a long time in the context of foundations and philosophy of physics. We won't face this general problem. Our goal here is rather to investigate the compatibility of the theoretical framework of microobjects with experimental data. In other words, how clearly and precisely fundamental physics can predict observational theory-laden results.

It has been shown how in the minimal version there are (at least) twenty-one free parameters in the standard model.[30] These parameters could be measured using essentially experimental results on cross sections, but in their determination the theory enters in many different ways: at the level of the approximate perturbative computations (or the model dependent computer estimations in nonperturbative QCD) by which we introduce a theoretical error affecting the cross-section evaluation and consequently the determination of the parameters; at the level of the theoretical calculations that are needed to exploit the technology of colliders and to improve the measurements of some crucial parameters of the experimental devices, like the luminosity. Every uncertainty we introduce in the error estimation due to our theoretical model can be a way to escape the fact that the theory is not the correct one. It is in fact the evidence of some discrepancy between experimental data and theory that can guide us to the need of a new theoretical framework. On the contrary, clear evidence of no discrepancy can confirm to us that the theory is essentially correct. Moreover, in the determination of such parameters (and in particular the running of the coupling constants), one has to deal with some measurements in which statistical errors and theoretical (model-dependent) errors are concurrent. Not always can we simply reduce statistically the errors without a clear full inclusion of the theory errors.

The greater is the number of free parameters of the theory, the greater is the number of "degrees of freedom" to adjust the theory to fit the experimental data. The present theory would seem to contain to much arbitrariness. But not all the twenty-one parameters enter in the same way in the theory. The masses of fundamental particles play a role in the theory that is less crucial than that of the coupling constants. These latter are only four: three for the electroweak sector (the charge e, the Fermi's constant G_E, and the mass of the Z^0) and one for the strong sector (the strong coupling α_s). Physicists would prefer to have a theory with only one general symmetry by which to obtain all the particles' masses (as the labels of some irreducible representation). Up to now only the direct measurement of masses of particles can be used in order to fix the values of the corresponding parameters.

But once the masses are fixed, the theory depends essentially only on four parameters. Thus there are good grounds for concluding that we have at our disposal a well-defined and coherent theoretical framework and quite accurate calculations, even if in no way definitive.

What physicists mean, in conclusion, by a particle? As far as experimental evidence is concerned, we could say that microphysical objects are "cross-sectional entities." We also suggest that this is the origin of the physicist's image of microobjects as something very similar to the material particles of classical physics.

Notes

This work was supported in part by Istituto Nazionale di Fisica Nucleare and by Dipartimento di Fisica Nucleare e Teorica, Pavia.

1. See N. N. Bogoliubov, A. A. Logunov, I. T. Todorov, *Introduction to Axiomatic Quantum Field Theory* (Reading, Mass.: W. A. Benjamin, 1975).
2. See E. P. Wigner, "On Unitary Representations of the Inhomogeneous Lorentz Group," *Annals of Mathematics* 40 (1939): 149–204.
3. Interactions among particles are of four types: gravitational, weak, electromagnetic, and strong interactions, in order of growing strength. At the present, in relativistic quantum field theory framework, the weak and electromagnetic interactions are low-energy manifestations of a single unified interaction called "electroweak interaction." Electroweak and strong interactions are treated together in the so-called standard model for elementary particles. What follows in the text will regard these three fundamental interactions.
4. The leptons are particles that are not affected by the strong interaction. There are three families of leptons containing each a negatively charged particle (electron, muon, and tau respectively) and a neutrino (electron neutrino, muon neutrino, and tau neutrino respectively). At present they appear to be truly elementary particles, in the sense they have no measured internal structure and are sometimes referred to as *pointlike particles*. If the electron can be thought as a pointlike particle in respect to its low-value mass, this fact could appear strange for the muon and the tau, which have masses comparable with proton mass (one-tenth and double of the mass of the proton, respectively).

   The quarks are objects different from usually called particles in many respects. First of all the magnitude of their electric charge is one-third or two-thirds of the electron's charge; then free quarks have never been observed in experimental devices. There are six quarks, set up in three families, each one grouping two quarks, one with $+2/3$ and the other with $-1/3$ electric charge (up and down, charm and strange, top and beauty are the names of the six quarks in the three families, respectively). Each quark exists under three varieties of "color" or "color charge"—whereas leptons have no "color"— hence the name of quantum chromodynamics (QCD) for the theory that describes their interactions. According to QCD, when a quark attempts to leave the interior of an hadron, it would cause new quark-antiquark pairs to be created (what happens when, for example, a meson made of a quark and an antiquark is smashed apart is very similar, mutatis mutandis, to the impossibility of isolating north and south poles by breaking in two a bar magnet). The quarks and antiquarks would then recombine so as to form a new hadron, the way in which QCD accounts for the impossibility to observe free quarks. Very energetic quarks would form a narrow spray of hadrons called "jet," at present the most direct experimental evidence of the existence of quarks.
5. Actually the number of "fundamental" particles can be counted as forty-five and not forty-eight. Since processes like beta-decay are observed to involve quarks and leptons with left-handed spins relative to their motions, we have

forty-five left-handed and forty-five right-handed fermion states in the gauge theory of electroweak interactions (there are no right-handed neutrinos or left-handed antineutrinos, if neutrinos are massless).

6. The gauge fields correspond to particles of spin-1 (or spin-2 in the case of the "hypothetical" quantum gravity); therefore they are bosons. These gauge bosons cannot all be massless to fit experimental data. Therefore, the symmetry cannot be exact or, in other words, symmetry is *spontaneously* broken, that is, broken in such a way to retain *renormalizability* of the theory—since renormalization is a consistent mathematical device to eliminate divergent quantities from the theory while pursuing calculations to arbitrarily high orders of approximation, it is a crucial property to select physical theories.

It is possible to mark a difference between matter fields and radiation fields? At the fundamental level (elementary constituents and processes) matter fields are fermions and radiation fields are bosons. This fact is one of the central points of the philosophical discussion about different content of individuality between matter constituents and radiation constituents. The fundamental constituents of matter (fermions) reveal more trace of individuality (on the ground of Pauli exclusion principle), whereas elementary interaction processes (bosons) are, in this respect, more problematical. For example, in the modal interpretation of quantum mechanics we can call for some individuality principle for fermions but not for bosons. This point has been analyzed especially by Bas van Fraassen. See, for instance, B. van Fraassen, *Quantum Mechanics: An Empiricist View* (Oxford: Oxford University Press, 1991), chap. 11. Cf. also G. Peruzzi, "Logical Anomalies of Quantum Objects. A Survey," *Foundations of Physics* 20 (1990): 337–352.

Supersymmetry, on the other hand, is a theoretical idea (at present without a satisfactory experimental evidence) that attempts to unify the treatment of particles with different spin. In its minimal version (the so called $N = 1$ supersymmetry), it requires every fundamental particle to have a partner, with the same electric charge and other properties like color but with spin differing by (minus) one-half. So the leptons and quarks are partnered by spin-0 sleptons and squarks, the photon, W, Z and gluon have spin one-half photino, wino, zino, and gluino partners. In the supersymmetry perspective the partition between matter fields and radiation field would, essentially, get lost.

7. Actually any relativistic theory of particles in interaction is a field theory. See L. D. Landau and E. M. Lifshitz, *Relativistic Quantum Theory* (Oxford: Pergamon Press, 1971), 29: "Every relativistic theory of particles must be a theory of systems having an infinite number of degrees of freedom. That is to say, any such theory of particles must be a field theory."

8. The quantum treatment of gravitational interaction is still an open question, but the Einstein field equations, by the light of the gauge principle, indicate that the basic process corresponds to the exchange of massless spin-2 gravitons.

9. As we know, there are three families of quarks and leptons in SM. Each family has a massless neutrino. One of the first experiments at LEP counted the number of apparently massless neutrino species and hence the number of families with which we should deal at the present stage. The shape of the cross section in electron-positron annihilation ($e^+ + e^- \rightarrow$ hadrons) in the region

of Z^0 peak is sensitive to the number of the neutrino species. One then can count the neutrinos produced in Z^0 decay despite the fact that they escape the detectors. The present number of massless neutrino species can now be given as 2.98 ± 0.027, that is, "there are" three families.

10. The space resolution achieving with the scattering of one particle from another is limited by the wavelength λ of their relative motion: $\lambda = 2\pi/k$ where k is the relative momentum of the two particles, which is proportional to approximately one-half the energy in the center-of-mass frame. So to probe small distances, we need large k or high energies in the center-of-mass frame.

11. On the grounds of the relativity statement, $E = mc^2$, a particle of mass m can be "created" only if there is enough energy available in the center-of-mass frame. Recent colliders $e^+ e^-$ can work at energies in the center-of-mass frame around 120 GeV, where the production of Z^0 becomes possible.

12. For a description of the main characters of a high-energy experiment, see A. Pickering, *Constructing Quarks* (Edinburgh: Edinburgh University Press, 1984), chap. 2, and for a more complete (and more technical) description see V. Barger and R. Phillips, *Collider Physics* (Reading, Mass.: Addison-Wesley, 1987).

13. See A. Pais, *Inward Bound: Of Matter and Forces in the Physical World* (Oxford: Oxford University Press, 1986), 2.

14. This statement is actually true if the target is sufficiently thin to exclude multiple scattering, otherwise the total target area will be less than the sum of the cross sections of the individual particles. We shall not enter in these subtleties and shall restrict our analysis neglecting multiple scattering.

15. This probability, again, is correct only for times so short that dP is small, which is a condition to prevent multiple scattering.

16. The differential cross section can be defined per unit of solid angle, and in this form is useful in experimental apparatus in which the scattered particles are counted with the aid of the detector: the number of particles scattered into the detector per unit time is $j\rho\sigma\, dx\, d\Omega$, where j is the incident current per unit area, and $d\Omega$ is the solid angle subtended by the detector at the target. From the measured value of this number, one can calculate σ (differential cross section per unit of solid angle), if j and ρ are known, that is, if the luminosity $\mathscr{L}$ is known.

17. A hard spherical particle of radius r_o has a potential that is zero everywhere for $r > r_o$ and it is infinite for $r < r_o$.

18. The Rutherford cross section for (elastic) scattering of an (nonrelativistic) electron of momentum p through an angle θ by a static point charge e is

$$\frac{d\sigma}{d\Omega_R} = \frac{m^2 \epsilon^2}{4 p^4 \sin^4(\theta/2)},$$

where $\epsilon = e^2/(4\pi\epsilon_0)$. If the same charge e is spread out in a spherically symmetric density distribution $e\rho(r)$, then the Rutherford cross section is replaced by

$$\frac{d\sigma}{d\Omega} = \frac{d\sigma}{d\Omega_R} G_E^2(q^2),$$

where q is the momentum transfer $\mathbf{q} = \mathbf{p} - \mathbf{p}'$, $\mathbf{p}$ and $\mathbf{p}'$ being the initial and final electron momenta, and G_E is called *form factor* and it is equal to the Fourier transform of the charge distribution with respect to q, a clear evidence of the size of the experimentally observed microobject. It can be proved that $G_E(0) = 1$ and $G_E(q^2)$ goes to zero as q^2 goes to infinity. Thus measurements of cross sections allow us to determine the form factor, that is, the Fourier transform of the charge distribution which has caused the scattering. We can obtain, in particular, the root-mean-square charge radius of the spherically charge distribution causing the scattering.

19. A common way is to assume that the potential can be expressed by some simple mathematical form containing a certain number of parameters (such as $V = Ce^{-r/r_o}/r^n$, where the parameters are C, r_o, and n), which can appropriately be chosen to fit experimental data.

20. The conservation of the number of particles in the nonrelativistic theory depends on the law of conservation of mass: the interactions do not affect the sum of the (rest) masses of the particles. In relativistic theory, however, only the total energy of the system is conserved, and consequently there is no law of conservation of mass.

21. Such systems, which integrate many different subdetectors in a single device, rely heavily, as we have noted, on fast electronics and computers to monitor and control the subdetectors, and to coordinate, classify and record the vast amount of information flowing in from different parts of apparatus.

22. If a particle is composite, we have peculiarities in the expression of the cross section. In the quantum domain the scattering of lepton (electron or muon) on hadron (proton or neutron) leads us to deep informations about the structure of the hadrons. In particular, the proton and the neutron are nonpointlike objects (in a sense very similar to the classical one) and they are composed of (three) pointlike entities. In the case of the elastic lepton scattering from the proton in addition to the electric form factor G_E associated with the charge distribution (cf. n. 18), there is also a magnetic form factor G_M associated with the magnetic moment distribution within the proton. The formulas for G_E and G_M are similar to the case of static charge and magnetic moment distribution for momentum transfers, which are much smaller than the proton mass so that the recoil energy of the proton is negligible. On the contrary the *static* interpretation of the form factors breaks down at high-momentum transfers, where the recoil energy of the proton cannot be neglected. We have also to take account of the magnetic moment of the electron, so the general cross-section formula becomes quite complicated. In any case the experimental data are in agreement with this nonpointlike behavior of the proton, which justifies the introduction of form factors.

The study of the inelastic lepton-nucleon (and, in general, lepton-hadron) scattering at high energies is of great importance, and it led to the first experimental evidence of pointlike components within the proton (initially called partons and then identified with quarks). The generalization of elastic differential cross section makes use of the *structure functions* (a generalization of the form factors), which are functions of two variables: the energy-momentum transfer squared Q^2 ($Q^2 = (\mathbf{p} - \mathbf{p}')^2 - (E - E')^2$, where the initial and final leptons have momenta and energies $(\mathbf{p}, E)$ and $(\mathbf{p}', E')$, respectively), and the dimensionless variable x, which, we may say, corresponds to the fraction of the nucleon momentum taken by the constituent which is observed. It can be shown that the decrease of the inelastic cross section with Q^2 is much more moderate at constant x. This type of behavior reminds us of the Rutherford experiments on the large-angle scattering of α particles by atoms, where the surprisingly large cross section led to the conclusion that the positive charge of the target atom is concentrated in a "pointlike" (with respect to the resolution pertaining the low energies of the Rutherford's α-particles) nucleus. The analogy suggests that the anomalously large cross section for deep inelastic lepton-nucleon scattering has its origin in the presence of "pointlike" nearly noninteracting constituents inside the nucleon. The results for the structure functions as a function of Q^2, for different values of x, are in good agreement with presently known values but not yet as precise. Certainly the comparison of experimental data with the theoretical results is in agreement with the quark structure of the hadrons.

23. Here the total cross section is the sum of the cross section for elastic scattering and that for inelastic reactions, where the latter is itself a sum over all possible inelastic processes allowed by conservation laws.

24. As early as the 1930s, proton and neutron, because of their striking similarity, were considered as two possible *states* of one and the same particle, the so-called *nucleon*, a name we used in what follows.

25. The nucleon resonances have the same internal quantum numbers of nucleons, and they are therefore interpreted as excited states of the nucleons. The peaks in the diagram of the cross section at low energies can be interpreted as the peaks showed by the Frank-Hertz experiment for the measure of energy of the excited states of an atom.

26. Cf. n. 22.

27. Ibid.

28. The reaction is $e^+ + e^- \rightarrow Z^0 \rightarrow$ quark + antiquark $\rightarrow$ hadrons.

29. A jet is qualitatively defined as a collimated spray of high-energy hadrons. For the purpose of performing accurate quantitative studies, one needs a precise definition of jet. Essentially, one has to specify how low-energy particles are assigned to jets in order to have a well-defined number of jets, which is the starting point for the computations of cross sections. We do not consider the problem in detail here but restrict ourselves only to the following considerations. The introduction of a *dimensionless resolution variable* allows us to formulate a definition by which we can classify the final-state particles in each event in a well-defined number of jets, a crucial step for the computation of the angular distribution of the jets. The main feature of this definition is that the corresponding iterative procedure of clustering type provides an unambiguous and exhaustive assignment of particles to jets. There are several jet clustering algorithms—with reference to e^+e^- annihilation at LEP—specified

by different definitions of the dimensionless resolution variables. Many theoretical and phenomenological investigations have been carried out in the past few years on this matter. There is no doubt that this is another example of the relevance of *mereology* in modern physics.

30. See P. Langacker, "Particle Physics Summary, Where Are We and Where Are We Going?," in J. Trân Thanh Vân, ed., *'92 Electroweak Interactions and Unified Theories* (Gif-sur-Yvette: Editions Frontières, 1992).

Bibliography

Accardi, L., and Fedullo, A. "On the Statistical Meaning of the Complex Numbers in Quantum Mechanics." *Lettere al Nuovo Cimento* 34 (1982): 161–172.

Adams, R. "Primitive Thisness and Primitive Identity." *Journal of Philosophy* 76 (1979): 5–26.

Aerts, D. "Description of Compound Physical Systems." In E. Beltrametti and B. van Fraassen, eds., *Current Issues in Quantum Logic*. New York: Plenum, 1981.

———. "The One and the Many." Ph.D. dissertation, Free University of Brussels, 1981.

———. "Description of Many Physical Entities without the Paradoxes Encountered in Quantum Mechanics." *Foundations of Physics* 12 (1982): 1131–1170.

———. "Classical Theories and Non Classical Theories as a Special Case of a More General Theory." *Journal of Mathematical Physics* 24 (1983): 2441–2453.

———. "How Do We Have to Change Quantum Mechanics in Order to Describe Separated Systems?" In S. Diner, D. Fargue, G. Lochak, and F. Selleri, eds., *The Wave-Particle Dualism*. Dordrecht: Reidel, 1984.

———. "The Physical Origin of the Einstein Podolsky Rosen Paradox." In G. Tarozzi and A. van der Merwe, eds., *Open Questions in Quantum Physics*. Dordrecht: Reidel, 1985.

———. "A Possible Explanation for the Probabilities of Quantum Mechanics and Example of a Macroscopical System That Violates Bell Inequalities." In P. Mittelstaedt and E. W. Stachow, eds., *Recent Developments in Quantum Logic*. Mannheim: Bibliographisches Institut, 1985.

———. "A Possible Explanation for the Probabilities of Quantum Mechanics." *Journal of Mathematical Physics* 27 (1986): 202–210.

———. "The Origin of the Non-Classical Character of the Quantum Probability Model." In A. Blaquiere, S. Diner, and G. Lochak, eds., *Information, Complexity, and Control in Quantum Physics*. Berlin: Springer, 1987.

———. "An Attempt to Imagine Parts of the Reality of the Microworld." In J. Mizerski, A. Posiewnik, J. Pykacz, and M. Zukowski, eds., *Problems in Quantum Physics*. Singapore: World Scientific, 1990.

———. "A Macroscopic Classical Laboratory Situation with Only Macroscopic Classical Entities Giving Rise to a Quantum Mechanical Probability Model." In L. Accardi, ed., *Quantum Probability and Related Topics*, vol. 6. Singapore: World Scientific, 1991.

———. "Quantum Structures Due to Fluctuations of the Measurement Situation." *International Journal of Theoretical Physics* 32 (1993): 2207–2220.

———. "Quantum Structures, Separated Physical Entities and Probability." *Foundations of Physics* 24 (1994): 1227–1258.

————. "Quantum Structures: An Attempt to Explain the Origin of Their Appearance in Nature." *International Journal of Theoretical Physics* 34 (1995): 1165–1186.

Aerts, D., and Aerts, S. "Applications of Quantum Statistics in Psychological Studies of Decision Processes." *Foundations of Science* 1 (1995): 85–97.

————. "Interactive Probability Models: From Quantum to Kolmogorovian." Preprint, CLEA, VUB, Brussels, 1995.

Aerts, D., Coecke, B., and Valckenborgh, F. "A Mechanistic Macroscopic Physical Entity with a Three Dimensional Hilbert Space Quantum Description." *Helvetica Physica Acta* 70 (1997): 793–814.

Aerts, D., and Durt, T. "Quantum, Classical and Intermediate: An Illustrative Example." *Foundations of Physics* 24 (1994): 1353–1368.

————. "Quantum, Classical and Intermediate: A Measurement Model." In T. Havyonen, ed., *Symposium on the Foundations of Physics*. Helsinki, 1994.

Aerts, D., Durt, T., Grib, A. A., Van Bogaert, B., and Zapatrin, R. R. "Quantum Structures in Macroscopic Reality." *International Journal of Theoretical Physics* 32 (1993): 489–498.

Aerts, D., Durt, T., and Van Bogaert, B. "A Physical Example of Quantum Fuzzy Sets, and the Classical Limit." *Tatra Montains Math. Publ.* 1 (1992): 5–15.

————. "Quantum Probability, the Classical Limit and Non-Locality." In T. Hyvonen, ed., *Symposium on the Foundations of Modern Physics*. Singapore: World Scientific, 1993.

Aerts, D., and Piron, C. "Physical Justification for Using the Antisymmetric Tensor Product." Preprint TENA. Free University of Brussels, 1979.

Aerts, D., and Reignier, J. "The Spin of a Quantum Entity and Problems of Non-Locality." In P. Lahti and P. Mittelstaedt, eds., *Symposium on the Foundations of Modern Physics, 1990*. Singapore: World Scientific, 1990.

————. "On the Problem of Non-Locality in Quantum Mechanics." *Helvetica Physica Acta* 64 (1991): 527–547.

Aerts, D., and Van Bogaert, B. "A Mechanical Classical Laboratory Situation with a Quantum Logic Structure." *International Journal of Theoretical Physics* 31 (1992): 1839–1848.

Aicardi, F., Borsellino, A., Ghirardi, G. C., and Grassi, R. "Dynamical Models for State-Vector Reduction: Do They Ensure That Measurements Have Outcomes?" *Foundations of Physics Letters* 4 (1991): 109–116.

Albert, D. Z. "On the Collapse of the Wave Function." In A. Miller, ed., *Sixty-Two Years of Uncertainty*. New York: Plenum, 1990.

————. *Quantum Mechanics and Experience*. Cambridge, Mass.: Harvard University Press, 1992.

Albert, D. Z., and Loewer, B. "Wanted Dead or Alive: Two Attempts to Solve Schrödinger's Paradox." In A. Fine, M. Forbes, and L. Wessels, eds., *PSA 1990*, vol. 1. East Lansing, Mich.: Philosophy of Science Association, 1990.

Albert, D. Z., and Vaidman, L. "On a Proposed Postulate of State-Reduction." *Physics Letters A* 139 (1989): 1–4.

Amstrong, D. M. *Universals and Scientific Realism*. 2 vols. Cambridge: Cambridge University Press, 1978.

Aspect, A., Grangier, P., and Roger, G. "Experimental Realization of Einstein-Podolsky-Rosen-Bohm *Gedankenexperiment*: A New Violation of Bell's Inequalities." *Physical Review Letters* 48 (1982): 91–94.

Bacon, J., Campbell, K., and Reinhardt, L., eds., *Ontology, Causality, and Mind: Essays in Honor of David Armstrong*. Cambridge: Cambridge University Press, 1993.

Barger, V., and Phillips, R. *Collider Physics*. Reading, Mass.: Addison-Wesley, 1987.

Barnette, R. L. "Does Quantum Mechanics Disprove the Principle of the Identity of Indiscernibles?" *Philosophy of Science* 45 (1978): 466–470.

Bell, J. S. "On the Einstein-Podolsky-Rosen Paradox." *Physics* 1 (1964): 195–200.

———. "Quantum Mechanics for Cosmologists." In C. Isham, R. Penrose, and D. Sciama, eds., *Quantum Gravity 2*. Oxford: Clarendon Press, 1981.

———. "Beables for Quantum Field Theory." CERN-TH. 4035/84, 1984. (Reprinted in J. S. Bell, *Speakable and Unspeakable in Quantum Mechanics*. Cambridge: Cambridge University Press, 1987.)

———. "Six Possible Worlds of Quantum Mechanics." In S. Allén, ed., *Proceedings of the Nobel Symposium 65: Possible Worlds in Arts and Sciences*. Stockholm: Nobel Foundation, 1986.

———. "Are There Quantum Jumps?" In C.E.W. Kilmister, ed., *Schrödinger-Centenary Celebration of a Polymath*. Cambridge: Cambridge University Press, 1987. (Reprinted in J. S. Bell, *Speakable and Unspeakable in Quantum Mechanics*. Cambridge: Cambridge University Press, 1987.)

———. "Against 'Measurement.' " In A. Miller, ed., *Sixty-Two Years of Uncertainty*. New York: Plenum, 1990.

Benatti, F., Ghirardi, G. C., Rimini, A., and Weber, T. "Quantum Mechanics with Spontaneous Localization and the Quantum Theory of Measurement." *Il Nuovo Cimento* 100 B (1987): 27–41.

Birkhoff, G., and von Neumann, J. "The Logic of Quantum Mechanics." *Annals of Mathematics* 37 (1936): 823–843.

Bogoliubov, N. N., Logunov, A. A., Todorov, I. T. *Introduction to Axiomatic Quantum Field Theory*. Reading, Mass.: W. A. Benjamin, 1975.

Bohm, D. *Wholeness and the Implicate Order*. London: Routledge, 1980.

Bohm, D., and Vigier, J. P. "Model of the Causal Interpretation of Quantum Theory in Terms of a Fluid with Irregular Fluctuations." *Physical Review* 96 (1954): 208–216.

Bohr, N. "The Quantum Postulate and the Recent Development of Atomic Theory." *Nature* 121 (1928): 580–590.

Born, M. *The Born-Einstein Letters*. New York: Walker, 1971.

Bose, S. N. "Plancks Gesetz und Lichtquanten-hypothese." *Zeitschrift für Physik* 26 (1924): 178–181.

Brown, C. *Leibniz and Strawson: A New Essay in Descriptive Metaphysics*. Munich: Philosophia Verlag, 1990.

Brown, H. "Bohm Trajectories and Their Detection in the Light of Neutron Interferometry." 3rd Annual Conference on the Foundations of Quantum Theory and Relativity, Cambridge, September 13–16, 1994.

Brown, H. R., and Harré, H., eds., *Philosophical Foundations of Quantum Field Theory*. Oxford: Clarendon Press, 1988.

Bub, J. "On the Completeness of Quantum Mechanics." In C. A. Hooker, ed., *Contemporary Research in the Foundations and Philosophy of Quantum Theory*. Dordrecht: Reidel, 1973.

Busch, P. "Unsharp Reality and Joint Measurements for Spin Observables." *Physical Review D* 33 (1986): 2253–2261.

———. "Unsharp Reality and the Question of Quantum Systems." In P. Lahti and P. Mittelstaedt, eds., *Symposium on the Foundations of Modern Physics, 1987*. Singapore: World Scientific, 1987.

———. "Macroscopic Quantum Systems and the Objectification Problem." In P. Lahti and P. Mittelstaedt, eds., *Symposium on the Foundations of Modern Physics, 1990*. Singapore: World Scientific, 1990.

Busch, P., and Lahti, P. "A Note on Quantum Theory, Complementarity, and Uncertainty." *Philosophy of Science* 52 (1985): 64–77.

Butterfield, J. "Interpretation and Identity in Quantum Theory." *Studies in the History and Philosophy of Science* 24 (1993): 443–476.

Butterfield, J., Fleming, G. N., Ghirardi, G. C., and Grassi, R. "Parameter Dependence in Dynamical Models for Statevector Reduction." *International Journal of Theoretical Physics* 32 (1993): 2287–2304.

Carnap, R. *Logical Foundations of Probability*. Chicago: University of Chicago Press, 1950.

Cassirer, E. *Determinismus und Indeterminismus in der modernen Physik*. Göteborg: Elanders Boktryckeri Aktiebolag, 1937.

———. "The Concept of Group and the Theory of Perception." *Philosophy and Phenomenological Research* 5 (1944): 1–35.

———. "Reflections on the Concept of Group and the Theory of Perception." In *Symbol, Myth and Culture. Essays and Lectures of Ernst Cassirer, 1935–1945*. New Haven: Yale University Press, 1979.

Castellani, E. "Sulla nozione di oggetto nella fisica classica e quantistica." In C. Cellucci, M. C. Di Maio, and G. Roncaglia, eds., *Logica e filosofia della scienza: problemi e prospettive*. (Proceedings Soc. Italiana di Logica e Filos. delle Scienze, Lucca, 1993.) Pisa: Edizioni ETS, 1994.

———. "Quantum Mechanics, Objects and Objectivity." In C. Garola and A. Rossi, eds., *The Foundations of Quantum Mechanics—Historical Analysis and Open Questions*. Dordrecht: Kluwer, 1995.

Cleland, C. "Space: An Abstract System of Non-Supervenient Relations." *Philosophical Studies* 46 (1984), 19–40.

Coburn, R. "Identity and Spatiotemporal Continuity." In M. K. Munitz, ed., *Identity and Individuation*. New York: New York University Press, 1971.

Coecke, B. "Generalization of the Proof on the Existence of Hidden Measurements to Experiments with an Infinite Set of Outcomes." Preprint TENA, Free University of Brussels, 1995.

———. "A Hidden Measurement Model for Pure and Mixed States of Quantum Physics in Euclidean Space." *International Journal of Theoretical Physics* 34 (1995): 1313–1320.

———. "A Hidden Measurement Representation for Quantum Entities Described by Finite Dimensional Complex Hilbert Spaces." Preprint TENA, Free University of Brussels, 1995.

Cortes, A. "Leibniz's Principle of the Identity of Indiscernibles: A False Principle." *Philosophy of Science* 43 (1976): 491–505.

Costantini, D. "The Relevance Quotient." *Erkenntnis* 14 (1979): 149–157.

Costantini, D., Galavotti, M. C., and Rosa, R. "A Rational Reconstruction of Elementary Particle Statistics." *Scientia* 117 (1982): 151–159.

———. "A Set of 'Ground Hypotheses' for Elementary Particle Statistics." *Il Nuovo Cimento* 74B (1983): 151–158.

Currie, D. G., Jordan, T. F., and Sudarshan, E.C.G. "Relativistic Invariance and Hamiltonian Theories of Interacting Particles." *Review of Modern Physics* 35 (1963): 350–375.

Cushing, J. T. "A Background Essay." In J. T. Cushing and E. McMullin, eds., *Philosophical Consequences of Quantum Theory: Reflections on Bell's Theorem*. Notre Dame, Ind.: University of Notre Dame Press, 1989.

Cushing, J. T., Delaney, C. F., and Gutting, G. M., eds., *Science and Reality: Recent Work in the Philosophy of Science*. Notre Dame, Ind.: University of Notre Dame Press, 1984.

da Costa, N.C.A., and French, S. "The Model Theoretic Approach in the Philosophy of Science." *Philosophy of Science* 57 (1990): 248–265.

da Costa, N.C.A., French, S., and Krause, D. "The Schrödinger Problem." In M. Bibtol and O. Darrigol, eds., *Erwin Schrödinger: Philosophy and the Birth of Quantum Mechanics*. Paris: Editions Frontières, 1992.

Dalla Chiara, M. L., and Toraldo di Francia, G. "Individuals, Properties and Truth in the E.P.R. Paradox." In P. Lahti and P. Mittelstaedt, eds., *Symposium on the Foundations of Modern Physics, 1985*. Singapore: World Scientific, 1985.

———. "Individuals, Kinds and Names in Physics." In E. Agazzi and M. Mondadori, eds., *Logica e Filosofia della Scienza, oggi*. (Proceedings Soc. Italiana di Logica e Filos. delle Scienze, San Gimignano, 1983.) Bologna: Clueb, 1986. (Reprinted in G. Corsi, M. L. Dalla Chiara, and G. C. Ghirardi, eds., *Bridging the Gap: Philosophy, Mathematics, and Physics*. Dordrecht: Kluwer, 1993.)

———. "Identity Questions from Quantum Theory." In K. Gavroglu, J. Stachel, and M. W. Wartofski, eds., *Physics, Philosophy and the Scientific Community*. Dordrecht: Kluwer, 1995.

Daniels, C. "Towards an Ontology of Numbers." Unpublished manuscript, 1980.

de Broglie, L. "Sur la possibilité de relier les phénomènes d'interference et de diffraction à la théorie des quanta de lumière." *Comptes Rendus* 183 (1926): 447–448.

de Muynck, W. "Distinguishable and Indistinguishable-Particle Descriptions of Systems of Identical Particles." *International Journal of Theoretical Physics* 14 (1975): 327–346.

D'Espagnat, B., and Klein, E. *Regards sur la matière*. Paris: Fayard, 1993.

Diosi, L. "Models for Universal Reduction of Macroscopic Quantum Fluctuations." *Physical Review A* 40 (1989): 1165–1174.

Dirac, P.A.M. "On the Theory of Quantum Mechanics." *Proceedings of the Royal Society of London,* ser. A 112 (1926): 661–677.

———. *The Principles of Quantum Mechanics*. 4th ed. Oxford: Oxford University Press, 1978. (Originally published in 1930.)

Dorling, J. "Probability, Information and Physics." Preprint. Department of History and Philosophy of Science, Chelsea College, University of London, 1978.

Einstein, A. "Quantentheorie des einatomigen idealen Gases." *Preussische Akad. der Wissenschaften (Phys.-math. Klasse) Sitzungsberichte* (= Berliner Berichte) 1924: 261–267, and 1925: 3–14.

Einstein, A., Podolski, B., and Rosen, N. "Can Quantum-Mechanical Description of Physical Reality Be Considered Complete?" *Physical Review* 47 (1935): 777–780. (Reprinted in J. Wheeler and W. Zurek, eds., *Quantum Theory and Measurement*. Princeton: Princeton University Press, 1983.)

Ellis, J., Mohanty, S., and Nanopoulos, D. V. "Quantum Gravity and the Collapse of the Wavefunction." *Physics Letters B* 221 (1989): 113–119.

Emch, G. G. *Mathematical and Conceptual Foundations of 20th Century Physics*. Amsterdam: North-Holland, 1984.

Evans, G. "Can There Be Vague Objects?" *Analysis* 38 (1978): 208. (Reprinted in *Collected Papers*. Oxford: Oxford University Press, 1985.)

Falkenburg, B. "The Analysis of Particle Tracks: A Case against Incommensurability." *Studies in History and Philosophy of Modern Physics* 27 (1966): 337–371.

Fermi, E. "Zur Quantelung des idealen einatomigen Gases." *Zeitschrift für Physik* 36 (1926): 902–912.

Fine, A. "Antinomies of Entanglement: The Puzzling Case of the Tangled Statistics." *Journal of Philosophy* 79 (1982): 733–747.

———. "Interpreting Science." In A. Fine and J. Leplin, eds., *PSA 1988*, vol. 2. East Lansing, Mich.: Philosophy of Science Association, 1989.

Fleming, G. "Lorentz Invariant State Reduction and Localization." In A. Fine and J. Leplin, eds., *PSA 1988*, vol. 2. East Lansing, Mich.: Philosophy of Science Association, 1989.

French, S. "First-Quantised Para-Particle Theory." *International Journal of Theoretical Physics* 26 (1987): 1141–1163.

———. "Identity and Individuality in Classical and Quantum Physics." *Australasian Journal of Philosophy* 67 (1989): 432–446.

———. "Individuality, Supervenience and Bell's Theorem." *Philosophical Studies* 55 (1989): 1–22.

———. "Why the Identity of Indiscernibles Is Not Contingently True Either." *Synthese* 78 (1989): 141–166.

———. "Hacking Away at the Identity of Indiscernibles: Possible Worlds and Einstein's Principle of Equivalence." *Journal of Philosophy* 92 (1995): 455–466.

French, S., and Krause, D. "The Logic of Quanta." In T. Y. Cao, ed., *Proceedings of the Boston Colloquium in the Philosophy of Science: A Historical Examination and Philosophical Reflections on the Foundations of Quantum Field Theory*. Cambridge: Cambridge University Press, forthcoming.

French, S., and Redhead, M. "Quantum Physics and the Identity of Indiscernibles." *British Journal for the Philosophy of Science* 39 (1988): 233–246.

Frenkel, A. "Spontaneous Localizations of the Wave Function and Classical Behavior." *Foundations of Physics* 20 (1990): 159–188.

Geach, P. T. "Identity." *Review of Metaphysics* 21 (1967): 3–12. (Reprinted in P. T. Geach, *Logic Matters*. Oxford: Blackwell, 1972.)

———. *Reference and Generality*. 3d ed. Ithaca, N.Y.: Cornell University Press, 1980.

Ghirardi, G. C., Grassi, R., and Benatti, F. "Describing the Macroscopic World: Closing the Circle with the Dynamical Reduction Program." *Foundations of Physics* 25 (1995): 5–38.

Ghirardi, G. C., Grassi, R., Butterfield, J., and Fleming, G. N. "Parameter Dependence and Outcome Dependence in Dynamical Models of State Vector Reduction." *Foundations of Physics* 23 (1993): 341–364.

Ghirardi, G. C., Grassi, R., and Pearle, P. "Relativistic Dynamical Reduction Models: General Framework and Examples." *Foundations of Physics* 20 (1990): 1271–1316.

———. "Relativistic Dynamical Reduction Models." In P. Lahti and P. Mittelstaedt, eds., *Symposium on the Foundations of Modern Physics, 1990*. Singapore: World Scientific, 1990.

Ghirardi, G. C., Grassi, R., and Rimini, A. "Continuous-Spontaneous-Reduction Model Involving Gravity." *Physical Review A* 42 (1990): 1057–1064.

Ghirardi, G. C., and Pearle, P. "Dynamical Reduction Theories: Changing Quantum Theory So the Statevector Represents Reality." In A. Fine, M. Forbes, and L. Wessels, eds., *PSA 1990*, vol. 2. East Lansing, Mich.: Philosophy of Science Association, 1991.

Ghirardi, G. C., Pearle, P., and Rimini, A. "Markov Processes in Hilbert Space and Continuous Spontaneous Localization of Systems of Identical Particles." *Physical Review A* 42 (1990): 78–89.

Ghirardi, G. C., and Rimini, A. "Old and New Ideas in the Theory of Quantum Measurement." In A. Miller, ed., *Sixty-Two Years of Uncertainty*. New York: Plenum, 1990.

Ghirardi, G. C., Rimini, A., and Weber, T. "Unified Dynamics for Microscopic and Macroscopic Systems." *Physical Review D* 34 (1986): 470–491.

Ginsberg, A. "Quantum Theory and Identity of Indiscernibles Revisited." *Philosophy of Science* 48 (1981): 487–491.

Gracia, J. J. *Individuality*. Albany, N.Y.: State University of New York Press, 1988.

Greenberg, O. W., and Messiah, A.M.L. "Symmetrization Postulate and Its Experimental Foundation." *Physical Review* 136 B (1964): 248–267.

Gudder, S. P. *Quantum Probability*. Boston: Academic Press, 1988.

Hacking, I. "The Identity of Indiscernibles." *Journal of Philosophy* 72 (1975): 249–256.

Hale, S. "Elementarity and Anti-Matter in Contemporary Physics." In A. Fine, M. Forbes, and L. Wessels, eds., *PSA 1990*, vol. 2. East Lansing, Mich.: Philosophy of Science Association, 1991.

Healey, R. *The Philosophy of Quantum Mechanics: An Interactive Interpretation*. Cambridge: Cambridge University Press, 1989.

Heisenberg, W. "Uber quantentheoretische Umdeutung kinematischer und mechanischer Beziehungen." *Zeitschrift für Physik* 33 (1925): 879–893.

———. "Schwankungserscheinungen und Quantenmechanik." *Zeitschrift für Physik* 40 (1926): 501–506.

———. *Die physikalische Prinzipien der Quantentheorie*. Leipzig: Hirzel, 1930.

Heller, M. *The Ontology of Physical Objects: Four-dimensional Hunks of Matter*. Cambridge: Cambridge University Press, 1990.

Hermann, G. "Die naturphilosophischen Grundlagen der Quantenmechanik." *Abhandlungen der Fries'schen Schule*, new ser. 6, vol. 2. Berlin: Verlag Öffentliches Leben, 1935.

Hirsch, E. "Essence and Identity." In M. K. Munitz, ed., *Identity and Individuation*. New York: New York University Press, 1971.

Howard, D. "Einstein on Locality and Separability." *Studies in History and Philosophy of Science* 16 (1985): 171–201.

———. "Holism, Separability and the Metaphysical Implications of the Bell Experiments." In J. T. Cushing and E. McMullin, eds., *Philosophical Consequences of Quantum Theory*. Notre Dame, Ind.: University of Notre Dame Press, 1989.

———. "A Peek Behind the Veil of Maya: Einstein, Schopenhauer, and the Historical Background of the Conception of Space as a Ground for the Individuation of Physical Systems." In J. Earman and J. D. Norton, eds., *The Cosmos of Science: Essays of Exploration*. Pittsburgh: University of Pittsburgh Press; Konstanz: Universitätsverlag, 1997.

Huggett, N. "What Are Quanta, and Why Does It Matter?" In D. Hull, M. Forbes, and R. M. Burian, eds., *PSA 1994*, vol. 2, East Lansing, Mich.: Philosophy of Science Association, 1995.

Huggett, N. "Identity, Quantum Mechanics, and Common Sense." *The Monist* 80 (1997): 118–130.

Jarrett, J. "On the Physical Significance of the Locality Conditions in the Bell Arguments." *Nous* 18 (1984): 569–589.

Jauch, J. M. *Foundations of Quantum Mechanics*. Reading, Mass.: Addison-Wesley, 1968.

Johnston, M. "Constitution Is Not Identity." *Mind* 101. 401 (1992): 89–105.

Jubian, M. *Ontonlogy, Modality, and the Fallacy of Reference*. Cambridge: Cambridge University Press, 1993.

Kant, I. *The Critique of Pure Reason*. Translated by N. Kemp Smith. London: Macmillan, 1980.

Kaplan, D. "How to Russell a Frege-Church." *Journal of Philosophy* 72 (1976): 716–729.

Karolyhazy, F. "Gravitation and Quantum Mechanics of Macroscopic Objects." *Il Nuovo Cimento A* 42 (1966): 390–402.

Kastler, A. "On the Historical Development of the Indistinguishability Concept for Microparticles." In A. van der Merwe, ed., *Old and New Questions in Physics, Cosmology, Philosophy, and Theoretical Biology*. New York: Plenum, 1983.

Komar, A. B. "Qualitative Features of Quantized Gravitation." *International Journal of Theoretical Physics* 2 (1969): 157–160.

Krause, D. "On a Quasi-Set Theory." *Notre Dame Journal of Formal Logic* 33 (1992): 402–411.

———. "Non-Reflexive Logics and the Foundations of Physics." In C. Cellucci, M. C. Di Maio, and G. Roncaglia, eds., *Logica e filosofia della scienza: problemi e prospettive*. (Proceedings Soc. Italiana di Logica e Filos. delle Scienze, Lucca 1993.) Pisa: Edizioni ETS, 1994.

Krause, D., and French, S. "A Formal Framework for Quantum Non-Individuality." *Synthese* 102 (1995): 195–214.

Kripke, S. *Naming and Necessity*. Oxford: Blackwell, 1980.

Ladyman, J. "Structural Realism and the Model-Theoretic Approach to Scientific Theories," *Studies in History and Philosophy of Modern Physics*, forthcoming.

Landau, L. D., and Lifshitz, E. M. *Relativistic Quantum Theory*. Oxford: Pergamon Press, 1971.

Langacker, P. "Particle Physics Summary, Where Are We and Where Are We Going?" In J. Trân Thanh Vân, ed., *'92 Electroweak Interactions and Unified Theories*. Gif-sur-Yvette: Editions Frontières, 1992.

Leggett, A. J. "Macroscopic Quantum Systems and the Quantum Theory of Measurement." *Progress of Theoretical Physics Supplements* 69 (1980): 80–100.

Lévy-Leblond, J.-M. "Galilei Group and Nonrelativistic Quantum Mechanics." *Journal of Mathematical Physics* 4 (1963): 776–788.

———. "Galilei Group and Galilean Invariance." In E. M. Loebl, ed., *Group Theory and Its Applications*, vol. 2. New York: Academic Press, 1971.

Lewis, D. "Survival and Identity." In A. Rorty, ed., *The Identities of Persons*. Berkeley: University of California Press, 1976. (Reprinted in D. Lewis, *Philosophical Papers*. Oxford: Oxford University Press, 1983.)

———. "Putnam's Paradox." *Australasian Journal of Philosophy* 62 (1984): 221–236.

———. *On the Plurality of Worlds*. Oxford: Blackwell, 1986.

———. "Vague Identity: Evans Misunderstood." *Analysis* 48 (1988): 128–130.

Lowe, E. J. "The Paradox of the 1,001 Cats." *Analysis* 42 (1982): 27–30.

Lucas, J. R. "The Nature of Things." *Presidential Address, British Society for the Philosophy of Science*, June 7, 1993.

Ludwig, G. *Foundations of Quantum Mechanics I*. Berlin: Springer, 1983.

———. *Foundations of Quantum Mechanics II*. Berlin: Springer, 1985.

Mach, E. *Erkenntnis und Irrtum*. Leipzig: J. A. Barth, 1905.

Maidens, A. "Particles and the Perversely Philosophical Schoolchild: Rigid Designation, Haecceitism, and Statistics." *Teorema* 17 (1998): 75–87.

———. "Trans-World Identity and Entities in Physics." Preprint.

Manin, Y. I. "Problem of Present Day Mathematics: I (Foundations)." In F. E. Browder, ed., *Proceedings of the Symposia in Pure Mathematics*, vol. 28. Providence, R.I.: American Mathematical Society, 1976.

Maudlin, T. *Quantum Non-Locality and Relativity*. Oxford: Blackwell, 1994.

Margenau, H. "The Exclusion Principle and Its Philosophical Importance." *Philosophy of Science* 11 (1944): 187–208.

———. *The Nature of Physical Reality*. New York: McGraw-Hill, 1950.

McCulloch, G. *The Game of the Name*. Oxford: Oxford University Press, 1989.

Mirman, R. "Experimental Meaning of the Concept of Identical Particles." *Il Nuovo Cimento* 18 B (1973): 110–122.

Misner, C. W., Thorne, K. S., and Wheeler, J. A. *Gravitation.* San Francisco: W. H. Freeman, 1973.

Mittelstaedt, P. *Philosophical Problems of Modern Physics.* Dordrecht: Reidel, 1976.

———. "Constituting, Naming and Identity in Quantum Logic." In P. Mittelstaedt and E. W. Stachow, eds., *Recent Development in Quantum Logic.* Mannheim: Bibliographisches Institut, 1985.

———. *Sprache und Realität in der modernen Physik.* Mannheim: Bibliographisches Institut-Wissenschaftsverlag, 1986.

———. "Unsharp Particle-Wave Duality in Double-Split Experiments." In F. Selleri, ed., *Wave-Particle Duality.* New York: Plenum, 1992.

———. "Constitution of Objects in Classical Mechanics and in Quantum Mechanics," *International Journal of Theoretical Physics* 34 (1995): 1615–1626.

Mittelstaedt, P., Prieur, A., and Schieder, R. "Unsharp Particle-Wave Duality in a Photon Split Beam Experiment." *Foundations of Physics* 19 (1987): 891–903.

Mondadori, F. "Reference, Essentialism and Modality in Leibniz's Metaphysics." *Studia Leibnitiana* 5 (1973): 74–101.

Newton, T. D., and Wigner, E. P. "Localized States for Elementary Systems." *Review of Modern Physics* 21 (1949): 400–406.

Noonan, H. W. "Constitution Is Identity." *Mind* 102. 405 (1993): 133–145.

Pais, A. *Inward Bound: Of Matter and Forces in the Physical World.* Oxford: Oxford University Press, 1986.

Parrini, P., ed., *Kant and Contemporary Epistemology.* Dordrecht: Kluwer, 1994.

Pauli, W. "Ueber die Zusammenhang des Abschlusses der Elektronengruppen im Atom mit der Komplexstruktur der Spektren." *Zeitschrift für Physik* 31 (1925): 765–783.

Pearle, P. "Combining Stochastic Dynamical State-Vector Reduction with Spontaneous Localization." *Physical Review A* 39 (1989): 227–239.

Pearle, P., and Squires, E. "Bound State Excitation, Nucleon Decay Experiments, and Models of Wave Function Collapse." *Physical Review Letters* 73 (1994): 1–5.

Penrose, R. "Gravity and State-Vector Reduction." In R. Penrose and C. J. Isham, eds., *Quantum Concepts in Space and Time.* Oxford: Clarendon Press, 1986.

Peruzzi, G. "Logical Anomalies of Quantum Objects. A Survey." *Foundations of Physics* 20 (1990): 337–352.

Pickering, A. *Constructing Quarks.* Edinburgh: Edinburgh University Press, 1984.

Piron, C. *Foundations of Quantum Physics.* Reading, Mass.: W. A. Benjamin, 1976.

Pitowski, I. *Quantum Probability–Quantum Logic.* Berlin: Springer, 1989.

Post, H. "Individuality and Physics." *Listener* 70 (1963): 534–537.

Putnam, H. "Truth and Convention: On Davidson's Refutation of Conceptual Relativism." *Dialectica* 41 (1987): 69–77.

Quine, W.V.O. *Word and Object.* Cambridge, Mass.: MIT Press, 1960.

———. "Whither Physical Objects?" In R. S. Cohen, P. K. Feyerabend, and M. W. Wartofsky, eds., *Essays in Memory of Jmre Lakatos.* Dordrecht: Reidel, 1976.

Randall, C., and Foulis, D. "The Operational Approach to Quantum Mechanics." In C. A. Hooker, ed., *Physical Theory as Logico-Operational Structure*. Dordrecht: Reidel, 1979.

Randall, C., and Foulis, D. "A Mathematical Language for Quantum Physics." In C. Gruber et al., eds., *Les fondements de la mécanique quantique*. Lausanne, 1983.

Redhead, M., and Teller, P. "Particles, Particle Labels, and Quanta: The Toll of Unacknowledged Metaphysics." *Foundations of Physics* 21 (1991): 43–62.

———. "Particle Labels and the Theory of Indistinguishable Particles in Quantum Mechanics." *British Journal for the Philosophy of Science* 43 (1992): 201–218.

Reichenbach, H. *Experience and Prediction*. Chicago: University of Chicago Press, 1938.

———. *Philosophical Foundations of Quantum Mechanics*. Berkeley: University of California Press, 1944.

———. *Elements of Symbolic Logic*. New York: Macmillan, 1947.

———. *The Theory of Probability*. Berkeley: University of California Press, 1949.

———. *The Direction of Time*. Edited by M. Reichenbach. Berkeley: University of California Press, 1991. (Originally published in 1956.)

———. *The Philosophy of Space and Time*. Translated by M. Reichenbach and J. Freund. New York: Dover, 1957. (Original German edition published in 1928.)

Resnick, M. "Between Mathematics and Physics." In A. Fine, M. Forbes, and L. Wessels, eds., *PSA 1990*, vol. 2. East Lansing, Mich.: Philosophy of Science Association, 1991.

Russell, B. *Human Knowledge: Its Scope and Limits*. London: Allen and Unwin, 1948.

Schrödinger, E. "Quantisiering als Eigenwertproblem." *Annalen der Physik* 79 (1926): 361–376.

———. "Die gegenwärtige Situation in der Quantenmechanik I–III." *Die Naturwissenschaften* 23 (1935): 807–812, 823–828, 844–849.

———. *Science and Humanism*. Cambridge: Cambridge University Press, 1952.

Segal, E. "Postulates for General Quantum Mechanics." *Annals of Mathematics* 48 (1947): 930–948.

Shapere, D. "Modern Physics and the Philosophy of Science." In A. Fine and J. Leplin, eds., *PSA 1988*, vol. 2. East Lansing, Mich.: Philosophy of Science Association, 1989.

———. "The Origin and Nature of Metaphysics." *Philosophical Topics* 18 (1990): 163–174.

Shimony, A. "Search for a Worldview Which Can Accommodate Our Knowledge of Microphysics." In J. T. Cushing and E. McMullin, eds., *Philosophical Consequences of Quantum Theory: Reflections on Bell's Theorem*. Notre Dame, Ind.: University of Notre Dame Press, 1989.

———. "Desiderata for a Modified Quantum Dynamics." In A. Fine, M. Forbes, and L. Wessels, eds., *PSA 1990*, vol. 2. East Lansing, Mich.: Philosophy of Science Association, 1991.

Simons, P. *Parts. A Study in Ontology*. Oxford: Clarendon Press, 1987.

Smart, J. J. C. "Space-Time and Individuals." In R. Rudner and I. Scheffler, eds., *Logic and Art*. Indianapolis: Bobbs-Merrill Company, 1972.

Stairs, A. "Sailing into the Charybdis: van Fraassen on Bell's Theorem." *Synthese* 61 (1984): 351–360.

Stolt, R. H., and Taylor, J. R. "Correspondence between the First- and Second-Quantized Theories of Paraparticles." *Nuclear Physics* 19 B (1970): 1–19.

Streater, R. F., and Wightman, A. S. *PCT, Spin and Statistics, and All That*. New York: W. A. Benjamin, 1964.

Strohmeyer, I. "Tragweite und Grenze der Transzendentalphilosophie zur Grundlegung der Quantenphysik." *Zeitschrift für allgemeine Wissenschaftstheorie* 18 (1987): 239–275.

Sudarshan, E., and Mehra, J. "Classical Statistical Mechanics of Identical Particles and Quantum Effects." *International Journal of Theoretical Physics* 3 (1970): 245–251.

Teller, P. "Quantum Physics, the Identity of Indiscernibles, and Some Unanswered Questions." *Philosophy of Science* 50 (1983): 309–319.

———. "Relational Holism and Quantum Mechanics." *British Journal for the Philosophy of Science* 37 (1986): 71–81.

———. "Relativity, Relational Holism and the Bell Inequalities." In J. T. Cushing and E. McMullin, eds., *Philosophical Consequences of Quantum Theory*. Notre Dame, Ind.: University of Notre Dame Press, 1989.

———. *An Interpretive Introduction to Quantum Field Theory*. Princeton: Princeton University Press, 1995.

Toraldo di Francia, G. *Le cose e i loro nomi*. Rome: Laterza, 1986.

Unger, P. "The Problem of the Many." *Midwest Studies in Philosophy* 5 (1980): 411–467.

Valckenborgh, F. "Closure Structures and the Theorem of Decomposition in Classical Components." In Proceedings of the 5th Winterschool on Measure Theory, Liptowski, 1995.

van Fraassen, B. "Singular Terms, Truth-Value Gaps' and Free Logic." *Journal of Philosophy* 63 (1966): 481–495.

———. "Probabilities and the Problem of Individuation." Presented at the American Philosophical Association, 1969. In S. Luckenbach, ed., *Probabilities, Problems and Paradoxes*. Encino, Calif.: Dickenson, 1972.

———. "Semantic Analysis of Quantum Logic." In C. A. Hooker, ed., *Contemporary Research in the Foundations and Philosophy of Quantum Theory*. Dordrecht: Reidel, 1973.

———. "The Einstein-Podolski-Rosen Paradox." *Synthese* 29 (1974): 291–309.

———. *The Scientific Image*. Oxford: Oxford University Press, 1980.

———. "A Modal Interpretation of Quantum Mechanics." In E. Beltrametti and B. van Fraassen, eds., *Current Issues in Quantum Logic*. New York: Plenum, 1981.

———. "The Charybdis of Realism: Epistemological Implications of Bell's Inequality." *Synthese* 5 (1982): 25–38.

———. "Quantification as an Act of Mind." *Journal of Philosophical Logic* 11 (1982): 343–369.

———. *An Introduction to the Philosophy of Time and Space.* New York: Columbia University Press, 1985. (Originally published in 1970.)

———. *Quantum Mechanics: An Empiricist View.* Oxford: Oxford University Press, 1991.

van Inwagen, P. *Material Beings.* Ithaca, N.Y.: Cornell University Press, 1990.

Varadarajan, V. S. *Geometry of Quantum Theory.* 2d ed. New York: Springer, 1985.

von Neumann, J. *Mathematical Foundations of Quantum Mechanics.* Princeton: Princeton University Press, 1955. (Original German edition published in 1932.)

von Weizsäcker, C. F. *Zum Weltbild der Physik.* Stuttgart: Hirzel, 1943.

Weyl, H. *Symmetry.* Princeton: Princeton University Press, 1982. (Originally published in 1952.)

Wiggins, D. *Identity and Spatio-temporal Continuity.* Oxford: Blackwell, 1967.

———. *Sameness and Substance.* Oxford: Blackwell, 1980.

Wigner, E. "Über nicht kombinierende Terme in der neueren Quantentheorie." *Zeitschrift für Physik* 40 (1927): 492–500.

———. "Über nicht kombinierende Terme in der neueren Quantentheorie. II Teil." *Zeitschrift für Physik* 40 (1927): 883–892.

———. "On Unitary Representations of the Inhomogeneous Lorentz Group." *Annals of Mathematics* 40 (1939): 149–204.

Zahar, E. G. "Poincaré's Structural Realism and His Logic of Discovery." Forthcoming.